T0062654

# Mobile SmartLife via Sensing, Localization, and Cloud Ecosystems

# Mobile SmartLife via Sensing, Localization, and Cloud Ecosystems

Kaikai Liu and Xiaolin Li

CRC Press
Taylor & Francis Group
Boca Raton London New York

CRC Press is an imprint of the
Taylor & Francis Group, an **informa** business

CRC Press
Taylor & Francis Group
6000 Broken Sound Parkway NW, Suite 300
Boca Raton, FL 33487-2742



Printed on acid-free paper
Version Date: 20171108

International Standard Book Number-13: 978-1-4987-3234-5 (Hardback)



**Library of Congress Cataloging-in-Publication Data**

Names: Liu, Kaikai, author. | Li, Xiaolin, author.
Title: Mobile smartlife via sensing, localization, and cloud ecosystems /
Kaikai Liu and Xiaolin Li.
Description: Boca Raton : CRC Press, 2017. | Includes bibliographical
references and index.
Identifiers: LCCN 2016043423| ISBN 9781498732345 (hbk) | ISBN 9781498732369
(ebk)
Subjects: LCSH: Ubiquitous computing. | Mobile computing. | Indoor
positioning systems (Wireless localization) | Cloud computing. | Sensor
networks.
Classification: LCC QA76.5915 .L56 2017 | DDC 004--dc23
LC record available at https://lccn.loc.gov/2016043423

**Visit the Taylor & Francis Web site at**
**http://www.taylorandfrancis.com**

**and the CRC Press Web site at**
**http://www.crcpress.com**

Dedicated to my beloved
and everyone who has helped me along the way

# Contents

# List of Figures

# List of Tables

# Preface

## Towards SmartLife Mobile Applications

Our lifestyle is changing dramatically with the ubiquity of mobile devices and network connectivity. By seamlessly collecting advanced data related to human activity, and providing responsive actions and services to users, developers can maximize the functionality of mobile devices, thereby improving livability, convenience and safety, and ultimately enabling a smart life. The spatial and contextual data, i.e., users' locations as well as their interaction with the cyber and physical world, has been a decisive driver for the ongoing trend towards smart and connected applications to date.

Context-aware location sensing is the cornerstone of this vision. Over the past few years, a broad variety of services have been targeted to revolutionize how people sense and interact with everyday objects and locations towards a smart life. For example, sensor networks provide realtime activity data for heating/air conditioning systems and fire and smoke detectors, GPS and WLAN systems provide way-finding and coarse-grained location services, RFID and short-range communication devices provide proximity detection and awareness. However, these separate and usually proprietary systems are far from satisfactory. The major metrics of these spatial enabling technologies, most notably accuracy, interoperability, and deployability performance, are far from satisfactory. Significant gaps exist in our understanding of how a scalable location sensing system design could meet a multitude of smart application demands. Moreover, no extensible and developer-friendly system frameworks are available for location and smart applications. Developers do not have any testbed or prototype system available for them to play with. Senior developers are reluctant to extend and debug their existing prototype infrastructures since the errors and software deficiencies are hard to identify in the distributed manner. Large-scale deployment is rarely available and hard to share with junior developers for partial or temporal development.

Our overarching goal is to develop an intellectual framework and a location infrastructure testbed to promote and guide the developers to realize the full-fledged smart applications for a smart lifestyle, to address societal challenges in local communities. We aim to offer a comprehensive book on a complete mobile system design for SmartLife applications. This book is structured to be a complete and updated guide for building the ecosystem step by step from the hardware to mobile apps, to the cloud processing and service backend. Beginners can start from the initial introductory and tutorial chapters while advanced readers can learn directly from the algorithms and prototype design. Practitioners can find inspiration for utilizing the proposed localization techniques in a variety of mobile applications including shopping map, indoor navigation, visitor guide, augmented reality, and location-based social-aided sensing/sourcing.

We have tested and verified the information in this book to the best of our ability, but you may find that features have changed (which may in fact resemble bugs). Please let us know about any errors you find, as well as your suggestions for future editions, by writing to the following address.

Please get in touch (kaikai.liu@sjsu.edu) if you know of services that are missing, or have other questions or suggestions.

I hope this book will give you a good head start and that you have fun in the process.

## How This Book Is Organized

The main body of this book is divided into three parts: 1) fine-grained smartphone indoor localization; 2) context-aware indoor augmented reality via sensing and vision tracking; 3) large-scale localization via crowds and social collaboration.

In the first part, we are focusing on fine-grained localization techniques. We design a complete indoor localization ecosystem (code-named Guoguo). The Guoguo ecosystem consists of a set of indoor positioning satellites (IPS), an app on a smartphone (iOS or Android) for opportunistic sensing (with IPS, WiFi, BLE, and GPS), and a cloud server for offloading heavy processing, location services, and content management. Without any hardware burden or attachment on a user's smartphone, we could achieve centimeter-level fine-grained smartphone indoor localization.

In the second part, we push the fine-grained indoor localization to context-aware indoor augmented reality via sensor fusion and vision tracking. To realize AR applications under various scales and dynamics, we propose a suite of algorithms for fine-grained AR view tracking, to improve the accuracy of attitude and displacement estimation, reduce the drift, eliminate the marker, and lower the computation cost. Instead of requiring extremely accurate absolute locations, we propose multi-modal solutions according to mobility levels without additional hardware requirements. Experimental results demonstrate significantly less error in projecting and tracking the AR view. These results are expected to make users excited to explore their surroundings with enriched content.

In the third part, we propose large-scale localization in social networks. Leveraging the ubiquitousness of "crowds" of sensor-rich smartphones, we design a cyber-physical mobile social network for smart life with the following salient features: crowd sensing/sourcing, crowd localizing, cooperative localization, connecting friends socially and physically, dynamic location-based check-in points, tracking nearby friends or team members or children/elders, and mission-oriented groups.

Finally, Chapter 12 summarizes this book and points out some future work.

## Acknowledgements

I would never have been able to finish this book without the guidance of my advisor and Ph.d committee members, help from colleagues, co-authors and friends, and support from my family. I would like to thank National Science Foundation (NSF) for the funding support under the grant number CNS 1637371. I would like to express my deepest gratitude to my Ph.d advisor, Dr. Xiaolin Andy Li, for his excellent assistance, encouragement, forward thinking, and providing me with an excellent research atmosphere that I can enjoy. I would like to thank Dr. Yuguang Fang, who let me experience the research in Wireless Networks Laboratory (WINET). I would also like to thank Dr. Dapeng Wu for patiently correcting my writing, and helping me to develop background in areas of wireless communication, network science, and computer vision. I would also like to thank my co-authors, Xinxin Liu, Ze Yu, Min Li, Qiuyuan Huang and Jiecong Wang for many valuable discussions.

Thanks for all the help I have received in writing and learning about this book. I regret for errors or inadequacies that may remain in this work, and the responsibility is entirely my own.

Last, and surely the most, I want to thank my family, for their love, selflessness, and sacrifice, not just this time, but so many times in my life. I would thank them for their encouragement and letting me join the adventure and the long lonely journey. I especially thank my wife Wenrong for being with me in places that are far away from home, and going through all the hardness in life together. I devote this book to all of you.

# Author Bios

**Dr. Kaikai Liu** is an assistant professor in the Department of Computer Engineering since August 2015. His research interests include Mobile and Cyber-Physical Systems (CPS), Smart and Intelligent Systems, Internet-of-Things (IoT), Software-Defined Computing and Networking. He has published over 20 peer-reviewed papers in journals and conference proceedings, 1 book, and holds 4 patents (licensed by three companies). He developed several prototype systems that can potentially improve peoples lives, for example, emergency communication system for smart city, Ultra-wideband communication and detection for search and rescue victims, indoor localization and navigation for the disabled. He is a recipient of the Outstanding Achievement Award at UF (four times), the Apple WWDC Student Scholarship (2013 and 2014), the Innovator Award from the Office of Technology Licensing at UF (2014), the Top Team Award at NSF I-Corps Winter Cohort (Bay area, 2015), the 2015 Gator Engineering Attribute Award for Creativity at UF, IEEE SWC 2017 Best Paper Award, IEEE SECON 2016 Best Paper Award, ACM SenSys 2016 Best Demo - Runner up, 2016 CoE Kordestani Endowed Research Professor, and 2017 CoE Research Professor Award. He served as the TPC member and technical reviewers for many IEEE/ACM conferences and journals.

**Dr. Xiaolin (Andy) Li** is a Full Professor and University Term Professor in Department of Electrical and Computer Engineering (ECE) and Department of Computer & Information Science & Engineering (CISE, affiliated) at University of Florida (UF). His research interests include Cloud Computing, Big Data, Deep Learning, Intelligent Platforms, HPC, and Security & Privacy for Health, Precision Medicine, CPS/IoT, Science, Engineering, and Business. He has published over 100 peer-reviewed papers in journals and conference proceedings, 5 books, and 4 patents (three licensees). His team has created many software systems and tools. He is the founding Director of Large-scale Intelligent Systems Laboratory (Li Lab) and the founding Director of National Science Foundation Center for Big Learning (CBL) with UF, CMU, U. Oregon, and UMKC. His research has been sponsored by National

Science Foundation (NSF), National Institutes of Health (NIH), Department of Homeland Security (DHS), and others. He was a faculty member (with early promotion and tenure) in the Computer Science Department at Oklahoma State University (OSU), a visiting professor at Nokia Research Center Beijing (NRC), a visiting scholar at University of Texas at Austin (UT), an Extreme Blue intern at IBM, a graduate research assistant at Rutgers University (RU), a research staff at Institute for Infocomm Research (I2R), a research scholar at National University of Singapore (NUS), and a research assistant at Zhejiang University (ZJU). He received a PhD degree in Computer Engineering from Rutgers University under the direction of Professor Manish Parashar. He is a recipient of the National Science Foundation CAREER Award in 2010, the Internet2 Innovative Application Award in 2013, NSF I-Corps Top Team Award (1 out of 24 teams, including Berkley, Harvard, and MIT) in 2015, Top Team (DeepBipolar) in the CAGI Challenge on detecting bipolar disorder in 2016, and best paper awards (IEEE ICMLA 2016, IEEE SECON 2016, ACM CAC 2013, and IEEE UbiSafe 2007).

# Chapter 1

# INTRODUCTION

**Kaikai Liu**

*Assistant Professor, San Jose State University*

**Xiaolin Li**

*Associate Professor, University of Florida*

## CONTENTS



## 1.1 Mobile SmartLife: An Overview

Everyone interacts with the world differently — guided by their own attitudes, their communities and families, and their smartphones with apps. Ever since smartphones became a virtual assistant for us to interact with the world on the move, mobile apps have never been proficient at helping people gather together virtually and informed of nearby and online updates and events. Mobile apps such as Google Maps and FourSquare have long used location data to help us sense and interact with our surroundings. These could be merely the tip of the iceberg for our needs of location sensing. Our physical world, disconnected from our devices, is waiting for smart connections to fill the gap. Researchers are excited about new technologies that could revolutionize how people sense and interact with everyday objects and locations.

As one of the two most important contexts (time and location), location is becoming a key entry for mobile Internet, which could navigate the elderly, disabled, and children to their destination safely; welcome a customer into a store with a personal message and recommendation; offer an in-store deal or deliver a coupon in the front of cereal boxes; deliver biographies or art commentary on a specific object when visitors are wandering through a museum; remind users of social events in proximity. Ubiquitous smartphone and location information are enabling these new mobile *smartlife* scenarios near a certain object or location towards mobile smart life.

## 1.2 Location Matters

As one of the two key components of a mobile context (time and location), localization has been the subject of extensive works ranging from algorithms, models, supporting technologies, to systems and applications. Current coarse-grained (room-level or meter-level) localization on a smartphone (based on GPS) has enabled a lot of mobile services, such as location-based services, maps, and navigation systems. Location information has infiltrated our everyday life in ways that we had not imagined at its start.

The vertical markets that are ripe for location have different needs for **accuracy**, **cost**, and **speed**. **Different Accuracy Requirements:** Meter-level (e.g., GPS with five-meter accuracy) localization is sufficient to navigate a car (meter-level footprint), on a street (several-meter footprint); but it is far from sufficient to navigate a user (foot-level footprint) in a library (with half-meter-wide isles and inch-level books) or a shopping mall (with inch-level items). With infrastructure already installed throughout the great indoors, the easiest solution may appear to be Wi-Fi triangulation. While the cost may be attractive, the accuracy is not precise enough for many apps. **Various Costs:** With infrastructure already installed throughout the great indoors, the easiest solution for shopping mall navigation may appear to be Wi-Fi fingerprinting approaches. While the cost may be attractive, the accuracy is barely sufficient for differentiating different stores. In-store navigation, e.g., shelf-to-shelf, requires high-density deployment of location anchor nodes, which increases the installation and operation cost significantly. **Speed in Terms of Latency or Refresh Rate:** Tens of seconds delay for outdoor navigation could lead to wrong turn decision for drivers. Second-level delay still makes it hard to get the complete moving trace of a basketball player for performance analysis and evaluation.

Various promising location applications that are emerging include indoor navigation, retail, advertising, manufacturing, asset tracking, gaming, intelligence, and public safety. The indoor location market will be more enormous than the outdoor, since we spend more than 80% of time indoors in our daily activities, e.g., working, shopping, eating, at the office, at home. However, these services are severely limited when applying existing localization solutions due to low accuracy, high cost, and low speed.

## 1.3 Application Scenarios

The ultimate aim of this work is to enable real-life applications that help people live independently and conveniently in a smart way. Specifically, this work narrows down the enabling techniques to three example applications: 1) helping the blind or other disabled to live independently; 2) accessing the information automatically via context-awareness; 3) locating lost children or other group members. These three applications require the assistance technique, safety guidance, and information recommendation, which pave the way to the future *smartlife* mobile applications.

### *1.3.1 Helping the Blind*

It is estimated that over one million persons in the U.S. are blind, and each year 50,000 more will become blind [77]. China has the world's top population of the blind, about 5 million. Studies show that blindness, AIDS and cancer are the top three fears of the people in the world [77].

The blind and visually impaired person faces two lifelong challenges: 1) accessing information; 2) navigating through space [118]. The second challenge, which we are addressing in this work, links one's ability to independently move through the world. The blind face a multitude of challenges every day that can prevent them from getting where they want to go, and even make the give up places that are essential to their life, e.g., schools, clinics, retail stores, and city facilities. The journey to public places is very daunting and leaves them stressed and anxious.

Helping the blind and visually impaired people back to the mainstream and employment has significant social importance for the whole society. Multiple approaches are already proposed to help the blind live independently like normal people. For example, the Americans with Disabilities Act (ADA) requires places of public accommodation to ensure that everyone regardless of disability has an equal opportunity to enjoy their services and facilities [3]. More specifically, some countries are requiring the construction of sidewalks for the blind in urban planning. Using Beijing as an example, it owns the longest sidewalk for the blind in the world, i.e., nearly 1600 km. The government and the whole society have a high motivation for creating an urban environment that is suitable for everyone, including the disabed. However, the blind sidewalk requires a vast investment in terms of construction and maintenance. In Los Angeles, for example, 4,600 of the city's 10,750 miles of sidewalks need some degree of repair at an estimated cost of $1.2 billion [2]. Even with the sidewalk in place, the coverage and utilization are still low. Studies show that most of the blind sidewalks cannot connect the most visited places for the blind, i.e., the hospital, the residence community, the school, and most of the indoor area [100]. People spend most of their time indoors, which is particularly true for the blind. However, most of the indoor places do not have sidewalks or other related accessibility assistance techniques for helping the blind.

Only using the sidewalk is not enough. With the advance of mobile technology, people are designing navigation devices or solutions for helping the blind to visit unknown public places. When visiting outdoor places, e.g., city streets, blind people can utilize existing GPS devices for navigation assistance. There have been many attempts to integrate GPS technology into a navigation-assistance system for blind and visually impaired people. Trekker Breeze handheld talking GPS [36], as an example, verbally announces the names of streets, intersections, and landmarks as blind people walk. It also provides information on all street intersections and stops when the blind are in a vehicle, which allows them to know where they are without missing the next stop. With the popularity of GPS-enabled smartphones, multiple solutions are proposed to leverage the general purpose device for navigation instead of expensive stand-alone devices. MIPsoft's BlindSquare is a GPS navigation software for iPhone and iPad. It uses Foursquare for points of interest (POI) and OpenStreetMap for street info [121]. However, these solutions are far from practical enough due to various constraints: 1) the routes in Trekker Breeze need to be recorded before they can be used, which limits the usability when blind and people visiting some unknown public places; 2) the error of GPS in city blocks may reach 28 meters for 95% of the time due to the so-called "Urban Canyon" effect [72]; 3) the nearby POI information is mostly static without dynamic traffic and road blockage information; 4) verbal navigation instructions are farless accurate than vision and maps due to positional ambiguity, insufficient interpretation, and missing fine-grained context.

Once blind people move indoors, the GPS navigation device carried will lose access to the required satellite signals, which becomes a nightmare for them especially in unknown places. No matter how good the systems are outdoors, their accuracy, coverage, and quality deteriorate significantly in small-scale indoor places. We spend more than 80% of our time indoors in our daily activities, e.g., working, shopping, eating, at the office, at home. Practical, robust, and accurate indoor location solutions are largely missing. In indoor places, technology that can help the blind navigate with foot-level accuracy is convenient. This has significant implications for finding the door, restroom, stair, entrance, and other key places. However, it is very hard to get the fine-grained location and sense the nearby objects in a room. Technologically, outdoor localization techniques cannot be directly moved indoors. Current GPS, cellular, and Wi-Fi-based localization techniques [6, 8, 15] are not reliable and fine-grained enough when it comes to pinpointing an individual's whereabouts, especially indoors. Despite significant efforts on indoor localization in both academia and industry in the past two decades, highly accurate and practical indoor localization remains an open problem. Alternative solutions require barcode or RFID to be strategically placed around buildings, and it is currently unrealistic to expect to find such systems installed in many places. Even with wide adoption, these technologies have their inherent limitations. Existing devices like the omniscanner could help the blind record the route and voice via UPC labels. This could add additional information for the blind to remember different places. However, scanning the UPC labels is not convenient. Using RFID could have more freedom, but the resolution is very low.

After the localization process, the blind's navigation system should start by establishing their location within the building, and on the map. It will require voice input for the destination or purpose, and then determines the best route or action to get them done, and guides them along it via verbal cues and feedback. However, indoor navigation and interaction are far more complex than outdoor GPS navigation via voice assistance, where the road is clear and predefined. There are so many obstacles, blockages, floors, and routes in indoor places, e.g., door, staircase, elevator, and restricted areas.

The limitation of physical freedom caused by vision loss has the greatest negative impact on human life. Our work aims to increase their confidence in visiting public place and taking new routes via our assistive technology. We propose mobile solutions to help *the blind* navigate, socialize, and explore the physical world better. It is a unique multi-modal eco-system with the following salient features: fine-grained step-by-step navigation, indoor navigation-specific voice interaction, and a participatory social network for assistance. With centimeter-level resolution for mapping their steps in the physical world, we hope our technology can improve their independence to a new level, and gradually help them to recover the hope of life and welcome the unknown world.

### 1.3.2 Context-Awareness Information Accessing

Locating and digitizing of the physical world has become the next battle for big brothers. Industry leaders are sketching out their forward vision for computing and how we relate to our physical world by playing around with context, especially locations. Major tech players are working in the context-awareness approaches to bring users even more information super quickly, without tedious manual typing and searching. Google Now and Apple Siri are all context-awareness examples. In addition to we you are talking, Google pushes context-awareness to the next level by proposing "Now on Tap" in Google I/O 2015.

Even when our confidence with smartphones as the "all-controlling" hub continues to gather steam, wearable devices, e.g., Google Glass and Smart Watches, are another revolution ready to shake things up. According to research conducted by PwC's Consumer Intelligence Series, 20% of the adults in the United States already own a wearable device [131]. It is the more complex sensing sphere where wearable technology can truly emerge as the driving force for future context-aware applications. For example, imagine if customers are visiting a mall or museum that is equipped with fine-grained localization and navigation techniques, all information displayed right through the wearable camera/device on their route at the right physical place. This augmented-reality (AR) method is a huge benefit for customers who can save on precious time in information accessing with hands free.

One key challenge in context-aware application is how to sense and interact with the real physical location in a fine-grained manner. To improve the location sensing accuracy of mobile device, Apple purchased WiFiSLAM in 2013, and Coherent Navigation in 2015. In 2013, Apple launched iBeacon, i.e., the Apple-certified version of a BLE beacon, which represents programming interface for proximity sensing for smartphone [5]. High-profile retailers such as Macy's and American Eagle Outfitters, along with major league baseball and the National Football League, are actively testing them for location-aware services around local navigation, augmented reality, retail recommendation, proximity social networking, and location-aware advertising. However, most of these solutions only rely on proximity detection, and provide services like pushing coupon, indoor check-in, far from the augmented reality way.

Existing augmented reality relies highly on the location sensing or computer vision techniques [7, 110]. However, both worlds have their own drawbacks and limitations. Computer-vision-based solutions are highly accurate and fine grained, but the distance and coverage are very small in addition to their high computational power requirement. Location-sensing-based solutions are lightweight, but only achieve coarse-grained resolution and only apply to position-of-interests (PoIs) at far distance (hundreds of meters away). How to balance the two worlds with **fine-grained location, low power consumption, and better coverage** is the motivation of our work on this part.

### 1.3.3 Locating Lost Children

Losing their beloved child is the worst nightmare of every parent. Sometimes after you turn around for just a few seconds, your child is gone when you turn back. If you are at home or some less-crowded small area, you probably can find your child in some corner or in the immediate vicinity. But for public areas, like shopping malls, streets, or even your child's favorite Disney World, it is hard to find your child in crowds when lost.

There are so many reasons for your child to get distracted and wander, then get lost. In places like Disney theme parks, there is simply so much to see, and so many people attend, especially for events like fast-paced parades. Even if your child doesn't typically wander, kidnapping could happen, making it even harder to find your child. Guarding a child in crowded places full of attractions is nontrivial; locating your lost child is mission impossible.

Normal approaches include writing your phone number on a shoe tag or sticker for your child, or going to your designated meeting place, if you have one or your child could do that. You also can look at the closest locations that are of interest, or go to the baby center to locate a lost child found by others. However, it is not efficient for this manually blind searching. Giving your "big kids" a cell phone could be a high-tech approach, but in most cases they are not old enough to carry one.

To find your child quickly if you are separated, lots of systems and approaches have been developed. One kind of approach is using a GPS locator that is installed on your child's shoes or clothes, e.g., Amber Alert GPS, PocketFinder, AT&T Family Locator [21]. This kind of device includes GPS and cellular communication modules. One problem for this kind locator is that it is high cost and bulky. GPS and cellular communication are all expensive and power hungry, especially when it works in continuous mode. Providing sufficient battery life for one day's use could result in a bulky and heavy device, and is not suitable for little kids to carry. For indoor places like castles and shopping malls, GPS may suffer significant performance degradation, or even not work due to the signal blockage.

Another category of approaches relies on devices with peer-to-peer communication capabilities. The transmitter and receiver pair, or smartphone and peripheral pair, is carried by parents and child, respectively. If the child goes out of the communication range or predetermined threshold, parents will get an alert. These kinds of approaches leverage the existing low-power communication standard and could be made with high efficiency in power, and portable in size, e.g., Toddler Tag Child Locator, Keeper 4.0, Chipolo [17]. A drawback of this approach lies in its lacking absolute location information, e.g., GPS location. It is impossible for parents to locate their child when the child goes out of the communication range.

It is very hard to balance between the coverage, accuracy, device complexity, and power consumption. Using simple and popular wearable devices for kids to carry is attractive; however, how to **achieve better accuracy and coverage of localization** without power-hungry GPS and cellular devices are the research problems that we need to solve.

## 1.4 Research Challenges

The three technical requirements for our envisioned application scenarios in Section 1.3 are: 1) smartphone localization and navigation technology with foot-level resolution; 2) indoor augmented reality with fine-grained location, low power consumption, and better coverage; 3) better accuracy and coverage of wearable device localization.

The biggest obstacle has been technical, i.e., fine-grained localization. Current coarse-grained (room-level or multi-meter-level) localization solutions can hardly meet the requirements of many smart life applications. For example, foot-level resolution is required for step-by-step navigation for the blind. Satellite-based localization, e.g., GPS, has been one of the most important technological advances of the last half century. No matter how good the systems get outdoors, their accuracy, coverage and quality deteriorate significantly in small-scale indoor places. We spend more than 80% of our time indoors in our daily activities, e.g., working, shopping, eating, at the office, at home. Practical, robust, and accurate indoor location solutions are largely missing. It is very hard for our smartphone to get the fine-grained location and sense the nearby object in a room. Technologically, outdoor localization techniques cannot be directly moved indoors. Current GPS, cellular, and Wi-Fi-based localization techniques [6, 8, 15] are not reliable and fine-grained enough when it comes to pinpointing an individual's whereabouts, especially indoors. Despite significant efforts on indoor localization in both academia and industry in the past two decades, highly accurate and practical smartphone-based indoor localization remains an open problem.

When pushing the fine-grained location awareness to context-awareness, there are two more pieces that are missing in addition to the vision tracking and fine-grained localization. These two pieces are: 1) **High-Accuracy Attitude Requirement:** Compared with almost invisible remote POIs, line-of-sight indoor AR views are more sensitive to the attitude estimation result. The attitude estimation error of the camera would introduce visible drift and bias between the "rendered" objects and the actual objects due to the short distance. 2) **Displacement Matters:** The displacement of a user's camera is ignored in outdoor AR applications, since its moving distance is far shorter than the POI distance. For indoor AR applications with short-range POIs, the movement even in a small hand shake would have significant impact for the screen location of the AR view. All these pieces need to be integrated together with the vision tracking and fine-grained localization approaches for lower computational power, higher tracking accuracy, and faster response.

For locating a lost child using low-power and miniature wearable devices without high cost cellular and GPS module. Although the required resolution is lower, the coverage requirement is significantly higher. Most of the environments are uncontrolled without any infrastructure. To solve the coverage and localization problems of wearable devices without infrastructure, we focus on the investigation of nearby "crowds" of smartphones for transparent ranging and locating via peer collaboration. State-of-the-art approaches leverage connection information to detect the presence of wearable devices in a specific region near to one participator. However, the resulting searching area of the obtained location resolution still makes it hard for parents to pinpoint their children in crowds, e.g., 20-meter peer-to-peer distance could exaggerate the initial location error surface to thousands of square meters.

## 1.5 Book Organization

This book surveys the state-of-the-art techniques, describes future mobile applications, and presents original technologies in this rapidly moving field. It represents a comprehensive and updated book in this field, covering a complete system design and implementation via sensing, localization, and a cloud ecosystem. The envisioned smartlife ecosystem is shown in Fig. 1.1. Using off-the-shelf smartphones with a developed cloud ecosystem (anchors, wearable tags, mobile apps, and cloud server), the proposed solution could revolutionize various smartlife applications such as indoor navigation, near-field advertising, augmented-reality, mobile education/campus/health and entertainment, and shopping/tour guides.

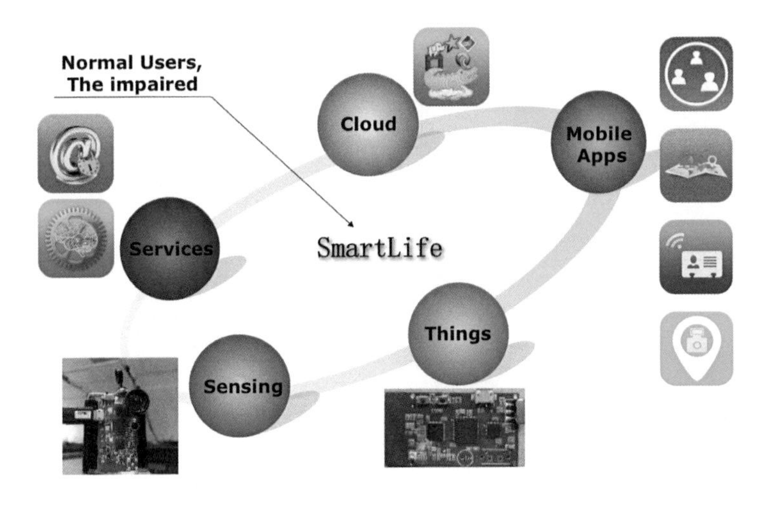


**Figure 1.1: Envisioned smartlife ecosystem.**

This book presents a prototype system to fill the long-lasting gap of smartphone-based indoor localization. This system consists of a constellation of low-complexity anchor networks that enable fine-grained localization with accuracy up to centimeter-level without any hardware/attachment burden on users' smartphones. A hybrid mobile social network is proposed for location-based social and crowd missions. Specific applications include social cooperative localization, locating your lost child via crowd sensing, and helping users coming, keeping, and working together.

# *Chapter 2*

# Overview of Mobile Systems

**Kaikai Liu**

*Assistant Professor, San Jose State University*

**Xiaolin Li**

*Associate Professor, University of Florida*

## CONTENTS



Modern smartphones and location-based services and apps are poised to transform our daily life. However, current smartphone-based localization solutions are limited mainly to the outdoors mostly missing practical, robust, and accurate indoor location solutions. Despite significant efforts on indoor localization in both academia and industry in the past two decades, highly accurate ones and practical smartphone-based indoor localization remains an open problem. To enable indoor location-based services (ILBS), there are several stringent requirements for an indoor localization system: highly accurate that can differentiate massive users' locations (foot-level); no additional hardware components or extensions on users' smartphones; scalable to massive concurrent users. Current GPS, Radio RSS (e.g. WiFi, Bluetooth, ZigBee), or Fingerprinting-based solutions can only achieve meter-level or room-level accuracy.

In this chapter, we provide an overview of location sensing systems and localization approaches. To address the challenges in existing systems, we propose a practical and accurate solution that fills the long-lasting gap of smartphone-based indoor localization. Specifically, we design and implement an indoor localization ecosystem called Guoguo. Guoguo consists of an anchor network with a coordination protocol to transmit modulated localization beacons using high-band acoustic signals, a realtime processing app in a smartphone, and a backend server for indoor contexts and location-based services. The overview architecture of our proposed system is given in this chapter. The detailed system design and implementation will be elaborated in the following chapters.

## 2.1 Overview of Location Sensing Techniques

Localization schemes can be classified into three categories of methods: angle-based, fingerprinting-based, and ranging-based. An angle-based approach relies on the directional antenna scan to achieve angle resolution. But a narrow-beam directional antenna is very expensive and unsuitable for consumer applications. Fingerprinting-based approaches [8, 15, 6] or other proximity approaches [78, 38, 18, 99] feature lowest complexity without any requirement for additional infrastructure. However, they achieve only room-level resolution and require the site-survey to build fingerprinting databases at the cost of intensive labor. A ranging-based approach is more straightforward. Measuring the radio signal strength (RSS) and time-of-arrival (TOA) are the two typical ranging solutions. Compared with TOA, RSS information is widely available and lots of work has been dedicated to using the RSS of WiFi, Bluetooth, or cellular signals, for indoor localization [127, 20, 44, 50]. However, the need for the prior information of the radio attenuation model, and the time-varying features of channel make these approaches only suitable for applications with low location accuracy needs.

From the measurement signal perspective, smartphone-based indoor localization could be summarized into the following categories: RSS-based ranging using propagation attenuation of radio (WiFi, cellular, and Bluetooth) [20, 44], fingerprinting-based matching using surround sound [6, 103] or radio profiles [114, 15, 19], and TOA-based ranging [53, 59, 74, 52, 133, 48]. Different applications may have different requirements as to the complexity, cost, and resolution. For fine-grained indoor localization application with high resolution requirements, TOA-based ranging is preferred.

TOA-based ranging is more accurate and robust. The Cramér-Rao Low Bound (CRLB) of TOA ranging is inversely propositional to the effective bandwidth of signal, it is also why ultra-wideband (UWB) signal received special attention for its high accuracy on ranging and localization. However, current UWB techniques are still under development and not available on current smartphones. Another option is to use the acoustic signal for TOA ranging due to its low transmission speed. MIT Cricket [85], Active Bat [113] and DOLPHIN [70] are well-known systems that use ultrasound for localization. However, normal smartphones cannot receive ultrasound. In addition, they require radio signal for synchronization, e.g., ZigBee for Cricket. Dependence on special devices and non-applicability on smartphones greatly limits their adoption in daily indoor activities.

Recent research on leveraging the ubiquitous microphone sensors in a smartphone introduces a convenient and low-complexity approach. A. Mandal et al. [62] used a PDA to transmit annoying 4kHz acoustic signal. C. Peng et al. [81] proposed to transmit low-attenuation 2-6kHz acoustic signal for better coverage, but their solution causes sound pollution due to the audible signal. H. Liu et al. [52] used a smartphone to transmit high-band audible sound, but the achieved ranging coverage is only three meters. Authors in [74] performed localization using desktop PCs and laptops, achieving location accuracy of several meters. In addition, these solutions are not scalable and feature low location update rates. Due to the use of two-way ranging and simple acoustic "Beep" signal, only one user can be handled and localized at a time. Adopting time-divided coordination among users could be a partial solution. However, random concurrent access patterns of many users in real situations makes it hard for their solutions to support multiple users.

## 2.2 Overview of TOA-Based Smartphone Localization Approaches

We divide existing TOA-based localization solutions into four different categories according to their operation mode, i.e., **two-way active mode** [74, 52, 133], **one-way active mode** [74, 62, 10, 34], **assisted passive mode** [85, 106, 107], and **no-assisted passive mode** [53, 48]. The meaning of "two-way" and "one-way" relates to the communication modes between the smartphone and the anchor node; "active" means the smartphone needs to transmit signals; "passive" means the smartphone only receives signals. The "assisted" mode represents approaches that use additional signal for the transmitter and receiver synchronization or timing comparison.

**Two-way active mode** is a typical way to solve the peer-to-peer synchronization problem during ranging. C. Peng et al. [81, 82] proposed a solution of self recording between peer-to-peer devices and achieved maximum operation distance around 4 meters. H. Liu et al. [52] utilized the acoustic ranging to assist the WiFi localization, and achieved $1 \sim 2$m accuracy with 3m operation distance and 7s latency. Two-way active mode involves the time-consuming process of transmitting and receiving, which limits the refresh rate. This is especially serious for acoustic signal. **One-way active mode** features a reduced transmission delay by relying on a synchronized anchor network. A. Mandal et al. [62] used PDA to transmit noticeable 4kHz acoustic signals to a WiFi-enabled sensor network for sound source localization. Nandakumar et al. [74] proposed DeafBeep, which locates a laptop or desktop with only speakers. They also proposed a hybrid RF-and-AR-based approach with meter-level accuracy and second-level delay. However, both two-way and one-way active mode suffer from limited scalability for multi-users, since signals from other users cause significant interference. Moreover, active mode requires the speaker in the smartphone to transmit the acoustic beacon, which covers a shorter range compared with the specialized speaker in the anchor node. Thus, one-way passive mode contributes to better user capacity and longer coverage.

MIT Cricket [85] is a classic system of **assisted passive mode** by using specialized devices (Zigbee radio assists simple unmodulated Ultrasound pulse). However, current radio modules (WiFi, cellular, and Bluetooth) embedded in a smartphone have random access delay with inadequate timing accuracy ($>$ 3ms) for ranging assistance. Mostafa et al. proposed RF-Beep and SpyLoc [106, 107] by leveraging the radio signal interface at the kernel-level of the smartphones for assisted ranging with $30 \sim 70$cm accuracy. However, they need to modify the off-the-shelf smartphone OS, i.e., Android or iOS, and the achieved accuracy is still insufficient for some fine-grained location applications.

The Guoguo and other systems [48] fall into the category of **non-assisted passive mode** localization. Lazik et al. [48] used smartphones as anchor nodes to generate Chirp signal for one-way ranging. However, the generation of Chirp signal requires devices containing DAC, e.g., smartphones. Low-complexity sensors cannot generate Chirps. Moreover, [48] rely on frequency-based $f$ modulation with long duration of signal in the time-domain ($t \propto 1/f$). Long time duration reduces the refresh rate of localization and increases power consumption. We utilize our own low-complexity anchors to transmit low duty cycle ($\approx$ 1.92%) sharp pulse by 2-PAM modulation. Our scheme reduces power emission, increases refresh rate, and resists interference.

To compare the performance of Guoguo with other existing indoor localization approaches, we summarize their key features in terms of principle, accuracy, cost, smartphone applicability and training requirements in Table 2.1. The multi-user refers to how many end users can be supported simultaneously. From Table 2.1, we observe that Guoguo achieves better balance in terms of cost and performance over other acoustic/ultrasound-based approaches. Compared with other low-cost approaches (mainly smartphone-based), e.g., RSS and fingerprinting-based approaches, Guoguo significantly outperforms other approaches in accuracy. Guoguo is outstanding as a practical smartphone-

**Table 2.1: Comparison of Guoguo with existing localization techniques**

| System | Signal Type | Technique | Accuracy | Cost | Multi User | Smart-phone | Training |
|---|---|---|---|---|---|---|---|
| Guoguo | Modulated acoustic signal (17-20kHz) | Passive ranging | High (6 $\sim$ 15cm) | Low | High | Yes | No |
| Cricket [85], RF-Beep, SpyLoc [106, 107] | Acoustic, Radio | Assisted Ranging | High (centi- meter) | Medium | High | Hard | No |
| Beep [62], BeepBeep [82] | Acoustic signal (Low-band) | Active-mode ranging | High (centi- meter) | Medium | Limited | Yes | No |
| Active Bat [113], DOLPHIN [70] | Ultrasound | Active-mode ranging | High (centi- meter) | Medium | Limited | No | No |
| Ultra wideband [49, 54] | Ultra wideband radio | Ranging-based | High | High | High | No | No |
| Radio RSS [20, 44] | WiFi, Bluetooth, or cellular signal | Ranging-based | Low | Lowest | High | Yes | Yes |
| Fingerprinting [6, 8, 10, 15] | WiFi, Cellular, Bluetooth, FM and ambient sound | Fingerprint matching | Low (room- level) | Lowest | High | Yes | Yes |
| Proximity [99, 18, 78] | RFID, NFC, and acoustic signal | Proximity-based | Lowest | Medium | High | Depends | No |

based solution with centimeter-level accuracy without special hardware extensions on users' phones.

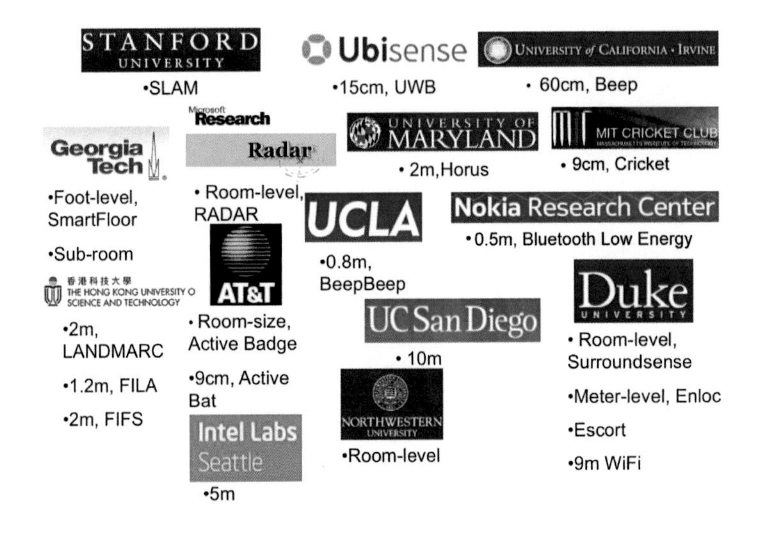


**Figure 2.1: Related localization systems from academia.**

## 2.3 Architecture of the Proposed Localization System

The architecture of the Guoguo localization system is shown in Fig. 2.3.

**Wireless Anchor Network:** We designed the anchor node from scratch and keep the BOM price as low as \$10 per node. The inter-node communication and synchronization have been implemented and achieved at a very highly accurate level. The transmitted beacon signal has been tuned to be unnoticeable to humans, with low-power and low-duty-cycle features. To enable passive sensing for multiple smartphone users, we designed the transmission waveform with wide-band modulation, and a transmission scheme of the acoustic beacon that follows the high-density pseudo-codes. We propose a *symbol-interleaved* beacon structure to overcome the drawback of the low transmission speed of acoustic signal and improve the *location update rate* [57].

**Smartphone Processing:** We perform preprocessing for the received acoustic beacon signal, e.g., adaptive filtering and wavelet decomposing, before signal detection. Transmit reference approach is utilized for the matched-filter estimation on the smartphone side [53]. The signal detection module is

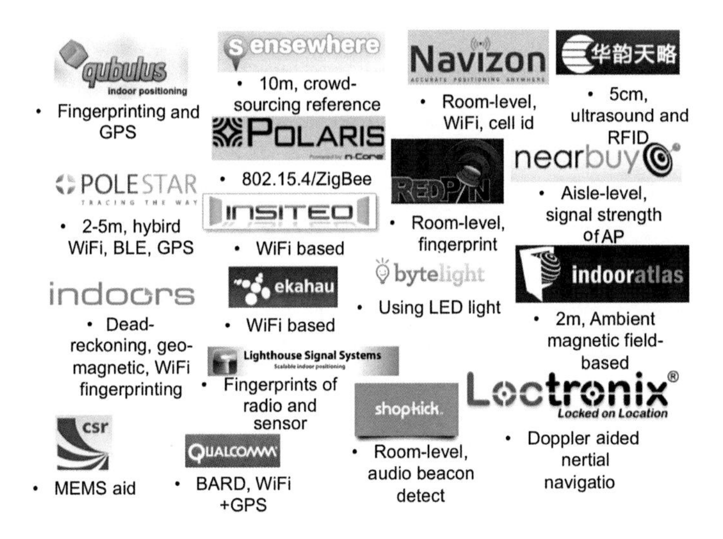


**Figure 2.2: Related localization systems from companies.**

designed with better robustness by applying cluster detection and spectrum matching to identify signal-like interference. The demodulated information bit could be used to extract the pseudo ID for each node. The synchronization is realized by tracking the demodulated signal in the time-domain, where the convergence of the tracking filter means the success of the synchronization.

We propose a dynamic TOA estimation scheme to obtain accurate ranging results by maximizing the TOA detection probability, along with a multiple-threshold backward approach to ensure the detection of TOA path. To minimize the NLOS bias effects, we propose approaches for NLOS mitigation.

**Server Processing:** To minimize the computation cost in a smartphone, we offload localization and tracking process into the back-end server. To mitigate the outliers and missing data in the ranging measurement, we propose track-before-localization for the ranging results of each station. We further propose semidefinite programming (SDP) for global optimal location estimation by leveraging the computational power of the server. To utilize the linear time varient feature of the unknown delay, we add delay-constraint into the location estimation for better robustness.

**Achieved Performance:** For the prototype of the Guoguo system, current experiment results demonstrated the promising features of our solution: low power and unnoticeable acoustic beacon, low-complexity anchor node (BOM price $<$ \$15 per node), high precision in anchor network synchronization ($< 10\mu s$), large coverage ($15 \sim 20$m), centimeter-level accuracy ($<$ 10cm), high refresh rate ($>$ 1Hz), robust under sound interference (up to 110dBA). When compared with commercial apps, the Guoguo app shows modest CPU utilization and network traffic (lower than Google Chrome) without inter-

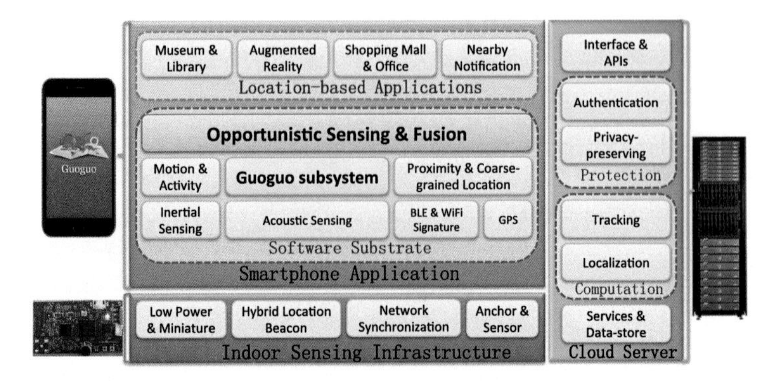


**Figure 2.3: System architecture.**

rupting normal phone usage. This practical and robust ecosystem promises enormous new possibilities for novel indoor location-based services and applications.

## 2.4 Hardware Design for Sensors, Anchor Nodes, and Wearables

### *2.4.1 Design Requirement and Overview*

**General Purpose Hardware.** Whether to opt for specialized or general purpose hardware solutions is the first decision we are faced with, and it's not an easy choice. Although both technologies go on different paths, they do have different economical features and must be evaluated carefully before implementation. The architecture of the specialized hardware could be very simple for one special application without considering the extension to other applications.

### *2.4.2 Hardware Design and Architecture*

Leveraging our wireless anchor network prototype, our indoor localization system consists of an app in a smartphone that adopts the non-assisted passive mode to avoid random-access interference and multi-user division issues in the system. A plurality of sensor nodes as preconfigured anchor constellation to broadcast acoustic beacon. Because of the non-assisted passive mode of the system, the location system can be highly scalable to support hundreds of users simultaneously in a large indoor space, e.g., museums, job fairs, and shopping centers. Moreover, the position information can be calculated locally on a smartphone without reporting to the third-party servers, making it privacy-proof if desired.

To meet our long-term objective, we designed the low-complexity anchor node from scratch using the TI MSP430 microcontroller and CC2533 Zigbee chip. An app on a smartphone (e.g., an iOS or Android mobile device) is designed to perform signal detection, ranging, and localization (the localization step can be optionally offloaded to the server). The architecture of the localization system is shown in Fig. 2.3.

We designed our own hardware as the low-complexity anchor node based on TI's MSP430 Microcontroller and the CC2533 Zigbee single-chip solution, and use a standard smartphone (e.g., Apple iOS platform) to perform signal detection, ranging and localization. The architecture of the localization system is shown in Fig. 2.3.

**Figure 2.4: The designed hardware node version 1.**

**Figure 2.5: The designed hardware node version 2.**

# 2.5 Cloud Architecture

To enable a full-fledged mobile sensing system, we provide cloud back-end for the sensing data from smartphones or other mobile devices. One key important requirement is data offloading and processing. To facilitate joint sensing data processing with high energy efficiency, we implemented the time-sensitive algorithms, e.g., data sensing, pre-processing, detection, TOA estimation, code matching, and NLOS mitigation, into the smartphone app. All the processing was put in the iOS Grand Central Dispatch (GCD) queue to enable concurrency with the mobile application without slowing down the smartphone's responsiveness. To ensure efficiency of the smartphone processing, we used the vDSP portion of the Accelerate framework in iOS. Other complex computation was executed in the server to minimize the computation cost in the smartphone. We designed a pub/sub framework based on the open-source Redis NoSQL server [92] as shown in Fig. 2.6. Once the smartphone publishes a preprocessed result to the server, the subscriber on the server side performs localization and returns the result to the smartphone asynchronously. Such configuration balanced the communication and computation cost. For our acoustic-based localization approach, the smartphone extracts ranging information from the audio raw data and greatly reduces communication overheads; the pub/sub server handles concurrency and serves multiple localization requests from smartphones. We leverage the scalable and robust features of the Redis server. Other features of our provided server include push notification support for social networks, big data processing back-end, and API servers for location-based services and user management. The details of the cloud backend will be elaborated in the following chapters.

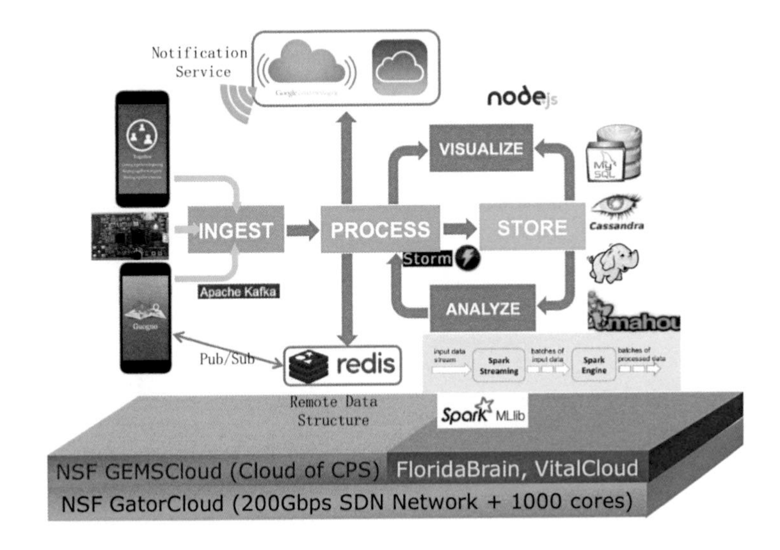


**Figure 2.6: The architecture of the cloud backend.**

# Chapter 3

# Acoustic Ranging and Communication

**Kaikai Liu**

*Assistant Professor, San Jose State University*

**Xiaolin Li**

*Associate Professor, University of Florida*

## CONTENTS

## 3.1 Introduction

Accurate indoor localization utilizing low-complexity devices promises a wide spectrum of applications, especially when the GPS signal is inaccessible due to the blockage of the satellite signal [80]. If sub-meter resolution can be achieved, it will fundamentally change and improve the way that current location-based services are delivered.

TOA estimation based on the communication channel can obtain the signal flight distance as pseudorange [80]. The ranging-based trilateration method utilizes these pseudoranges to calculate location information. For location-aware techniques, communication and accurate TOA estimation capabilities are two important prerequisites. Mainly, two categories of approaches have been proposed to solve this problem. The first type of solutions utilizes impulse radio ultra-wideband (UWB) technique for indoor TOA-based ranging, because ranging precision directly depends on the bandwidth of the operating signal. Using UWB signal has attracted significant research interest and has become a standard as IEEE 802.15.4a. However, the full-digital coherent IR-UWB system requires the bandwidth of several GHz to guarantee sub-meter ranging accuracy, which increases the overall hardware cost and processing power dramatically [130]. Some techniques, e.g., energy receiver and finite-resolution digital receiver, have been proposed to lower the overall complexity [130, 54]. Although much lower complexity can be achieved, it is still very expensive and requires additional special hardware.

The second kind of solutions utilizes the ultrasound signal to perform accurate ranging. Compared with electromagnetic signal used in a UWB device, the aerial acoustic signal is more pervasive and can achieve ranging accuracy with much lower hardware cost. Due to slower transmission speed of acoustic signal, even several KHz signal bandwidth can result in centimeter-level ranging accuracy. Yang et al. [128] used the acoustic approach to detect the position of a phone using car speakers. However, they only need to detect the relative region in a car. The Cricket localization system developed by MIT [85] using ultrasound for ranging *achieved* centimeter-level accuracy. They used the radio signal for synchronization and performed inter-node ranging by using their developed devices. The requirements of the dedicated device impede its widespread adoption by ordinary users.

In this chapter, we propose to directly utilize the existing hardware of consumers to achieve accurate ranging, i.e., the microphone sensor. We address the problem of ranging and communication by using the aerial acoustic signal over the microphone channel. Our scheme helps users achieve indoor localization by using their smartphones. Even the pre-placed anchor node providing the ranging beacon signal can be implemented with very low cost, i.e., a small speaker with a microcontroller is sufficient. We designed a joint symbol detection and TOA estimation method to achieve robust and accurate ranging results. We derived the TOA threshold by maximizing the TOA right detection probability that can be adaptively tuned in different environments. We also propose a dynamic demodulation method based on direct frequency discrimination and amplitude matching with transmit reference to address a worse communication channel condition than UWB and ultrasound signal. The experimental results show that our proposed TOA and communication scheme achieve very good mean-square error and bit-error rate with high probability.

## 3.2 System Architecture

To facilitate the indoor localization, we introduce an acoustic ranging and communication technique that leverages the existing low-complex anchor node infrastructure and microphone sensor in a mobile target (MT), e.g., a smartphone. An anchor node sends its unique acoustic beacon signal periodically. The beacon signal contains the ID or position information of its transmitter. On the receiver side, the microphone searches and captures the existing beacons and performs TOA estimation and communication decisions. After receiving beacon information from more than 3 anchor nodes, the target position can be determined by combing the information from the TOA estimation process and anchor positions. The simplified architecture of our audible-band acoustic localization system is shown in Fig. 3.1. In this paper, we focus on the two important prerequisites of localization, i.e., ranging and communication, and present in detail design goals, the signal modeling, symbol synchronization, TOA estimation, and communication decision.

## 3.3 Transmitter Signal Design and Modeling

### 3.3.1 Signal Modeling

Driven by the specific design goal of only using the microphone on the receiver side, we need to choose an appropriate transmitter signal band to match the capabilities of a user's smartphone. The acoustic signal band of the microphone is very limited; the typical band of a microphone is in the audible range,

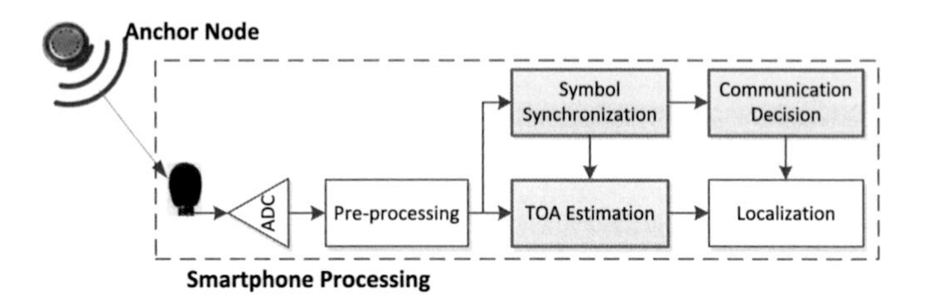


**Figure 3.1: The system architecture.**

i.e., 200Hz–20KHz. To reduce interference between beacon signals and daily environmental noises, we choose the high frequency side of 17KHz–20KHz as the operating band. By keeping the power spectrum density (PSD) lower than the perception level of human ears, and selecting an appropriate band-pass speaker to minimize the useless low-frequency part, the acoustic beacon can be unnoticeable even in very short distances.

The transmitted signal can be modeled as

$$g_t(t) = \sqrt{\varepsilon} \sum_{j=0}^{N_p-1} \sum_{i=0}^{N_s-1} g_{i,j}(t - jT_p - iT_s) \tag{3.1}$$

where $g_{i,j}(t)$ is the transmitted signal for the $i$th symbol in the $j$th period; $\varepsilon$ is the signal energy; $T_p$ is the beacon period, with total beacon numbers of $N_p$; $T_s$ is the symbol duration; $N_s$ is the symbol number, each information bit with values of $\pm 1$. Some spaces are left between beacon periods to avoid the inter-beacon interference, i.e., $N_s \times T_s < T_p$. The information bits are carried by $g_{i,j}(t)$.

The acoustic beacon signal is captured by the microphone and converted to the electrical domain after propagation through the free space with distortion. Passing through the analog-to-digital convertor (ADC), the received signal will be digitalized as $r(k)$ with the sampling frequency of $F_s$, and $k \in [1, \ldots, N_k]$, where $N_k = T_s \times F_s$. The digital version of the transmitted signal is $g_{i,j}(k)$; every symbol contains $N_k$ sampling points. The digital result that we get is

$$r_{i,j}(k) = \sum_{l=0}^{\xi_{i,j}-1} A_{i,j}^l \cdot g_t(k - k_{i,j}^l) + n_{i,j}(k) \tag{3.2}$$

where $\xi_{i,j}$ is the total number of propagation paths, with $A_{i,j}^l$, $k_{i,j}^l = \tau_{i,j}^l \times F_s$ representing the digital version of the multi-path delay. The term $n_{i,j}(t)$ is independent white Gaussian noise in the $i$th symbol of $j$th period. $N_b = T_p \times F_s$ means the number of sampling points in one beacon period.

### 3.3.2 Acoustic Beacon Generation

The generation of acoustic beacon relies on a plurality of sensor nodes with low-complexity and low-power consumption. To trade off between the goal of making the acoustic signal unnoticeable and the constraint of the microphone bandwidth, we choose 17kHz-20kHz as the operating band.

To lower the complexity of the anchor node, a low-cost microcontroller (MSP430) is utilized to drive an audio chip with configurable gain. The audio chip generates the acoustic beacon signal and performs 2-PAM (pulse amplitude modulation) to carry the information bit $p_j$. We use low-duty-cycle transient acoustic pulse for high timing resolution and low-power transmission (below the perception level of the human ear). The Zigbee chip connected to the microcontroller enables the wireless synchronization among $M$ anchor nodes.

### 3.3.3 Signal and Interference

The environmental sound noise and interference is uncontrollable, making the acoustic beacon signal hard to detect in this crowded band. Fig. 3.2(a) shows the sound spectrum in a normal environment without acoustic beacon signal; Fig. 3.2(b) shows the spectrum of the same environment with acoustic beacon signal generated. From the high frequency side of Fig. 3.2(b), we see the acoustic signal working in the band near 17 $\sim$ 20kHz. Fig. 3.2(c) shows the energy spectrum density (ESD) of the received acoustic beacon under interference, e.g., human talking, or video/music. From Fig. 3.2(c), we know that normal sound interference has a very strong frequency component even in the high frequency band, where our generated acoustic beacon has been completely submerged. Such high dynamic and wideband interference poses stringent challenges for acoustic localization systems.

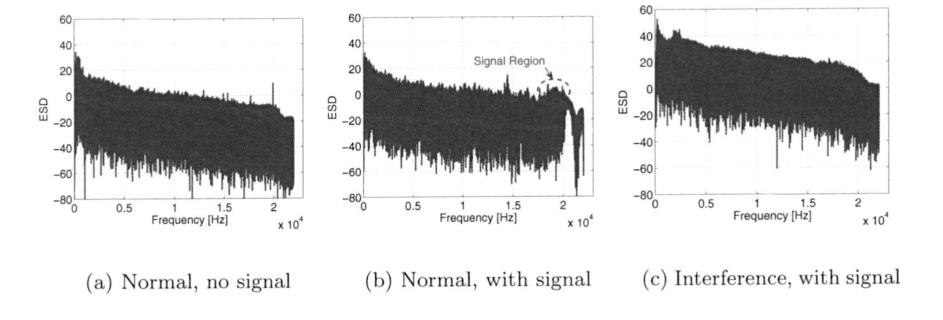


(a) Normal, no signal (b) Normal, with signal (c) Interference, with signal

**Figure 3.2: Energy spectrum density for three different cases: (a) no acoustic beacon signal, (b) with acoustic beacon, and (c) with sound interference and acoustic beacon.**

### 3.3.4 Realtime Filtering and Wavelet Denoising

To filter out out-band interference and noise, a frame-based finite impulse response (FIR) filter is used for its low complexity and high stability. To minimize spectrum leakage, we utilize an overlap-and-add approach for calibrating specific boundary results.

Filtering before detecting the beacon signal under noisy background is necessary, but not a perfect solution. Fig. 3.2(c) shows that the acoustic signal is still submerged by interference and noise; even out-band frequency components can be perfectly filtered out.

Wavelet approaches that utilize the localizing and concentrating properties of the wavelet transform allow effective filtering of noise from a signal. For example, some signal's energy could concentrate in a small number of wavelet coefficients, allowing denoise by thresholding wavelet coefficient. The rationale is that our generated beacon signal is local in time-domain and also local in the frequency-domain. Wavelets, concentrated in time, offer an ideal means to process signal that is localized in both time and frequency, whereas Fourier transform is localized only in frequency. Using the combined time and frequency locality, wavelet could provide better anti-noising and anti-interference performance.

The assumption used in wavelet transform need the whole batch of signal. Processing signal in a batch-based form is incompatible with the realtime feature of a localization and tracking system. Performing wavelet for small segments of signal and using overlap-and-add as in frame-based FIR is impossible. Unlike the linear process of FIR, wavelet performs non-linear processing, directly using overlapping segments of signal after wavelet transform introduces significant errors.

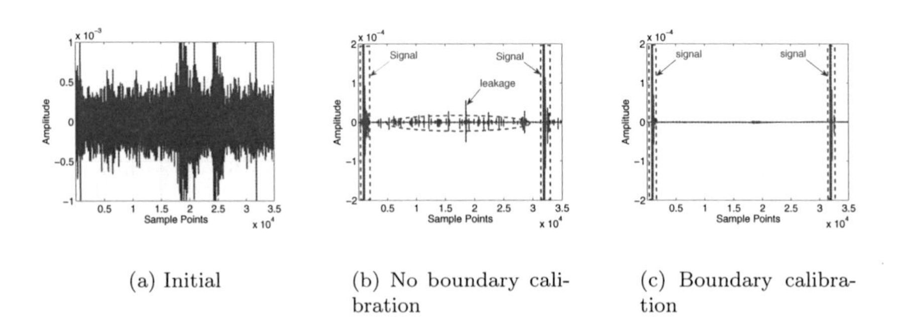


(a) Initial

(b) No boundary calibration

(c) Boundary calibration

**Figure 3.3: Time-domain waveform: (a) system input, (b) after normal FIR and wavelet processing, (c) after FIR and wavelet processing with boundary calibration.**

We adopt border extension and calibration for each frame as proposed in [90]. Fig. 3.3(a) shows the initial received acoustic signal in a normal indoor environment. After performing the frame-based FIR and wavelet denoising, the signal could be easily captured as shown in Fig. 3.3(b) and Fig. 3.3(c) with significant improvement in signal to noise ratio (SNR). As shown in Fig. 3.3(b), the spectrum leakage could cause small pulses and may result in high false detection rate. To avoid the spectrum leakage, we further propose frame-based realtime processing with the boundary calibration. The enhanced result is shown in Fig. 3.3(c).

# 3.4 Acoustic Receiver Design

## 3.4.1 Acoustic Receiver Signal Modeling

After propagation through the free space and distortion, the acoustic beacon signal captured by the microphone is converted to the electrical domain. The acoustic signal received can be represented as

$$\begin{aligned} r(t) =& \sqrt{\varepsilon} \sum_{j=0}^{N_p-1} \sum_{i=0}^{N_s-1} \sum_{l=0}^{\xi_{i,j}-1} A_{i,j}^l \cdot g_{i,j}(t - \tau_{i,j}^l - jT_p - iT_s) \\ &+ I_{i,j}(t) + n_{i,j}(t) \end{aligned} \tag{3.3}$$

where $\xi_{i,j}$ is the total number of propagation paths, with $A_{i,j}^l$ and $\tau_{i,j}^l$ as the amplitude and delay of the $l$-th path in the $i$th symbol of $j$th period, respectively. The term $n_{i,j}(t)$ is independent white Gaussian noise process in the $i$th symbol of $j$th period with spectral density of $N_0/2$; $I_{i,j}(t)$ is the frequency interference caused by the environmental sound noise.

After being received, the beacon signal of (3.3), $r(t)$ will be digitalized as $r(k)$ with the sampling frequency of $F_s$, and $k \in N_k$, where $N_k$ denotes the sampling points in every symbol, and can be written as $N_k = T_s \times F_s$. The digital version of the transmitted signal is $g_{i,j}(k)$, while $k_{i,j}^l = \tau_{i,j}^l \times F_s$ represents the digital version of the multi-path delay. $N_b = T_p \times F_s$ means the number of sampling points in one beacon period. $n_{i,j}(k)$ is the digital representation of Gaussian noise $n_{i,j}(t)$.

## 3.4.2 Hypothesis Test for Symbol Synchronization

The symbol synchronization is a prerequisite for ranging with the duty of detecting the signal start region. To better illustrate the symbol synchronization theoretically, we can rewrite the model $r_{i,j}(k)$ as $r_j(k) = \sqrt{\varepsilon}A_j^0 \cdot g_j(k - k_j^\tau) + n_j(k), i = 0$ in the $j$-th beacon period estimation, where $k_j^\tau$ is the TOA sampling point. For notation convenience, we define vector $\mathbf{r}_{j,k} = [r_j(k) \cdots]$, $\mathbf{s}_{j,k} = [g_j(k - k_j^\tau) \cdots]$, $\mathbf{n}_{j,k} = [n_j(k) \cdots]$. Assuming the length of signal under observation is $M_o$, the signal length equals the beacon duration, i.e.,

$M_p = N_k N_s$. The two conditions of the hypothesis for detecting the signal are

$$\begin{cases} H_0: \ \mathbf{r}_{j,k} = \mathbf{n}_{j,k} \quad k = 1 \cdots M_o \\ H_1: \begin{cases} \mathbf{r}_{j,k} = \sqrt{\varepsilon}\mathbf{s}_{j,k} + \mathbf{n}_{j,k} \quad k = k_j^{\mathcal{T}} \cdots k_j^{\mathcal{T}} + M_p - 1 \\ \mathbf{r}_{j,k} = \mathbf{n}_{j,k} \quad k = 1 \cdots k_j^{\mathcal{T}} - 1, k_j^{\mathcal{T}} + M_p \cdots M_o \end{cases} \end{cases} \tag{3.4}$$

where the signal is only presented with length of $M_p$ in $H_1$ condition with its starting point as $k_j^{\mathcal{T}}$. The symbol synchronization process is to detect the signal region in the noise background, i.e., detect $H_1$ condition out of $H_0$, while the TOA estimation is to detect the first path of signal and its delay $k_j^{\mathcal{T}}$.

To detect the signal region ($H_1$ condition), the receiver structure can be obtained by generalized likelihood ratio test (GLRT) as

$$\Lambda(\mathbf{r}_{j,k}) = p(\mathbf{r}_{j,k}|H_1, k_j^{\mathcal{T}})/p(\mathbf{r}_{j,k}|H_0) \tag{3.5}$$

where $p(\mathbf{r}_{j,k}|H_1, k_j^{\mathcal{T}})$ is the likelihood function of the received data vector conditioned on the maximum likelihood (ML) estimate of the TOA sample $k_j^{\mathcal{T}}$. The decision variable can be written as

$$\mathbf{z}_{j,i} = \arg\max_k \left| \frac{1}{M_s} \sum_{k=iN_k}^{iN_k+M_s} (\mathbf{r}_{j,k} \cdot \mathbf{t}_{j,k}) \right|, k \in \{1, M_o\} \tag{3.6}$$

where the $|\cdot|$ operation in (3.6) is due to the unknown sign of the received signal, $M_s$ is the length of signal under detection. $i = \lfloor k/N_k \rfloor$; and $\mathbf{t}_{j,k}$ is the template of the received signal term. (3.6) is performed for every $N_k$ points to speed up the detection. $\mathbf{t}_{j,k} = \mathbf{s}_{j,k}$ when the signal term $\mathbf{s}_{j,k}$ is known. If $\mathbf{s}_{j,k}$ is unknown, $\mathbf{t}_{j,k}$ can be simplified as a rectangular signal for non-coherent detection. The value of $M_s$ can be chosen as $M_s = (1/f_0) \times F_s$, i.e., covering the whole period for the highest frequency component. Denoting the detection threshold as $\eta_{syn}$, the symbol synchronization process can be written as

$$\hat{i} = \arg\min_i \{(\mathbf{z}_{j,i} > \eta_{syn}) \& (\mathbf{z}_{j,i+N_s-1} > \eta_{syn})\} \tag{3.7}$$

where the first term $(\mathbf{z}_{j,i} > \eta_{syn})$ indicates that the first symbol of a beacon signal is detected; the length of a beacon period is $N_s$; & is the "and" operation. If and only if the first symbol and the length of the beacon period is detected and validated, the symbol synchronization can be asserted.

### 3.4.3 Symbol Synchronization

The first path of the multipath signal $(\tau_{i,j}^l, i = 0, l = 0)$ in one beacon is often called the TOA path, which can be used to characterize the line-of-sight distance [31]. To lower the overall system complexity and exploit the similarity feature of the symbol synchronization (SS) and TOA, we use the

Neyman Pearson (NP) criterion in SS to detect the signal region with a fixed false-alarm rate. We then use the result of SS to improve the reliability and accuracy of the TOA estimation, and perform TC-based TOA estimation by maximizing our derived right detection probability.

To detect the $i$-th symbol that the beacon signal starts, we choose to extract continuous $M_s$ points in every $N_k$ interval to speed up the detection; $N_k$ interval equals to the symbol rate $T_s$. Due to the bandpass properties of the received signal, we need sufficient length of samples to cover the whole period of the high frequency signal, e.g., choosing $M_s$ to cover the whole period for the highest frequency component that $M_s = \lceil (1/f_0) \times F_s \rceil$. The decision process for symbol detection can be expressed as

$$\hat{i}_s = \min_i \left( \frac{1}{M_s} \sum_{k=iN_k}^{(iN_k+M_s)} |r_{i,j}(k)| > \hat{\eta}_{syn} \right) \tag{3.8}$$

where $\min(\cdot)$ is the function that selects the first $i$ that the decision vector crossing the threshold; $\hat{\eta}_{syn}$ is the threshold for synchronization; $|\cdot|$ is the absolute function to extract the amplitude information. For simplicity, we define decision vector as $z_i = \frac{1}{M_s} \sum_{k=iN_k}^{(iN_k+M_s)} |r_{i,j}(k)|$.

An important parameter involved in (3.8) is the synchronization threshold $\hat{\eta}_{syn}$. To determine this parameter, hypothesis tests can be used to minimize error detection probability. The process of (3.8) is to detect the signal from the noise component $n_{i,j}(k)$ with variance of $\sigma^2$. Since each individual sample $r_{i,j}(k)$ is a Gaussian random variable, the first moment of $r_{i,j}(k)$ when signal is present ($H_1$ condition) can be written as $\mathbf{E}(r_{i,j}(k); H_1) = \sqrt{\varepsilon}\mathbf{E}_l(A^l_{i,j}) \triangleq \sqrt{\varepsilon}\hat{A} = \mu$, $\mathbf{Var}(r_{i,j}(k); H_1) = \sigma^2$, where $\hat{A}$ is the estimated mean value of the multi-path amplitude $(A^l_{i,j})$. In the noise region where signal is not present ($H_0$ condition), the statistical parameters of $r_{i,j}(k)$ are $\mathbf{E}(r_{i,j}(k); H_0) = 0$, $\mathbf{Var}(r_{i,j}(k); H_0) = \sigma^2$. The decision vector $z_i$ is the mean absolute value of $|r_{i,j}(k)|$, and has a folded normal distribution. The probability density function (PDF) of $z_i$ is given by

$$\begin{aligned} f(z_i; \hat{A}, \sigma) = & \frac{1}{\sqrt{2\pi\sigma^2}} \exp\left( -\frac{(-z_i - \sqrt{\varepsilon}\hat{A})^2}{2\sigma^2} \right) \\ & + \frac{1}{\sqrt{2\pi\sigma^2}} \exp\left( -\frac{(z_i - \sqrt{\varepsilon}\hat{A})^2}{2\sigma^2} \right) \end{aligned} \tag{3.9}$$

where $z_i \geq 0$. The moment can be simplified under $H_0$ condition that $\hat{A} = 0$, as $\mathbf{E}(z_i) = \sqrt{2/\pi}\sigma$, $\mathbf{Var}(z_i) = (\pi - 2)/\pi\sigma^2$.

After performing the absolute function $|\cdot|$ in (3.8), the probability that the sample crosses the threshold $\hat{\eta}_{syn}$ in the noise region ($H_0$ condition) can be shown as

$$P_{fa}(|r_{i,j}(k)|) = \Pr(|r_{i,j}(k)| > \hat{\eta}_{syn}; H_0) = 2Q\left(\hat{\eta}_{syn}/\sigma\right) \tag{3.10}$$

where $\sigma$ is an unknown parameter; $Q(x)$ is the Gaussian Q function. $P_{fa}(|r_{i,j}(k)|)$ is the false alarm rate that detected the wrong sampling point. To estimate $\sigma$, one method is directly using $\hat{\sigma} = \sqrt{\mathbf{E}(r_{i,j}(k))^2}$ in the noise region. However, using the first moment of the folded normal distribution can lead to a more simplified method as $\hat{\sigma} = \sqrt{\pi/2}\mathbf{E}(z_i; H_0)$, where it does not need to perform complex $(\cdot)^2$ and $\sqrt{(\cdot)}$ in estimation and is really suitable for real-time situations.

With the unknown a prior information, using the Neyman Pearson (NP) criterion can achieve optimal performance by maintaining a constant false-alarm rate $\gamma_{fa}$. Thus, the detection threshold ($\hat{\eta}_{syn}$) can be determined by (3.10) as

$$\hat{\eta}_{syn} = \sqrt{\pi/2}\mathbf{E}(z_i; H_0)Q^{-1}(P_{fa}(|r_{i,j}(k)|)/2) \tag{3.11}$$

where $P_{fa}(|r_{i,j}(k)|)$ is the constant false alarm rate that should be pre-set according to application demands; high requirement on $P_{fa}(|r_{i,j}(k)|)$ is often achieved at the sacrifice of detection probability. In the first step detection, $P_{fa}(|r_{i,j}(k)|)$ should be set slightly higher to ensure no information loss, while the second step in TOA estimation can help keep the overall $P_{fa}(|r_{i,j}(k)|)$ at a lower level.

With the threshold ($\hat{\eta}_{syn}$) available, the probability that the correct signal is detected in the signal region ($H_1$ condition) is given by

$$\begin{aligned} P_d(|r_{i,0}(k)|) &= \Pr(|r_{i,j}(k)| > \hat{\eta}_{syn}; H_1) \\ &= Q\left(\frac{(\hat{\eta}_{syn} - \sqrt{\varepsilon}\hat{A})}{\hat{\sigma}}\right) + Q\left(\frac{(\hat{\eta}_{syn} + \sqrt{\varepsilon}\hat{A})}{\hat{\sigma}}\right) \end{aligned} \tag{3.12}$$

where $\sqrt{\varepsilon}\hat{A}$ is the signal parameter that can be estimated in the signal region as $\sqrt{\varepsilon}\hat{A} = \mathbf{E}(r_{i,j}(k); H_1)$. By using the detection threshold (3.11) with a fixed false alarm rate in (3.8), the signal start region ($\hat{i}_s$) can be detected at the probability of (3.12) and the communication synchronization can be asserted.

## 3.5 TOA Estimation

### *3.5.1 Spectrum Matching*

To eliminate the large error that misdetected the interference as signal during the symbol synchronization process, distinct features of the signal and the interference should be extracted. However, the acoustic waveform is directly generated by low-complexity anchor nodes; sophisticated waveform design or pulse compression approaches are not applicable.

The common way to differentiate two signals is to perform matching by utilizing their time-domain or frequency-domain feature. However, the remaining interferences after the preprocessing could be very similar to the normal acoustic signal. The reason is that the interferences were sharpened similar to the beacon by the FIR filter. It is also the reason that in-band interference is hard to mitigate.

Our solution is to calculate the energy spectral density (ESD) of the square root of the acoustic signal power. By comparing the feature of the ESD of received signal to the pre-stored feature of the beacon, interferences with different ESD features could be mitigated. The rationale is that the spectral differences would be enlarged when performing non-linear transform. The calculation of the square root of the signal power is a kind of nonlinear process by which the spectrum of the initial signal is changed. If the initial spectrum of the interference looks similar to the signal and is hard to differentiate, the transformed spectrum has a decreased probability of similarity.

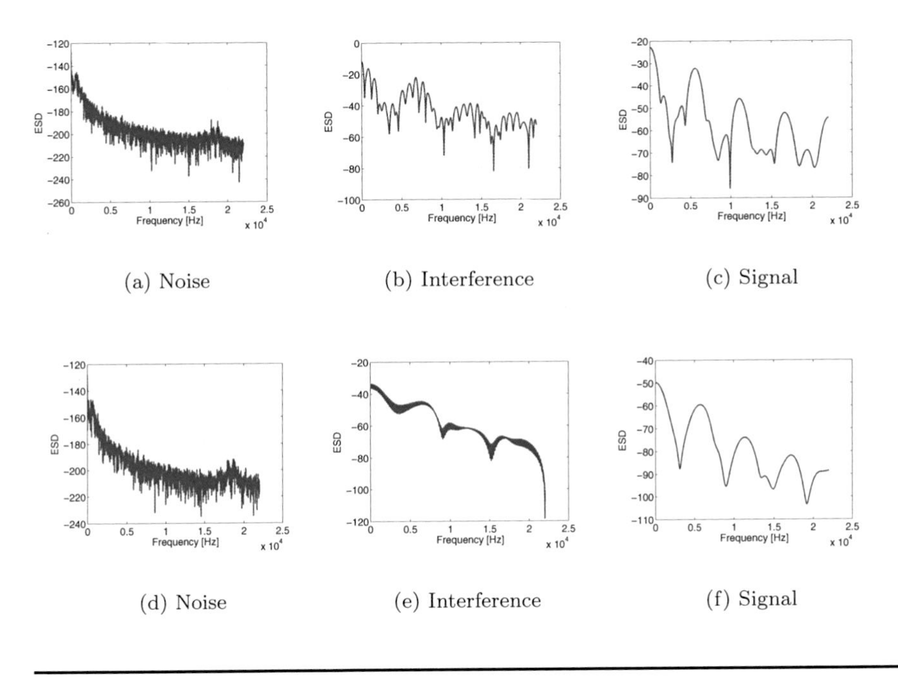


(a) Noise (b) Interference (c) Signal

(d) Noise (e) Interference (f) Signal

**Figure 3.4: The energy spectral density for the cases of (a) pure noises, (b) interference, and (c) signal.**

Fig. 3.4 lists the examples of the energy spectrum density (ESD) for three kinds of signals: noise, interference, and beacon signal. From Fig. 3.4, we know that the ESD of the signal features three peaks in the center region that differs the noise and interference. Detecting the three peaks (local extrema) by adjusting the kurtosis, slope threshold, and amplitude threshold can be an effective noise and interference mitigation approach. The solution that we propose here is not perfect for solving the problem of in-band interference mitigation, but works surprisingly well in real situations with carefully tuned parameters of thresholds.

### 3.5.2 Cluster Detection

When considering interference, one symbol duration could contain multiple pulses and at most one of them is the real signal. If the wrong pulse is detected as the signal, significant ranging error would be introduced. To address this problem, we utilize the high timing resolution property of the low-duty-cycle acoustic beacon, i.e., the interference and the signal are separable in the time-domain for most of the cases. First, detect the existence of multiple pulses; second, cluster the signal regions by the detection result; and finally, perform the spectrum matching for each cluster. In this way, individual pulses could be examined and interference mitigated, making the detection probability of the signal significantly improved.

Fig. 3.5(a) shows an example of two pulses in one frame duration. Initially, no distinct feature exists in the time-domain, and it is hard for the TOA estimation module to determine which one is the true signal.

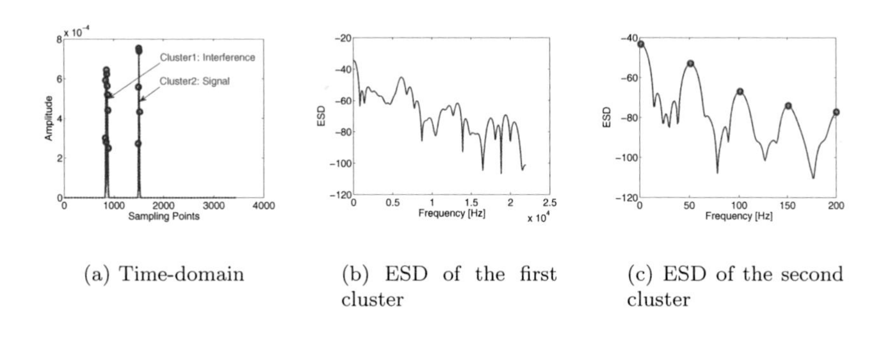


(a) Time-domain

(b) ESD of the first cluster

(c) ESD of the second cluster

**Figure 3.5: (a) Initial time-domain waveform of two clusters, (b) ESD of the first cluster, and (c) ESD of the second cluster.**

We propose to perform feature extraction, then cluster feature values detected in the frame using the $K$-nearest approach, where $K$ could be chosen as the length of transmitted pulse duration. In Fig. 3.5(a), two clusters without overlap are obtained. In the next step, spectral clustering could be utilized for each cluster. Fig. 3.5(b) and Fig. 3.5(c) are the ESD features for the two clusters during the spectrum matching process. By matching ESD, the second cluster is identified as the true TOA path.

### 3.5.3 TOA Estimation Procedure

When $\hat{i}_s$ is obtained in (3.8), we perform precise detection for the TOA path in the symbol region near $\hat{i}_s$. Jump-back and search-forward (JBSF) [49] is a method that is suitable for precise TOA searching after coarse detection. We define $z_k = |r_{i_s,j}(i_s N_k - J_b, \ldots, i_s N_k + M_s)|$ as the subtracted decision sequence for TOA estimation, and $J_b$ is the number of sampling points that jumped back in JBSF; the length of $z_k$ is $M_k = J_b + M_s + 1$. The decision process in this part can be written as

$$\hat{\tau}_j^{TOA} = \begin{cases} T_s \cdot [\min(k|z_k > \hat{\eta}_{TOA}) + \hat{i}_s N_k - J_b] - \frac{1}{2}T_s & k < M_k \\ \text{Re-estimate } \hat{i}_s & k \geq M_k \end{cases} \quad (3.13)$$

where $\hat{\eta}_{TOA}$ is the precise TOA estimation threshold; $\hat{i}_s$ obtained in the previous detection stage indicates the symbol period; $J_b$ can be set the same as the step size in the first step detection; i.e., $J_b = N_k$. The second part of (3.13) means that we need to re-estimate the signal region when the decision process cannot obtain a TOA value in $z_k$ and $k \geq M_k$, that is, no sampling point crossed the threshold. This problem is often due to the false detection in symbol synchronization; a suitable way to solve this problem is to re-estimate $\hat{i}_s$. From such a process, we know that the overall false alarm rate in SS (3.8) can be lowered in TOA estimation.

### 3.5.4 Maximizing the Right Detection Probability

In (3.13), the TOA threshold $\hat{\eta}_{TOA}$ should be determined before the decision process. The decision vector $z_k$ also follows the folded normal distribution in (3.9) and with the same statistical parameters. Then the probability that sample $z_k$ crosses the threshold in the noise and signal region can be calculated as in (3.10) and (3.12); shown as $P_{fa}(z_k)$ and $P_d(z_k)$, respectively. The probability that detected the right TOA path can be denoted as $P_{rt}$, the two kinds of TOA estimation errors are the early detection $P_{ed}$ and late detection $P_{ld}$. $P_{ed}$ is caused by selecting the incorrect crossing samples earlier than the true TOA path due to noise interference; $P_{ld}$ is the probability is that the TOA estimator missed the true TOA path, and detected a wrong sample in the signal region later due to channel fading. Assume the true TOA path $k_j^{TOA}$ is uniformly distributed in the region of the estimation area $(i_s N_k - J_b, \ldots, i_s N_k + M_s)$ with total length of $M_k = J_b + M_s + 1$. The probability that the true TOA path in any sampling point is $p_{toa} = 1/M_k$. The early detection probability that detected one point in the noise region is $P_{ed}(k_j^{TOA}) = 1-(1-P_{fa}(z_k))^{k_j^{TOA}}$, which means at least one sample crossed the threshold before the TOA path. The late detection $P_{ld}(k_j^{TOA}) = (1 - P_d(z_k))(1 - P_{fa}(z_k))^{k_j^{TOA}}$ shows the TOA point is missed. The right detection probability is $P_{rt}(k_j^{TOA}) = P_d(z_k)(1 - P_{fa}(z_k))^{k_j^{TOA}}$, where $\sum (P_{ed}(k_j^{TOA}) + P_{rt}(k_j^{TOA}) + P_{fa}(k_j^{TOA})) = 1$. Since $k_j^{TOA}$ is uniformly distributed with probability $p_{toa}$, the expectation for the error estimation probability over $k_j^{TOA}$ is

$$\begin{aligned} P_{err}^{TOA} &= \mathbf{E}_{k_j^{TOA}} \left( P_{ed}(k_j^{TOA}) + P_{fa}(k_j^{TOA}) ) \right) \\ &= \sum_{k_j^{TOA}=0}^{M_k} \left( \frac{1}{M_k} [1 - (1 - P_{fa}(z_k))^{k_j^{TOA}}] \right) \\ &+ \sum_{k_j^{TOA}=0}^{M_k} \left( \frac{1}{M_k} [(1 - P_d(z_k))(1 - P_{fa}(z_k))^{k_j^{TOA}}] \right) \\ &= 1 - \frac{P_d(z_k)}{M_k P_{fa}(z_k)} [1 - (1 - P_{fa}(z_k))_k^M] \end{aligned} \tag{3.14}$$

where the last part of (3.14) is the correct detection probability $P_{rt}^{TOA} = P_d(z_k)[1 - (1 - P_{fa}(z_k))_k^M]/M_k P_{fa}(z_k)$. Performing Taylor expansion of $(1 - P_{fa}(z_k))_k^M$, we have $(1 - P_{fa}(z_k))_k^M \approx 1 - M_k P_{fa}(z_k) + C_{M_k}^2 (P_{fa}(z_k))^2$. Then $P_{rt}^{TOA}$ can be simplified as

$$\begin{aligned} P_{rt}^{TOA} &\approx \frac{P_d(z_k)}{M_k P_{fa}(z_k)} (M_k P_{fa}(z_k) - C_{M_k}^2 (P_{fa}(z_k))^2) \\ &= P_d(z_k)[1 - \frac{1}{2}(M_k - 1) P_{fa}(z_k)] \end{aligned} \tag{3.15}$$

From (3.15), we know that increasing $P_d(z_k)$ or decreasing $P_{fa}(z_k)$ of the single sampling point can contribute a better $P_{rt}^{TOA}$. Small $M_k$ also helps to improve the performance in that less sampling points are detected.

To achieve a better TOA estimation scheme, selecting an appropriate TOA threshold by maximizing $P_{rt}^{TOA}$ provides a feasible way. Since $P_d(z_k)$ and $P_{fa}(z_k)$ is a function of $\hat{\eta}_{TOA}$, $P_{rt}^{TOA}$ can be written as $P_{rt}^{TOA}(\hat{\eta}_{TOA})$. The maximum value of $P_{rt}^{TOA}$ can be achieved when $\hat{\eta}_{TOA} = \hat{\eta}_{TOA}^{oth}$, as

$$\hat{\eta}_{TOA}^{oth} = \arg \max_{\hat{\eta}_{TOA}} (P_{rt}^{TOA}(\hat{\eta}_{TOA})) \tag{3.16}$$

For simplification, we can use the Maclaurin series of the error function $\text{erf}(x) = \frac{2}{\sqrt{\pi}} \sum_{n=0}^{\infty} [(-1)^n x^{2n+1}]/[n!(2n+1)]$ to express the approximate function of $Q(x)$ as

$$\begin{aligned} Q(x) &\approx \frac{1}{2} - \frac{1}{2} \text{erf}(x/\sqrt{2}) \\ &= \frac{1}{2} - \frac{1}{\sqrt{\pi}} (x/\sqrt{2} - \frac{(x/\sqrt{2})^3}{3} + \ldots) \end{aligned} \tag{3.17}$$

when $x$ is very small; even the first-order term of (3.17) can well represent $Q(x)$ as $Q(x) \approx 1/2 - x/\sqrt{2\pi}$. Let's define $z = \hat{\eta}_{syn}/\hat{\sigma}$ and $s = \sqrt{\varepsilon}\hat{A}/\hat{\sigma}$.

If you only consider $s > 0$, then $P_d(z_k)$ and $P_{fa}(z_k)$ can be simplified as $P_d(z_k) \approx 1/2 - (z-s)/\sqrt{2\pi}$ and $P_{fa}(z_k) \approx 1 - 2z/\sqrt{2\pi}$.

By substituting $P_d(z_k)$ and $P_{fa}(z_k)$ into (3.15), and defining $w = \frac{1}{2}(M_k - 1)$, $P_{rt}^{TOA}$ can be written as

$$\begin{aligned} P_{rt}^{TOA}(z) &= [1/2 - (z-s)/\sqrt{2\pi}][1 - w(1 - 2z/\sqrt{2\pi})] \\ &= -\frac{w}{\pi}z^2 + (\frac{2w-1}{2\sqrt{2\pi}} + \frac{ws}{\pi})z + \frac{1-w}{\sqrt{2\pi}}s + \frac{1-w}{2} \end{aligned} \tag{3.18}$$

where (3.18) is a convex function with its maximum achieved when $z = \frac{s}{2} + \frac{\sqrt{\pi}}{2\sqrt{2}}(1 - \frac{1}{2w})$. Then, $\hat{\eta}_{TOA}^{oth}$ can be written as

$$\hat{\eta}_{TOA}^{oth} = \frac{1}{2}\sqrt{\varepsilon}\hat{A} + \frac{\sqrt{\pi}}{2\sqrt{2}}(1 - \frac{1}{M_k - 1})\hat{\sigma} \tag{3.19}$$

where $\sqrt{\varepsilon}\hat{A}$ and $\hat{\sigma}$ can be estimated the same as in (3.11). By using (5.10) in TOA estimation of (3.13), the optimized TOA performance can be achieved under the criterion of maximum right detection probability as shown in (3.16).

## 3.6 Acoustic Symbol Demodulation

### 3.6.1 Challenges of Audible-Band Communication

For frequency-modulated signal, we can model $g_{i,j}(k) = \sqrt{\varepsilon}\cos(2\pi f_d k + \phi), d = 0, 1$, where $\phi$ is the unknown phase information between the local template and received carrier wave, $f_d = f_1$ represents the symbol "1"; $f_d = f_0$ represents the symbol "0." Using an audible band microphone receiver poses two additional challenges for communication decisions. The first challenge is hardware limitation. Conventional receivers often perform orthogonal down-conversion to the baseband, and such a process can estimate both the frequency and phase information; i.e., jointly using sin and cos to estimate the $\phi$. The reason that we cannot use orthogonal demodulation is due to the limitation of our hardware structure. In an audio system, the acoustic signal captured by a microphone is directly sampled by the analog digital converter (ADC) at the frequency of $F_s$. When the signal is sampled to the digital domain, there is no orthogonal down-conversion process for the typical processing unit. Such a process is quite simple since the microphone receiver is designed for normal voice and music, which is already in the baseband. With no down-conversion and frequency discriminator circuit, the conventional FSK demodulation method is not suitable. For such a reason, we need to design a new communication decision scheme to derive the information bits carried by the acoustic signal. The second challenge is caused by the poor channel condition of audible band acoustic signal compared with microwave and ultrasound signal. Assuming the channel coherent bandwidth is $B_{co}$, the system bandwidth $B = 1.48\text{KHz}$ is $B \ll B_{co}$ for microwave signal. Such a channel condition can be assumed as flat for $f_1$ and $f_0$. However, 1.48KHz should be viewed as wideband for acoustic signal, i.e., $B > B_{co}$. Thus the channel-fading effect should be considered for better demodulation performance.

### *3.6.2 Communication Demodulation*

Assume the received multi-path frequency-modulated signal is $r_{i,j}(k) = \sum_{l=0}^{\xi_{i,j}-1} \sqrt{\varepsilon}\cos(2\pi f_b k + \phi + \varphi_l(f_b)), b = 0, 1$; $\varphi_l(f_b)$ is a function of $f_b$ due to the frequency selective channel where $f_0$ and $f_1$ may suffer different fading; $f_0$ and $f_1$ represent the symbol "0" and "1." Construct the local correlation template $v_0(k) = \cos(2\pi f_0 k)$ and $v_1(k) = \cos(2\pi f_1 k)$ to demodulate the symbol "0" and "1." Then, use the constructed local templates to perform correlation and information extraction. For symbol representation, the correlation process can be shown as

$$\begin{aligned} a_{i,j}^0 &= \mathbb{E}_k\{g_{i,j}(k) \cdot v_0(k)\} + w_{i,j}^0, \\ a_{i,j}^1 &= \mathbb{E}_k\{g_{i,j}(k) \cdot v_1(k)\} + w_{i,j}^1 \end{aligned} \tag{3.20}$$

where $w_{i,j}^b = \mathbb{E}\{n_{i,j}(k) \cdot v_b(k)\}$; $b = 0, 1$; $k \in [k_j^{toa} + iN_k, k_j^{toa} + (i+1)N_k]$, $i = 0, \ldots, N_s - 1$. Define $\mathbf{s}_F^b = [\mathbb{E}_k\{g_{i,j}(k) \cdot v_0(k)\}, \mathbb{E}_k\{g_{i,j}(k) \cdot v_1(k)\}]^T$, $\mathbf{a} = [a_{i,j}^0, a_{i,j}^1]^T$, and $\mathbf{w} = [w_{i,j}^0, w_{i,j}^1]$. The process of $\mathbb{E}(\cdot)$ performs filtering and averaging. Use symbols "0," for example. When the signal is presented with two multi-paths ($\xi_{i,j} = 2$), frequency demodulation of $\mathbf{s}_F^0$ can be rewritten as

$$\begin{aligned} &\mathbf{s}_F^0 \\ &= \sup_{\phi} \frac{1}{\lceil F_s/B \rceil} \sum_{k=1}^{\lceil F_s/B \rceil} \sum_{l=0}^{\xi_{i,j}-1} \sqrt{\varepsilon}\cos(2\pi\Delta f \cdot k + \phi + \varphi_l(f_b)) \\ &= \sup_{\phi} \operatorname*{mean}_{k}[2\sqrt{\varepsilon}\cos(2\pi\Delta f \cdot k + \phi + \varphi_+(f_b))\cos(\varphi_-(f_b))] \end{aligned} \tag{3.21}$$

where $\varphi_+(f_b) = (\varphi_0(f_b) + \varphi_1(f_b))/2$, $\varphi_-(f_b) = (\varphi_0(f_b) - \varphi_1(f_b))/2$ for $l = 0, 1$; $\Delta f = f_b - f_0$; the bandwidth is $B = (f_0 - f_1)$. $\sup_\phi$ calculates the super-bound of the function with the parameter of $\phi$, i.e., the envelope. When symbol "0" is transmitted ($f_b = f_0, \Delta f = 0$), we have

$$\begin{aligned} \mathbf{s}_F^0(0) &= \sup_{\phi} 2\sqrt{\varepsilon}\cos(2\pi k 0 + \phi + \varphi_+(f_b))\cos(\varphi_-(f_b)) \\ &= 2\sqrt{\varepsilon}\cos(\varphi_-(f_b)) \end{aligned} \tag{3.22}$$

When the opposite symbol is transmitted ($f_b = f_1, \Delta f = B$), (3.21) can be shown as

$$\mathbf{s}_F^1(0) = \sup_{\phi} 2\sqrt{\varepsilon}\cos(2\pi k B + \phi + \varphi_+(f_b))\cos(\varphi_-(f_b)) \tag{3.23}$$

From (3.22) and (3.23), we know that decision vector $\mathbf{s}_F^0 = [2\sqrt{\varepsilon}\cos(\varphi_-(f_0)), 0]^T$ when symbol "0" is transmitted; $\mathbf{s}_F^1 = [0, 2\sqrt{\varepsilon}\cos(\varphi_-(f_1))]^T$ when symbol "1" is transmitted. With the equal prior probability $P(b = 0, 1) = 1/2$, the maximum-likelihood (ML) is the optimal decision rule, resulting in

$$\hat{d}^{ml} = sgn[log\frac{P(\mathbf{a}|b=1)}{P(\mathbf{a}|b=0)}] = sgn[\Lambda] \tag{3.24}$$

where $P(\mathbf{a}|b = 1)$ is the likelihood function of the received data vector $\mathbf{s}_F$ conditioned on the symbol "1" transmitted; $P(\mathbf{a}|b = 0)$ is for symbol "0." $\Lambda$ is the log-likelihood ratio (LLR).

For the frequency demodulation of (3.20), we define the decision vector as $\mathbf{y}$. The decision process can be written as

$$\mathbf{y} = \mathbf{a}(1) - \mathbf{a}(0) = \mathbf{d}_\delta + \mathbf{w}_F(1) - \mathbf{w}_F(0) \gtrless_0^1 = \eta_d \tag{3.25}$$

where $\mathbf{d}_\delta = \mathbf{s}_F^1 - \mathbf{s}_F^0$. From (3.22) and (3.23), we can calculate $\mathbf{d}_\delta = [2\sqrt{\varepsilon}\cos(\varphi_-(f_1)), -2\sqrt{\varepsilon}\cos(\varphi_-(f_0))]^T$, where $\eta_d$ is the communication decision threshold. When $\mathbf{y}$ is larger than the threshold, we can declare the information bit "1" is transmitted, and vice versa.

### 3.6.3 Dynamic Demodulation with Transmit Reference

Only using FSK for demodulation may suffer significant performance loss due to the multi-path fading effect. To deal with such a problem, we should utilize additional difference in symbol "0" and "1" to facilitate demodulation. From $r_{i,j}(k)$, we know that $f_0$ and $f_1$ have the same amplitude when the channel is flat enough, i.e., $\varphi_l(f_0) = \varphi_l(f_1)$. Conventional FSK is constant-amplitude demodulation for narrow bandwidth signal, but $B = 1.48KHz$ can be considered wideband for acoustic signal. These frequency-selective phenomena do cause some problems when using FSK demodulation, but allow the signal amplitude to be different for $f_0$ and $f_1$. Utilizing the amplitude difference can help to compensate the attenuation when $\varphi_-(f_b) \approx \pi/2$. Considering the envelope of $r_{i,j}(k)$, and only calculating two multi-paths, we have

$$\begin{aligned}
\mathbf{s}_A(b) &= \sup_{\phi} \sum_{l=0}^{\xi_{i,j}-1} \sqrt{\varepsilon}\cos(2\pi f_b k + \phi + \varphi_l(f_b)) \\
&= \sup_{\phi} 2\sqrt{\varepsilon}\cos(2\pi k f_b + \phi + \varphi_+(f_b))\cos(\varphi_-(f_b)) \\
&= 2\sqrt{\varepsilon}\cos(\varphi_-(f_b))
\end{aligned} \tag{3.26}$$

Since $\varphi_-(f_1) \neq \varphi_-(f_0)$, the amplitude of symbol "1" and "0" is also different, i.e., $\mathbf{s}_A(0) \neq \mathbf{s}_A(1)$. If the FSK result for one symbol is equal to zero, we can assume its probability is $p_F = \Pr\{\varphi_-(f_0) \approx \pi/2\}$. Then, the probability that another symbol also has $\varphi_-(f_1) \approx \pi/2$ is $(p_F)^2$, and $(p_F)^2 \ll$

$p_F < 1$. For such a reason, we can assume that the $\varphi_-(f_b)$ for symbol "1" and "0" is different and can be used for detection when the FSK result of one symbol shows significant small-scale fading.

Assuming the equal prior probability $P(b = 0, 1) = 1/2$, and the ML is the optimal decision rule, has

$$\hat{d}^{ml} = sgn[log\frac{P(\mathbf{a}|b=1)}{P(\mathbf{a}|b=0)}] = sgn[\Lambda] \tag{3.27}$$

where $P(\mathbf{a}|b = 1)$ is the likelihood function of the received data vector conditioned on the symbol "1" transmitted; $P(\mathbf{a}|b = 0)$ is for symbol "0." $\Lambda$ is the log-likelihood ratio (LLR).

For the frequency demodulation of (3.34), we define the decision vector as $\mathbf{y}_F$, the decision process can be written as

$$\mathbf{y}_F = \mathbf{s}_F(1) - \mathbf{s}_F(0) = \pm\sqrt{\varepsilon} + \mathbf{w}_F(1) - \mathbf{w}_F(0) \gtrless_0^1 0 \tag{3.28}$$

when $\mathbf{y}_f$ is larger than "0," then we can declare the information bit "1" is transmitted, and vice versa.

For the amplitude detection when one symbol shows significant small-scale fading, the difference between the received signal amplitude and the reference value can be used as the decision vector, as

$$\bar{\mathbf{s}}_A = \left[|\mathbf{s}_A(b) - s_A^{ref}(0)|, |\mathbf{s}_A(b) - s_A^{ref}(1)|\right]^T \tag{3.29}$$

where $s_A^{ref}(0)$ and $s_A^{ref}(1)$ served as prior information are the known amplitude for symbol "0" and "1." These two parameters can be obtained by using transmit reference (TR), e.g., transmit two bit of "1" and "0" at the beginning of the beacon period. At the receive side, we can assume that these TR bits suffer the same attenuation as other bits in the same beacon period. Calculating the amplitude of TR bit can provide $s_A^{ref}(0)$ and $s_A^{ref}(1)$ in amplitude demodulation of (3.37). Such a TR scheme is a simplified method for channel estimation, and can be used as a fingerprint to characterize the channel effect to the received information bit. From (3.26), we can know that $s_A^{ref}(b)$ is also equal to $s_A^{ref}(b) = 2\sqrt{\varepsilon}\cos(\varphi_-(f_b))$.

By comparing the amplitude difference in (3.37), the symbol that is most likely transmitted will be declared. Substituting (3.26) into (3.37), we have $\bar{\mathbf{s}}_A = [\sqrt{\varepsilon_\delta}, 0]^T$ when the information bit $b = 1$; $\bar{\mathbf{s}}_A = [0, \sqrt{\varepsilon_\delta}]^T$ when $b = 0$, where $\sqrt{\varepsilon_\delta} = 2\sqrt{\varepsilon}|\cos(\varphi_-(f_1)) - \cos(\varphi_-(f_0))|$. We define the decision vector as $\mathbf{y}_A$, then we have the decision process as

$$\mathbf{y}_A = \bar{\mathbf{s}}_A(1) - \bar{\mathbf{s}}_A(0) = \pm\sqrt{\varepsilon_\delta} + \mathbf{w}_A(1) - \mathbf{w}_A(0) \gtrless_0^1 0 \tag{3.30}$$

where $\mathbf{w}_A$ is the noise vector when symbol "1" or "0" is transmitted.

Combining (3.28) and (3.30) together, we can obtain a joint ASK/FSK decision rule as

$$\mathbf{y}_{joint} = \frac{\mathbf{s}_F(1) - \mathbf{s}_F(0)}{\sqrt{\varepsilon}}(1-\mu) + \frac{\bar{\mathbf{s}}_A(1) - \bar{\mathbf{s}}_A(0)}{\sqrt{\varepsilon_\delta}}\mu \gtrless_0^1 0 \tag{3.31}$$

where $\mu$ is the weighting coefficient. The joint demodulation of (3.38) is only needed when one symbol suffers significant attenuation; otherwise, using the amplitude in demodulation with no envelope distinction may provide negative effects. Thus, we should set $\mu$ according to

$$\mu = \begin{cases} \frac{|s_A^{ref}(0)-s_A^{ref}(1)|}{(s_A^{ref}(0)+s_A^{ref}(1))}, & \xi \neq 0 \\ 0, & \xi \approx 0 \end{cases} \tag{3.32}$$

where $\xi = (s_A^{ref}(0) - s_{noise}) \cdot (s_A^{ref}(1) - s_{noise})$, $s_{noise} = \sqrt{2/\pi}\hat{\sigma}$ is the amplitude value of the noise and can be estimated by using $\hat{\sigma} = \sqrt{\pi/2}\mathbf{E}(z_i; H_0)$. (3.32) shows that when one of the symbols is seriously attenuated, joint demodulation should be used to prevent performance degradation; when the amplitude of symbols does not show difference, only using frequency demodulation is sufficient and should set $\mu$ in (3.38).

The decision error can be written as

$$\begin{aligned} P_{error} &= P(b=0)P(\mathbf{y} > 0|b=0) + P(d=1)P(\mathbf{y} < 0|b=1) \\ &= \frac{1}{2}\int_0^{\infty} \frac{1}{\sqrt{2\pi\sigma^2}} \exp(-(x+\sqrt{E_b})^2/(2\sigma^2))dx \\ &+ \frac{1}{2}\int_{-\infty}^{0} \frac{1}{\sqrt{2\pi\sigma^2}} \exp(-(x-\sqrt{E_b})^2/(2\sigma^2))dx \\ &= \hat{Q}(-\sqrt{E_b/N_b}) \end{aligned} \tag{3.33}$$

where the $\sigma^2 = N_0$ in (3.40) is noise variance of the decision vector.

After the symbol synchronization and TOA estimation, we can perform symbol demodulation for every $N_k$ points from $k_j^{TOA}$ to obtain the information bit. For the frequency modulated signal of $g_{i,j}(k)$, we can model $g_{i,j}(k) = \sqrt{\varepsilon}\cos(2\pi f_d k + \phi), d = 0, 1$, where $\phi$ is the fixed unknown phase information between the local template and received carrier wave, $f_d = f_1$ represents the symbol "1"; $f_d = f_0$ represents symbol "0."

One feasible way to demodulate the information is to construct the local correlation template $v_0(k) = \cos(2\pi f_0 k)$ and $v_1(k) = \cos(2\pi f_1 k)$ for symbol "0" and "1," respectively. The constructed local template is used to perform correlation and information extraction. For symbol representation, the decision vector can be shown as $\mathbf{s}_F = [\mathbb{E}_k\{g_{i,j}(k) \cdot v_0(k)\}, \mathbb{E}_k\{g_{i,j}(k) \cdot v_1(k)\}]^T$, the process of $\mathbb{E}(\cdot)$ performs filtering and averaging. Using symbol "0" for example, and assuming the bandwidth is $B = (f_0 - f_1)$, then

$$\begin{aligned} \mathbf{s}_F(0) &= \mathbb{E}_k\{\sqrt{\varepsilon}\cos(2\pi f_d k + \phi)\cos(2\pi f_0 k)\} \\ &= \sup_{\phi} \frac{1}{\lceil F_s/B \rceil} \sum_{k=1}^{\lceil F_s/B \rceil} \sqrt{\varepsilon}\cos(2\pi(f_d - f_0)k + \phi) \end{aligned} \tag{3.34}$$

where the high frequency part $f_d + f_0$ has been filtered out; $\lceil \cdot \rceil$ is to calculate the minimum integer that is larger than the input value; $\sup_\phi$ is used to calculate the super-bound of the function with the parameter of $\phi$, i.e., calculate the envelope of the input signal. When $f_d = f_0$, then (3.34) can be written as

$$\mathbf{s}_F(0) = \sup_\phi \sqrt{\varepsilon}\cos(2\pi k0 + \phi) = \sqrt{\varepsilon} \tag{3.35}$$

When $f_d \neq f_0$, the result of (3.34) is

$$\begin{aligned} \mathbf{s}_F(0) &= \sup_\phi \frac{1}{\lceil F_s/B \rceil} \sum_{k=1}^{\lceil F_s/B \rceil} \sqrt{\varepsilon}\cos(2\pi Bk + \phi) \\ &= \sup_\phi \mathop{\mathbb{E}}_{\theta\in[0\sim 2\pi]} [\cos(\theta + \phi)] \approx 0 \end{aligned} \tag{3.36}$$

where $\mathbb{E}_\theta[\cdot]$ approximates to obtain the mean value of $cos(\cdot)$ in a whole period, which is equal to zero and irrelevant to the phase $\phi$. From (3.35) and (3.36), we can know that the same local template can obtain distinct values (0 and $\sqrt{\varepsilon}$) when the input signal is different. Using these distinct values will map the input signal to the information domain. Define the decision vector for frequency demodulation as $\mathbf{s}_F$, and $\mathbf{s}_F = [\sqrt{\varepsilon}, 0]^T$ when the information bit $b = 0$; $\mathbf{s}_F = [0, \sqrt{\varepsilon}]^T$ when $b = 1$, $\varepsilon$ is the transmitted bit energy.

Only using frequency for demodulation may suffer significant performance loss due to the multi-path fading effect. For microwave signal, the channel can be assumed as flat for several KHz, but not for acoustic signal. Frequency $f_1$ and $f_0$ may suffer different attenuation; such a channel fading effect should be considered for better demodulation performance. To deal with such problems, the amplitude difference of symbols "1" and "0" can help to compensate the performance loss when we perform joint frequency and amplitude detection.

For the amplitude detection when one symbol shows significant small-scale fading, the difference between the received signal amplitude and the reference value can be used as the decision vector, as

$$\bar{\mathbf{s}}_A = \left[ |\mathbf{s}_A(b) - s_A^{ref}(0)|, |\mathbf{s}_A(b) - s_A^{ref}(1)| \right]^T \tag{3.37}$$

where $s_A^{ref}(0)$ and $s_A^{ref}(1)$ are the prior information of amplitude for symbol "0" and "1," $\mathbf{s}_A(b)$ is the amplitude of the received signal. The parameters of $s_A^{ref}(0)$ and $s_A^{ref}(1)$ can be obtained by using transmit reference (TR), e.g., transmit two bit of "1" and "0" at the beginning of the beacon period. At the receive side, we can assume that these TR bits suffer the same attenuation as other bits in the same beacon period. Calculating the amplitude of TR bit provides $s_A^{ref}(0)$ and $s_A^{ref}(1)$ in amplitude demodulation of (3.37). Such a TR scheme is a simplified method for channel estimation, and is used as fingerprint to characterize the channel effect. For the decision vector, and $\bar{\mathbf{s}}_A = [\sqrt{\varepsilon_\delta}, 0]^T$ when the information bit $b = 0$; $\bar{\mathbf{s}}_A = [0, \sqrt{\varepsilon_\delta}]^T$ when $b = 1$, $\varepsilon_\delta$ is the energy difference.

We can obtain the dynamic decision rule as

$$\mathbf{y}_{joint} = \frac{\mathbf{s}_F(1) - \mathbf{s}_F(0)}{\sqrt{\varepsilon}}(1-\mu) + \frac{\overline{\mathbf{s}}_A(1) - \overline{\mathbf{s}}_A(0)}{\sqrt{\varepsilon_\delta}}\mu \gtrless_0^1 0 \tag{3.38}$$

where $\mu$ is the weighting coefficient. The joint demodulation of (3.38) is only needed when one symbol suffers significant attenuation; otherwise, using the amplitude in demodulation with no envelope distinction may provide negative effects. Thus, we set $\mu$ according to

$$\mu = \begin{cases} \frac{\left|s_A^{ref}(0) - s_A^{ref}(1)\right|}{(s_A^{ref}(0) + s_A^{ref}(1))}, & \xi \neq 0 \\ 0, & \xi \approx 0 \end{cases} \tag{3.39}$$

where $\xi = (s_A^{ref}(0) - s_{noise}) \cdot (s_A^{ref}(1) - s_{noise})$, $s_{noise} = \sqrt{2/\pi}\hat{\sigma}$ is the amplitude value of the folded noise. (3.39) shows that when one of the symbols is seriously attenuated, joint demodulation should be used.

If the decision vector $\mathbf{y}_{joint}$ is larger than 0, symbol "1" can be declared; otherwise symbol "0" is detected. After demodulating the information bit carried by the acoustic beacon signal, we can decode these bits to obtain the position information of the anchor node for the localization purpose. The decision error can be written as

$$\begin{aligned} P_{error} &= P(d=0)P(\mathbf{y} > 0|d=0) + P(d=1)P(\mathbf{y} < 0|d=1) \\ &\frac{1}{2}\int_0^\infty \frac{1}{\sqrt{2\pi\sigma^2}} exp(-(x+\sqrt{E_b})^2/(2\sigma^2))dx \\ &+ \frac{1}{2}\int_{-\infty}^0 \frac{1}{\sqrt{2\pi\sigma^2}} exp(-(x-\sqrt{E_b})^2/(2\sigma^2))dx \\ &= \hat{Q}(-\sqrt{E_b/N_b}) \end{aligned} \tag{3.40}$$

where the $\sigma^2 = N_0$ in (3.40) is noise variance of the decision vector.

# 3.7 Performance Evaluation

## 3.7.1 Experiment Setup

We conducted the measurement in an office environment to test the signal-to-noise ratio (SNR), bit-error-rate (BER), and TOA normalized mean square error (NMSE) at different communication distances to evaluate the system performance of TOA estimation and communication. We moved the anchor node away from the microphone from 0.254m to 7.366m, and conducted measurement for every 0.1m with more than 400s sampling data acquired in each measurement, $N_p = 400$. The beacon period is $T_p = 0.9710s$, while symbol duration is $T_s = 0.0205s$. The symbol number in one beacon period is $N_s = 17$, with 15 information bits as the unique ID of each anchor, and 2 bits ("1" and

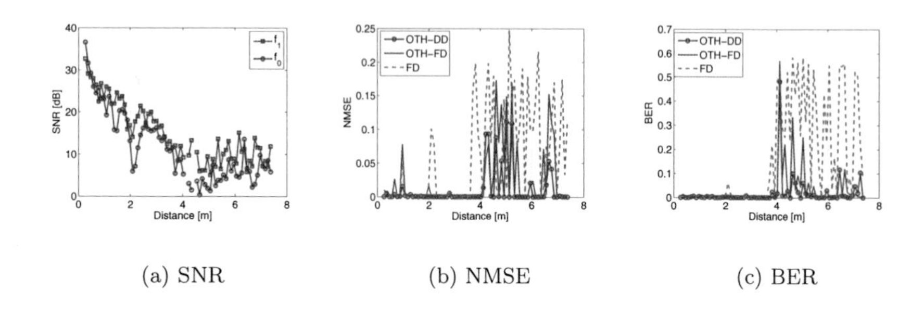


(a) SNR (b) NMSE (c) BER

**Figure 3.6: The SNR (a), NMSE (b) and BER (c) measurement results with respect to the distance.**

"0") for transmit reference. The sampling rate is $F_s = 44.1KHz$, with the symbols modulated in the frequency of $f_1 = 17.72KHz$ and $f_2 = 19.2KHz$.

To observe the signal attenuation with distance change, we measure the SNR value of symbols "1" and "0" with respect to the communication distance. The empirical SNR calculation equation used is $SNR(b) = 20\log_{10}[(S(b)/s_{noise})]$, where $S(b)$ is the signal mean absolute value for information bit $b$, and $s_{noise} = \sqrt{2/\pi}\hat{\sigma}$ is the mean absolute value of noise.

For the evaluation of the TOA estimation and communication performance, three different methods are evaluated and compared. The first one uses the conventional two-step TOA estimation without the threshold optimization and frequency demodulation ("FD") method; the second one uses the optimized threshold in TOA estimation, called "OTH-FD"; the last one uses both the TOA optimized threshold and our proposed dynamic demodulation (DD), called "OTH-DD."

The normalized mean square error (NMSE) is used in this paper to characterize the accuracy of the TOA result by using $NMSE = [(\hat{\tau}^{TOA} - \tau^{TOA})/\tau^{TOA}]^2$. We measure the BER value by comparing the demodulated information bit $(\hat{b}_i)$ to the real transmitted data $(b_i)$ by $BER = \sum_{i=0}^{N_s-1} |\hat{b}_i - b_i|/N_s$.

### 3.7.2 Experimental Results

The measurement results of SNR degradation vs. distance are shown in Fig. 3.6(a). The results show that the small-scale fading of the acoustic signal is strong, while $f_0$ and $f_1$ suffered different attenuation. For most distances, the attenuation of the frequency $f_0$ is stronger than $f_1$ due to its slightly higher frequency.

The results of NMSE with respect to distance are shown in Fig. 3.6(b). The x-coordinate is the distance between the transmitter and the receiver, and the y-coordinate is the measured NMSE. For smaller distances ($< 4.4m$), the

**Table 3.1: Performance comparison with respect to different methods under specific probability**

| Methods | Metrics | 70% | 80% |
|---|---|---|---|
| OTH-DD | NMSE | 0.0005 | 0.0031 |
| | Range Error (m) | 0.0864 | 0.5270 |
| | BER | 0.0055 | 0.0099 |
| OTH-FD | NMSE | 0.0041 | 0.0259 |
| | Range Error (m) | 0.69 | 4.4 |
| | BER | 0.0083 | 0.0152 |
| FD | NMSE | 0.0812 | 0.1142 |
| | Range Error (m) | 13.8107 | 19.4062 |
| | BER | 0.2303 | 0.4267 |

NMSE values for the three methods are all very low, showing that the TOA estimation is very accurate when SNR is strong. For larger distances ($> 4.4m$), the estimation error is increased due to the attenuation of the signal. Using our proposed optimized TOA threshold, "OTH-FD" and "OTH-DD" achieved better performance than the "FD" case. While using dynamic demodulation, some erroneous TOA estimation results can be identified and filtered by its BER value, thus making the TOA estimation accuracy of "OTH-DD" slightly better than "OTH-FD" due to its better BER performance.

The BER experiment results are shown in Fig. 3.6(c), with its y-coordinate as the measured BER. Using our optimized TOA threshold, "OTH-FD" and "OTH-DD" achieve better performance than the "FD" case, while using dynamic demodulation can even lower the error rate as shown in "OTH-DD."

For Fig. 3.6(b) and Fig. 3.6(c), the most interesting point is near $4.11m$; the performance of TOA estimation and communication for all the three methods are really worse at that point. From the echo in that distance, it shows that the waveform has been seriously attenuated both for symbol "1" and "0." There may be several multi-path signals arrived at for the receiver, with negative phase and canceling with each other. Such a dead zone may affect the final position result, but using more redundant anchor nodes can compensate the performance loss.

To evaluate the performance at specific probability, we write the maximum TOA estimation error and BER under a given probability as shown in Table 3.1. The metric of range error is calculated by NMSE to characterize the ranging accuracy. This table illustrates that 70% or 80% of the total results are less than the given value, e.g., the value 0.0005 means 70% of the NMSE results are less than 0.0005. From Table 3.1, we know that the maximum range error is less than 0.0864m with 70% probability when using "OTH-DD," such ranging accuracy is much higher than other existing schemes based on pervasive hardware; it is sufficient to guarantee a precise indoor localization result.

### 3.7.3 Communication and Ranging Results

The metric for assessing the performance of communication is bit-error-rate (BER); the metric for ranging accuracy is the estimation variance defined by $\sigma_{range} = c\sqrt{\mathrm{E}[(\hat{\tau}^{TOA} - \tau^{TOA})]^2}$, where $c$ is the speed of acoustic signal. To test the communication and ranging performance under different operating distances, we conducted experiments for measuring the BER and variance when putting the smartphone at different distances from $2.18m \sim 7.26m$. The BER and variance results obtained by detecting the beacon signal for one anchor node are shown in Table 3.2. From Table 3.2, we know that some distance region, e.g., near 5.2m, has high BER and variance due to the blockage or interference of the beacon signal. When the distance between anchor node and smartphone reaches $7.26m$, the ranging and communication results are still acceptable. Such a result demonstrates that the operating distance of our proposed system is sufficient for indoor localization.

**Table 3.2: BER and ranging NMSE results in different distances**

| Distance(m) | 2.1844 | 3.2004 | 4.2164 | 5.2324 | 6.2484 | 7.2644 |
|---|---|---|---|---|---|---|
| BER | 0.0015 | 0 | 0.0016 | 0.0202 | 0.0020 | 0.0095 |
| Variance(m) | 0.021 | 0.016 | 0.035 | 4.637 | 0.015 | 0.067 |

**Table 3.3: BER and ranging variance results for 4 anchor nodes when a smartphone is placed in Env1$(1.68, 1.02)m$ and Env2$(4.6, 1.03)m$**

| Index | Metrics | m=1 | m=2 | m=3 | m=4 |
|---|---|---|---|---|---|
| Env1 | BER | 0.0049 | 0.0040 | 0.0011 | 0.0015 |
| | Range Variance (m) | 0.0521 | 0.9597 | 0.0747 | 0.5004 |
| Env2 | BER | 0.0043 | 0.0029 | 0.0031 | 0.0029 |
| | Range Variance (m) | 0.9433 | 0.6917 | 0.2488 | 1.2249 |

To test the BER and variance in real localization scenarios, the communication and ranging results from 4 anchor nodes when a smartphone is placed in two positions are shown in Table 3.3. From Table 3.3, we know that the communication performance is sufficient for localization, with the largest BER less than 0.49%. The ranging variance from different anchor nodes is hard to compare due to the various propagation blockages or interference. The average ranging variance in Env2 is slightly larger than in Env1.

To better evaluate the relative ranging performance of $\hat{r}_m$ with the reference node of $f = 1$, we calculate the CDF of the measured relative distance from node 2 to 1 ("RD2-1"), node 3 to 1 ("RD3-1"), and node 4 to 1 ("RD4-

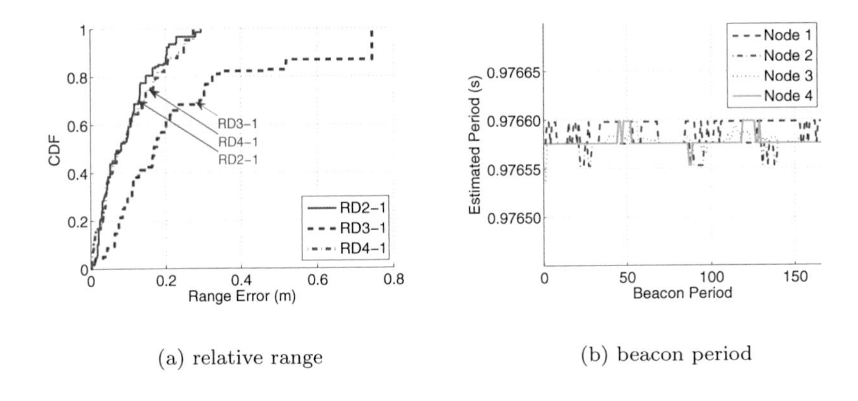


(a) relative range

(b) beacon period

**Figure 3.7: The relative range results (a) and beacon period estimation results (b) when a smartphone is placed in Env1**$(1.68, 1.02)m$.

1") as shown in Fig. 3.7(a). The results show that the ranging result of node 4 is less accurate than those for nodes 2 and 3. If choosing 80% probability as standard, the accuracy of relative distance of "RD2-1" and "RD3-1" is near 0.17 meters, while the "RD4-1" case can achieve accuracy of 0.32 meters in Env1.

The beacon period $T_p$ in (4.4) is another important parameter that needs to be estimated before the localization process. Due to the imperfect clock used in hardware, different anchor nodes may have slightly different offsets. The estimated period from four anchor nodes is shown in Fig. 3.7(b), with very small noise variance. Estimating $T_p$ periodically in real-time is a suitable approach to compensate the offset and improve accuracy.

# 3.8 Conclusion

We developed a TOA estimation and communication scheme that utilizes pervasive microphone channels and low-complex anchor nodes. Using the acoustic signal for ranging and localization can achieve high accuracy with no additional hardware requirement for users. To facilitate the ranging and communication under the restriction of the speaker/microphone pair, an optimized TOA estimation method and a dynamic demodulation scheme have been proposed. The experiment results show that the TOA ranging accuracy can be achieved within 0.0864 meters with 70% probability; the communication BER can be less than 0.55% for 70% distances in the range of $0.254 \sim 7.366$ meters. Such results of ranging and communication are sufficient for accurate indoor localization applications.

# *Chapter 4*

# Localization Algorithms

**Kaikai Liu**

*Assistant Professor, San Jose State University*

**Xiaolin Li**

*Associate Professor, University of Florida*

## CONTENTS

# 4.1 Trilateration via Relative Distance

## 4.1.1 Relative Distance from Anchor Nodes

In our design mode for localization, the microphone in the mobile phone only receives beacons broadcasted by the anchor nodes passively. Such a broadcast mode can ease the network design and support infinite users at the same time. The drawback is that two-way ranging cannot be performed. Thus, the unknown delay needs to be solved during the estimation process.

The obtained ranging results in a passive ranging process are the pseudo-ranges with unknown delay. Using the pseudo-range in localization, the timing delay should be resolved in positioning estimation. For 2-D localization, the three unknown parameters are coordinates (x, y) and the timing delay ($\delta_b$). To discriminate between different beacon signals, every anchor node modulates its unique pseudo-code in the beacon signal. For the multi-anchor operation, we choose the time division multiplexing (TDM) technique. One anchor serves as the control node and sends synchronization commands and tokens to other nodes in the network periodically. The period of the control node is also the beacon period ($T_p$) of the whole network, where $T_s \times N_s < T_p$. It shows that every beacon period has some free space in case of inter-beacon interference. Assuming there are a total of $M$ anchor nodes with each index as $m$, the beacon information will be repeated for every $M$ beacons in the $j$-th symbol, have $m \equiv j(\mod M)$ and $j = m + \lfloor j/M \rfloor \times M$, $g_{i,j}(k) = g_{i,m}(k)$. The TOA value $\hat{\tau}_m^{TOA}$ is obtained from the $m$th anchor node periodically. For one round beacon, the TOA value from the $m$-th anchor can be written as

$$\hat{\tau}_m^{toa} = \delta_b + (m-1)T_p + t_m \tag{4.1}$$

where $m \in [1, \ldots, M]$, and $\delta_b$ is the unknown beginning time. $T_p$ is the period time; $t_m$ is the flight time of the beacon signal to the microphone. The real distance $r_m$ can be represented by $t_m$ as

$$\hat{r}_m = ct_m = c(\hat{\tau}_m^{toa} - \delta_b - (m-1)T_p) \tag{4.2}$$

where $\delta_b + (m-1)T_p$, different from beacon to beacon, is the unknown delay. To minimize the effect of $\delta_b$, we substrate the same distance from every measurement to obtain a relative distance value. By selecting one anchor node with minimum ranging variance as reference, i.e., $m = f$. By setting $\hat{r}_f = 0$, the relative distance of other nodes to this reference point can be written as

$$\hat{r}_m = [\hat{\tau}_m^{toa} - \hat{\tau}_f^{toa} - (m-f)T_p]c + n_m - n_f \tag{4.3}$$

In (4.3), $n_m$ and $n_f$ are the measurement noise for $\hat{\tau}_m^{toa}$ and $\hat{\tau}_f^{toa}$. $T_p$ is the preset beacon period with the known initial value. For the distributed system, physical clocks are not synchronized between the anchor nodes and the microphone. To improve the accuracy of ranging estimation in (4.3), $T_p$ should be updated when a new beacon period is received. Define $m_t = \lfloor j/M \rfloor$ as period index parameter to label the total $\lfloor N_p/M \rfloor$ times beacon round. For $m_t$-th round, $T_p$ can be estimated by

$$\hat{T}_p(m_t) = \frac{1}{M}\left(\sum_{m=1}^{M} \hat{\tau}^{toa}_{m+M(m_t+1)} - \hat{\tau}^{toa}_{m+M(m_t)}\right) \tag{4.4}$$

when only consider one round time that the position result is calculated, we can simplify $\hat{T}_p = \hat{T}_p(m_t)$ in the following analysis.

### 4.1.2 Joint Estimation of the Position and Unknown Bias

With the measured distance from $M$ anchor nodes, trilateration can be performed to localize the position of the microphone. Assume the real position of the microphone is $\mathbf{p} = (x, y)$. The real distance from the microphone to the anchor nodes can be assumed as $r_m = \sqrt{(x - x_m)^2 + (y - y_m)^2}$. The observation equation for the pseudorange $\hat{r}_m$ is

$$\hat{r}_m = \sqrt{(x - x_m)^2 + (y - y_m)^2} + \delta + \widetilde{n}_m \tag{4.5}$$

where $m = 1, \ldots, M$, $\hat{r}_f = 0$; $\hat{r}_m$ is obtained by (4.3). $\delta$ is the unknown fixed delay for every anchor node, which has been modified to $\delta = -r_f$. $\widetilde{n}_m = n_m - n_f$ is the error of the pseudorange in (4.3). $n_m$ is a i.i.d Gaussian random parameter with noise variance of $\sigma_m^2$. The variance of $\widetilde{n}_m$ is $\sigma_m^2 + \sigma_f^2$ due to the substraction made in (4.3). In vector notations, (4.5) can be expressed as

$$\hat{\mathbf{r}} = \mathbf{f}(x, y, \delta) + \widetilde{\mathbf{n}} \tag{4.6}$$

where $\hat{\mathbf{r}} = [\hat{r}_1, \cdots, \hat{r}_M]^T$, $\mathbf{f}(x, y, \delta) = [r_1 + \delta, \cdots, r_m + \delta]^T$, and $\widetilde{\mathbf{n}} = [n_1, \cdots, n_m]^T$.

The localization process via the traditional maximum *a posteriori* (MAP) estimator can be written as

$$\hat{\theta}_{MAP} = \arg\max_{\theta} P(\hat{\mathbf{r}}|\theta) g(\theta) \tag{4.7}$$

where $g(\theta)$ is the probability density function (pdf) of parameter $\theta$; $P(\hat{\mathbf{r}}|\theta)$ is the pdf of the measurement vector $\hat{\mathbf{r}}$ conditioned on $\theta$. The conditional pdf of $\hat{\mathbf{r}}$ can be expressed as

$$P(\hat{\mathbf{r}}|\theta) = \frac{1}{\sqrt{2\pi \det(\sigma)}} \exp\left(-\frac{\mathbf{Q}^T \sigma^{-1} \mathbf{Q}}{2}\right) \tag{4.8}$$

where $\mathbf{Q} = (\hat{\mathbf{r}} - \mathbf{f}(x, y, \delta))$. $\sigma$ is the covariance matrix for $\hat{\mathbf{r}}$ as $\sigma = \{\mathrm{Cov}(\hat{r}_i, \hat{r}_j)\}$,

$i, j = 1, \ldots, M$. $\det(\sigma)$ calculates the determinant of $\sigma$. For the Gaussian noise component in (4.23), it has $\mathbf{n} \sim \mathcal{N}(0, \sigma)$ with

$$\sigma = \begin{bmatrix} \sigma_m^2 + \sigma_f^2 & \sigma_f^2 & \cdots & \sigma_f^2 \\ \sigma_f^2 & \sigma_m^2 + \sigma_f^2 & \ddots & \vdots \\ \vdots & \ddots & \ddots & \sigma_f^2 \\ \sigma_f^2 & \cdots & \sigma_f^2 & \sigma_m^2 + \sigma_f^2 \end{bmatrix} \tag{4.9}$$

After obtaining the conditional pdf of $P(\hat{\mathbf{r}}|\theta)$, we should know the distribution $\theta$ before performing MAP estimation.

From $\mathbf{f}(x, y, \delta)$, we define the vector of unknown parameters as $\theta = [x \quad y \quad \delta]^T$. The localization purpose is to estimate $(x, y)$ from measurements. Different approaches like Bayesian or ML estimation techniques can be applied depending on the prior information about $\theta$. If the prior probability distribution of $\theta$ is known, a maximum *a posteriori* (MAP) estimator can be applied. For the case where the distribution of the unknown parameter $\theta$ is unknown, uniform distribution can be assumed. Then, MAP can be simplified to a maximum log-likelihood (ML) estimator as

$$\hat{\theta}_{ML} = \arg\max_{\theta} \log P(\hat{\mathbf{r}}|\theta) \tag{4.10}$$

where $P(\hat{\mathbf{r}}|\theta)$ is the pdf of the measurement vector $\hat{\mathbf{r}}$ conditioned on $\theta$. The conditional pdf of $\hat{\mathbf{r}}$ can be expressed as

$$P(\hat{\mathbf{r}}|\theta) = \frac{1}{\sqrt{2\pi \det(\sigma)}} \exp\left(-\frac{\mathbf{Q}^T \sigma^{-1} \mathbf{Q}}{2}\right) \tag{4.11}$$

where $\mathbf{Q} = (\hat{\mathbf{r}} - \mathbf{f}(x, y, \delta))$. $\sigma$ is the covariance matrix for $\hat{\mathbf{r}}$ as $\sigma = \{\mathrm{Cov}(\hat{r}_i, \hat{r}_j)\}$, $i, j = 1, \ldots, M$. $\det(\sigma)$ calculates the determinant of $\sigma$ with

$$\sigma = \begin{bmatrix} \sigma_m^2 + \sigma_f^2 & \sigma_f^2 & \cdots & \sigma_f^2 \\ \sigma_f^2 & \sigma_m^2 + \sigma_f^2 & \ddots & \vdots \\ \vdots & \ddots & \ddots & \sigma_f^2 \\ \sigma_f^2 & \cdots & \sigma_f^2 & \sigma_m^2 + \sigma_f^2 \end{bmatrix} \tag{4.12}$$

For the Gaussian noise component in (4.23), it has $\tilde{\mathbf{n}} \sim \mathcal{N}(0, \sigma)$, where $\sigma$ is the error covariance matrix with diagonal entries of $\sigma_m^2 + \sigma_f^2$, other elements of $\sigma_f^2$. Using uniform distribution in ML estimation is sort of the worst-case scenario, while achieving optimal performance under worst-case instead of global optimal is practical and reasonable.

The result of (4.24) can be achieved by searching over possible parameters that maximize the log-likelihood. For real system implementation, such a searching process is very computationally intensive and can drain the battery life of your smartphone quickly. To make this problem even simpler, we can assume that all the measurements are independent. Such a special case is understandable in that all the pseudoranges for anchors are measured independently and in different beacon periods. With such an assumption, the noise distribution with zero mean and a known covariance matrix (4.24) can be simplified as

$$\hat{\theta}_{ML} = \arg\min_{\theta} (\hat{\mathbf{r}} - \mathbf{f}(x, y, \delta))^T \sigma^{-1} (\hat{\mathbf{r}} - \mathbf{f}(x, y, \delta)) \tag{4.13}$$

where $(\hat{\mathbf{r}} - \mathbf{f}(x, y, \delta))$ represents the estimation error, and $\sigma^{-1}$ can be represented as the weighting coefficient for each independent measurement. (4.26) is also the form of non-linear least-squares (NLS) estimator for which steepest descent, Gauss-Newton, and Taylor series based methods can be used to solve the problem [30]. These kinds of methods are all complicated and require a good initial value in calculation to avoid converging to the local minima of (4.26).

With the TOA value from $M$ anchor nodes to the reference point obtained, we can perform trilateration to localize the position of the microphone. Assume the real position of the microphone is $\mathbf{p} = (x, y)$; each anchor node has its known position as $\mathbf{p}_m = (x_m, y_m)$, and $m \in [1, \ldots, M]$. The real distance from the microphone to the anchor node can be assumed as $r_m = \sqrt{(x - x_m)^2 + (y - y_m)^2}$. Assume the measured distance between the anchor node to the smartphone is $\hat{r}_m$, and can be shown as

$$\hat{r}_m = c\hat{\tau}_m^{TOA} = r_m + c\delta_t, m = 1, \ldots, M \tag{4.14}$$

where the $c$ is the speed of the acoustic signal, $\delta_t$ is the unknown fixed timing delay for every anchor node. For simplicity, we set $\delta = c\delta_t$, and every measurement from the anchor node has nearly the same $\delta$. In real systems, $\delta$ may be a very large number of value, and different from beacon to beacon. To minimize the effect of $\delta$, we can substrate the same distance from every measurement to obtain a relative distance value. By selecting a reference point of $m = f$, and setting $\delta = -r_f$, we have $\hat{r}_f = 0$. Then the relative distance of other node to this reference point can be written as

$$d_m(f) = [\hat{\tau}_m^{TOA} - \hat{\tau}_f^{TOA} - (m - f)T_p]c \tag{4.15}$$

where $m \neq f, \{m, f\} \in [1, \ldots, M]$.

If the reference anchor is $f$, then the relative distance obtained in (4.15) can be expressed in terms of $\hat{r}_m$ as $d_m(f) = \hat{r}_m - \hat{r}_f$. Choose one anchor as the reference node $m = f$, and other equations in (4.17) subsidize the reference measurement.

Rather than solving (4.26) directly, we use the properties of (4.3) to cancel out the nonlinear term. In (4.3), we only care about the relative distance to the reference point, with the fixed delay of $\delta = -r_f$. Such a process can avoid the non-linear term in (4.26). To better illustrate this process, we can substitute $r_m$ into (4.5), and squaring both sides of (4.5) yields

$$(x - x_m)^2 + (y - y_m)^2 = (\hat{r}_m - \delta - \widetilde{n}_m)^2, m = 1, \ldots, M \tag{4.16}$$

where $\widetilde{n}_m = n_m - n_f$. Simplifying both sides of (4.16) gives

$$\begin{aligned} &2x_m x + 2y_m y - 2\delta\hat{r}_m + \delta^2 - x^2 - y^2 - 2(\hat{r}_m - \delta)\widetilde{n}_m \\ &= x_m^2 + y_m^2 - \hat{r}_m^2 \end{aligned} \tag{4.17}$$

where $\widetilde{n}_m^2 = 0$ has been eliminated in (4.17) due to independent properties of the noise. $\delta^2 - x^2 - y^2$ is the non-linear term for the unknown parameter $\theta$, and the same for $m = 1, \ldots, M$. By utilizing the reference point $(f)$, i.e., $\hat{r}_f = r_f + \delta = 0 + n_f$, the simplification can be made as

$$\delta^2 - x^2 - y^2 = x_f^2 - 2xx_f + y_f^2 - 2y_f y + 2n_f\delta \tag{4.18}$$

The non-linear term in (4.17) can be canceled out by subsidizing (4.18) into (4.17) as

$$\begin{aligned} &[(x_m^2 + y_m^2 - \hat{r}_m^2) - (x_f^2 + y_f^2 - \hat{r}_f^2)] + 2\hat{r}_m\widetilde{n}_m - 2\delta n_m \\ &= 2(x_m - x_f)x + 2(y_m - y_f)y - 2\delta(\hat{r}_m - \hat{r}_f) \end{aligned} \tag{4.19}$$

Equation (5.15) can be expressed in matrix form as

$$\mathbf{A}\theta = \nu + \mathbf{p_n} \tag{4.20}$$

where $\mathbf{A} = \begin{bmatrix} x_m - x_f & y_m - y_f & \hat{r}_f - \hat{r}_m \\ \cdots & & \\ \cdots & & \cdots \end{bmatrix}$, $\theta = \begin{bmatrix} x \\ y \\ \delta \end{bmatrix}$ and $\nu = \frac{1}{2}[(x_m^2 + y_m^2 - \hat{r}_m^2) - (x_f^2 + y_f^2 - \hat{r}_f^2)]_{(M-1)\times 3}$, $m = 1, \ldots, M$ when $m \neq f$. $\hat{r}_m$ is the measured distance obtained from (4.3) and $\hat{r}_f = 0$. $\mathbf{p_n} = 2\hat{r}_m\widetilde{n}_m - 2\delta n_m$ is the noise term with its first moment as $\mathrm{E}(\mathbf{p_n}) = 0$ and the covariance matrix of vector $\mathbf{p_n}$ as $\widetilde{\sigma} = \mathrm{Cov}(\mathbf{p_n})$. The diagonal elements of the covariance matrix are $2\hat{r}_m(\sigma_m^2 + \sigma_f^2) + 2\delta\sigma_m^2$, with other elements in the matrix as $2\hat{r}_m\sigma_f^2$. Note that $\delta$ is not available in practice; it can be set to 0 at the first snapshot of localization to simplify the solution. For the following snapshots, $\delta$ can be replaced by the calculated value to improve accuracy.

Then (4.26) can be written as a least-square (LS) problem [30], as $\min_\theta \|\mathbf{A}\theta - \nu\|_2$. It is equivalent to find $\theta$, which minimizes the sum squares of $M$ independent error vector with $\|\mathbf{e}\|^2 = (\nu - \mathbf{A}\theta)^T\widetilde{\sigma}^{-1}(\nu - \mathbf{A}\theta)$. By minimizing this quadratic function, we can obtain the solution of

$$\hat{\theta} = (\mathbf{A}^T\widetilde{\sigma}^{-1}\mathbf{A})^{-1}\mathbf{A}^T\widetilde{\sigma}^{-1}\nu \tag{4.21}$$

Assuming all the noise variances are identical and replacing $\sigma_m^2 = 1$, the common part $2\hat{r}_m$ can be ignored. In such a case, $\widetilde{\sigma}$ is equivalent to $\sigma$. The diagonal element of $\widetilde{\sigma}^{-1}$ is $(M-2)/(M-1)$ with other elements of $1/(M-1)$. The estimated position of the target can be obtained by selecting the first two parameters $(x, y)$ of $\hat{\theta}$. The obtained $\delta$ in each result of $\hat{\theta}$ can be used as a constraint such that $\delta \approx r_f = \sqrt{(x - x_f)^2 + (y - y_f)^2}$. By calculating $r_f$ and using the obtained $(x, y)$, the difference between the estimated $\delta$ can be shown as $\hat{e} = \sqrt{(\hat{\theta}(1) - x_f)^2 + (\hat{\theta}(2) - y_f)^2} - \hat{\theta}(3)$. Small $\hat{e}$ indicates good position results. Such delay-constraint (DC) can be used as the self-evaluation of the position results and can filter out some incorrect estimated positions.

# 4.2 Mobile Phone Localization via Semidefinite Programming

## 4.2.1 *Localization Measurement Model*

Assume the real position of a mobile phone is $\mathbf{y} \in \mathbb{R}^d$, i.e., the 2-D coordinate $(d = 2)$ of $\mathbf{y}$ is $\mathbf{y} = [x, y]^T$. Denote the known anchor position vector as $\mathbf{x}_m \in \mathbb{R}^d$, where $m$ is the anchor index with total $M$ anchors. Using $d = 2$ as an example, each element of $\mathbf{x}_m$ is a 2-D coordinate as $[x_m, y_m]^T$, $m = 1, \ldots, M$. The objective of localization is to estimate $\mathbf{y}$ from distance measurements of $\hat{\mathrm{r}}_m$, where $\hat{\mathrm{r}}_m$ denotes the measured distance from the target $(\mathbf{y})$ to anchor node $m$. The real distance $r_m$ from the mobile phone to the $m$-th anchor node can be written as $r_m = ||\mathbf{y} - \mathbf{x}_m||_2$, where $|| \cdot ||_2$ calculates the 2-norm and obtains the Euclidean distance. Then, the distance measurement from the $m$-th anchor can be written as

$$\hat{\mathrm{r}}_m = ||\mathbf{y} - \mathbf{x}_m||_2 + \delta_r + n_m \tag{4.22}$$

where $\delta_r$ is the unknown fixed bias for the whole anchor network due to the unknown access time of the mobile phone. The random access time of the mobile phone to the anchor network is unavoidable. Using the appropriate algorithm to estimate $\delta_r$ during the localization process is necessary. In vector notations, (4.22) can be expressed as

$$\hat{\mathbf{r}} = \mathbf{f}(\theta) + \mathbf{n} \tag{4.23}$$

where $\hat{\mathbf{r}} = [\hat{r}_1, \cdots, \hat{r}_M]^T$, $\mathbf{n} = [n_1, \cdots, n_m]^T$. We define the unknown parameter vector as $\theta = [\mathbf{y}, \delta_r]^T$, then $\mathbf{f}(\theta) = [r_1 + \delta_r, \cdots, r_m + \delta_r]^T$. The localization process is to estimate $\theta$ by using approaches like Bayesian or maximum-likelihood (ML) estimation techniques.

To estimate the unknown parameter $\theta$, maximum *a posteriori* (MAP) estimator is optimal when the prior probability distribution $g(\theta)$ is known. For most cases, the distribution $g(\theta)$ of the unknown parameter $\theta$ is unknown. Using uniform distribution instead of the $g(\theta)$ is like achieving optimal performance under the worst-case scenario. Such simplification is practical and reasonable, and can simplify the MAP estimator to the maximum likelihood (ML) estimator as

$$\hat{\theta}_{ML} = \arg\max_{\theta} \log P(\hat{\mathbf{r}}|\theta) \tag{4.24}$$

where $P(\hat{\mathbf{r}}|\theta)$ is the pdf of the measurement vector $\hat{\mathbf{r}}$ conditioned on unknown parameter $\theta$. The conditional pdf of $\hat{\mathbf{r}}$ can be expressed as

$$P(\hat{\mathbf{r}}|\theta) = \frac{1}{\sqrt{2\pi \det(\sigma)}} \exp\left(-\frac{\mathbf{Q}^T\sigma^{-1}\mathbf{Q}}{2}\right) \tag{4.25}$$

where $\mathbf{Q} = (\hat{\mathbf{r}} - \mathbf{f}(\theta))$. $\sigma$ is the covariance matrix for $\hat{\mathbf{r}}$ as $\sigma = \{\text{Cov}(\hat{r}_i, \hat{r}_j)\}$, $i, j = 1, \ldots, M$. $\det(\sigma)$ calculates the determinant of $\sigma$. For the Gaussian noise component in (4.23), it has $\mathbf{n} \sim \mathcal{N}(0, \sigma)$.

The result of (4.24) can be achieved by searching over possible parameters that maximize the log-likelihood. For noise distribution with zero mean and a known covariance matrix, (4.24) can be simplified as

$$\hat{\theta}_{ML} = \arg\min_{\theta} (\hat{\mathbf{r}} - \mathbf{f}(\theta))^T\sigma^{-1}(\hat{\mathbf{r}} - \mathbf{f}(\theta)) \tag{4.26}$$

where $(\hat{\mathbf{r}} - \mathbf{f}(\theta))$ represents the estimation error, and $\sigma^{-1}$ can be represented as the weighting coefficient for each independent measurement. (4.26) is also the form of a non-linear least-squares (NLS) estimator. The steepest descent, Gauss-Newton, and Taylor-series-based method can be used to solve the problem. These kinds of methods require a good initial value in calculation to avoid converging to the local minima of (4.26), or need to calculate the computational complex of the matrix inverse operation [64].

### 4.2.2 Min-Max Criterion

To prevent the algorithm from converging to the local optimal values, the concept of relaxation onto convex sets has been proposed and demonstrated as tight bound to the initial *non-convex* problem [117]. Among existing relaxation criteria, using *minimax* approximation and semidefinite relaxation can find the global minimum value without the "inside convex hull" requirement [68].

The solution of (4.26) by using the NLS estimator is a *nonconvex* optimization problem, while a semidefinite programming (SDP) technique can be used to relax the initial *nonconvex* problem into a convex one, and has been proven to have a tight bound to the original problem. To utilize the SDP relaxation, we can modify the problem formulation by rewriting (4.23) into $\hat{\mathbf{r}} - \delta_r = ||\mathbf{y} - \mathbf{x}_m||_2 + \mathbf{n}$. Performing square operation in both sides will lead to

$$(\hat{\mathbf{r}} - \delta_r)^T\sigma^{-1}(\hat{\mathbf{r}} - \delta_r) = (||\mathbf{y} - \mathbf{x}_m||_2 + \mathbf{n})^2 \tag{4.27}$$

where the right side of (4.27) will lead to $||\mathbf{y} - \mathbf{x}_m||_2^2 + 2\mathbf{n}^T||\mathbf{y} - \mathbf{x}_m||_2 + \mathbf{n}^T\mathbf{n}$. $\mathbf{n}$ is the variance vector of the ranging error. Assuming every ranging measurement is independent, we will have $\mathbf{n}^T\mathbf{n} = 0$; $2\mathbf{n}||\mathbf{y} - \mathbf{x}_m||_2$ will be the new noise term as $\mathbf{n}'$. By adopting the *min-max* criterion [68, 98], (4.27) can be formulated as

$$\mathbf{y} = \arg\min_{\mathbf{y}} \max_{m=1,\ldots,M} \underbrace{\left| ||\mathbf{y} - \mathbf{x}_m||_2^2 - (\hat{\mathbf{r}} - \delta_r)^T\sigma^{-1}(\hat{\mathbf{r}} - \delta_r) \right|}_{\xi} \tag{4.28}$$

where the term $\xi$ can be viewed as the residual error. (4.28) calculates $\mathbf{y}$ by minimizing the maximum residual error. Compared with (4.26), (4.28) remains nonconvex, but it is comfortable for the following semidefinite relaxations.

The first term in (4.28) can be written into a matrix form of

$$\begin{aligned}||\mathbf{y}-\mathbf{x}_m||_2^2 &= \begin{bmatrix} \mathbf{y}^T & 1 \end{bmatrix} \begin{bmatrix} \mathbf{I}_d & -\mathbf{x}_m \\ -\mathbf{x}_m^T & \mathbf{x}_m^T\mathbf{x}_m \end{bmatrix} \begin{bmatrix} \mathbf{y} \\ 1 \end{bmatrix} \\ &= \text{trace}\left\{ \begin{bmatrix} \mathbf{y} \\ 1 \end{bmatrix} \begin{bmatrix} \mathbf{y}^T & 1 \end{bmatrix} \begin{bmatrix} \mathbf{I}_d & -\mathbf{x}_m \\ -\mathbf{x}_m^T & \mathbf{x}_m^T\mathbf{x}_m \end{bmatrix} \right\} \\ &= \text{trace}\left\{ \begin{bmatrix} \mathbf{Y} & \mathbf{y} \\ \mathbf{y}^T & 1 \end{bmatrix} \begin{bmatrix} \mathbf{I}_d & -\mathbf{x}_m \\ -\mathbf{x}_m^T & \mathbf{x}_m^T\mathbf{x}_m \end{bmatrix} \right\}\end{aligned} \tag{4.29}$$

where $\mathbf{Y} = \mathbf{y}\mathbf{y}^T$, trace$\{\cdot\}$ calculates the trace of the matrix, and $\mathbf{I}_d$ is an identity matrix of order $d$. Using the same process in (4.29), the second term in (4.28) can be written into

$$\begin{aligned}&(\hat{\mathbf{r}}-\delta_r)^T\sigma^{-1}(\hat{\mathbf{r}}-\delta_r) \\ &= \text{trace}\left\{ \begin{bmatrix} \delta & \delta_r \\ \delta_r^T & 1 \end{bmatrix} \begin{bmatrix} \sigma^{-1}\mathbf{I}_d & -\sigma^{-1}\hat{\mathbf{r}} \\ -\hat{\mathbf{r}}^T\sigma^{-1} & \hat{\mathbf{r}}^T\sigma^{-1}\hat{\mathbf{r}} \end{bmatrix} \right\}\end{aligned} \tag{4.30}$$

where $\delta = \delta_r\delta_r^T$.

### 4.2.3 Semidefinite Programming

The objective function of $\xi$ can be converted to minimize $\epsilon$ at the constraint of an inequality expression $-\epsilon < \xi < \epsilon$, while $\xi$ can be written as the form of (4.29) and (4.30). The constraints form of (4.29) and (4.30) are convex, but the equality constraints of $\mathbf{Y} = \mathbf{y}\mathbf{y}^T$ and $\delta = \delta_r\delta_r^T$ are nonconvex. Using semidefinite relaxation, these two equalities can be relaxed to inequality constraints of $\mathbf{Y} \succeq \mathbf{y}\mathbf{y}^T$ and $\delta \succeq \delta_r\delta_r^T$, respectively. The matrix form of these two equalities is

$$\begin{bmatrix} \mathbf{Y} & \mathbf{y} \\ \mathbf{y}^T & 1 \end{bmatrix} \succeq 0, \quad \begin{bmatrix} \delta & \delta_r \\ \delta_r^T & 1 \end{bmatrix} \succeq 0 \tag{4.31}$$

where $\succeq$ means a positive definite (semidefinite) matrix, which is different from $\geq$.

Accordingly, the initial problem of (4.26) can be relaxed to a semidefinite programming form as

$$\begin{aligned}&\min_{\{\mathbf{y},\mathbf{Y},\delta_r,\delta\}} \epsilon \\ \text{s.t.} \quad & -\epsilon < \text{trace}\left\{ \begin{bmatrix} \mathbf{Y} & \mathbf{y} \\ \mathbf{y}^T & 1 \end{bmatrix} \begin{bmatrix} \mathbf{I}_d & -\mathbf{x}_m \\ -\mathbf{x}_m^T & \mathbf{x}_m^T\mathbf{x}_m \end{bmatrix} \right\} \\ & - \text{trace}\left\{ \begin{bmatrix} \delta & \delta_r \\ \delta_r^T & 1 \end{bmatrix} \begin{bmatrix} \sigma^{-1}\mathbf{I}_d & -\sigma^{-1}\hat{\mathbf{r}} \\ -\hat{\mathbf{r}}^T\sigma^{-1} & \hat{\mathbf{r}}^T\sigma^{-1}\hat{\mathbf{r}} \end{bmatrix} \right\} < \epsilon, \\ & m = 1, \ldots, M, \\ & \begin{bmatrix} \mathbf{Y} & \mathbf{y} \\ \mathbf{y}^T & 1 \end{bmatrix} \succeq 0, \quad \begin{bmatrix} \delta & \delta_r \\ \delta_r^T & 1 \end{bmatrix} \succeq 0\end{aligned}$$

The mobile phone position $\mathbf{y}$ can be extracted from the optimal solution of $\{\mathbf{y}, \mathbf{Y}, \delta_r, \delta\}$. The SDP problem can be solved by some standard convex optimization packages, e.g., the SeDuMi and CVX package [28].

# 4.3 Performance Bound and Anchor Network Coverage

## 4.3.1 *Fisher Information Matrix*

The measured distance in (4.23) is a set of stochastic variables $\hat{\mathbf{r}} = [\hat{r}_1 \cdots \hat{r}_M]^T$. The probability density function of $\hat{\mathbf{r}}$ can be expressed by using $P(\hat{\mathbf{r}}; \theta)$ instead of the conditional form of $P(\hat{\mathbf{r}}|\theta)$. For simplicity, we use the *joint log-probability density function* $q(\hat{\mathbf{r}}; \theta) = \log P(\hat{\mathbf{r}}; \theta)$ for the following analysis.

Fisher information is often used as a metric to compare the performance of different estimation schemes and investigate the theoretical upper bound (extreme performance) for a particular algorithm. The Fisher score vector $s_\theta$ of the stochastic variable $\hat{\mathbf{r}}$ can be defined as the partial derivatives of $q(\hat{\mathbf{r}}; \theta)$, with respect to the parameter $\theta$, as

$$s_\theta = \frac{\partial q(\hat{\mathbf{r}}; \theta)}{\partial \theta} = \frac{\partial \mathbf{f}^T(\theta)}{\partial \theta} \sigma^{-1} (\hat{\mathbf{r}} - \mathbf{f}(\theta)) \tag{4.32}$$

where the expectation of $s_\theta$ is $\mathrm{E}(s_\theta) = 0$ under suitable regularity conditions [109]. Then the covariance matrix of $s_\theta$ yields $\mathrm{Cov}(s_\theta, s_\theta) = \mathrm{E}\left[(s_\theta - \mathrm{E}(s_\theta))(s_\theta - \mathrm{E}(s_\theta))^T\right] = \mathrm{E}[s_\theta s_\theta^T]$. The *Fisher information matrix* of the observations $\hat{\mathbf{r}}$ is described by

$$\begin{aligned}
\mathbf{J}_\theta = \mathrm{E}[s_\theta s_\theta^T] &= \mathrm{E}\left[\frac{\partial q(\hat{\mathbf{r}}; \theta)}{\partial \theta} \frac{\partial q(\hat{\mathbf{r}}; \theta)}{\partial \theta^T}\right] \\
&= \mathrm{E}\left[\frac{\partial \mathbf{f}^T(\theta)}{\partial \theta} \sigma^{-1} (\hat{\mathbf{r}} - \mathbf{f}(\theta))(\hat{\mathbf{r}} - \mathbf{f}(\theta))^T \sigma^{-1} \frac{\partial \mathbf{f}(\theta)}{\partial \theta}\right] \\
&= \frac{\partial \mathbf{f}^T(\theta)}{\partial \theta} \sigma^{-1} \frac{\partial \mathbf{f}(\theta)}{\partial \theta}
\end{aligned} \tag{4.33}$$

where $\mathbf{J}_\theta$ is actually a covariance matrix and positive semidefinite. It is positive definite if and only if the elements of $s_\theta$ are linearly independent random variables.

## 4.3.2 *Cramer-Rao Low Bound*

To evaluate the position accuracy, the Cramér-Rao low bound (CRLB) is often used as a theoretical optimal value from any unbiased estimator. The CRLB can be written as the reciprocal of the Fisher information. For an estimate of $\hat{\theta}$ obtained, we have the CRLB as

$$\mathrm{E}_r\{(\hat{\theta} - \theta)(\hat{\theta} - \theta)^T\} \succeq \mathbf{J}_\theta^{-1} \tag{4.34}$$

where (4.34) is in the form of a covariance matrix, the right part of (4.34) is $CRLB = \mathbf{J}_\theta^{-1}$, and this means the variance of the estimated parameter could not be lower than the CRLB under the given estimator. The term $\mathbf{A} \succeq \mathbf{B}$ expresses that the difference $\mathbf{A} - \mathbf{B}$ of the real symmetric matrices $\mathbf{A}$ and $\mathbf{B}$ is positive semidefinite. $\succeq$ will be reduced to $\geq$ if $\theta$ is a scalar [11].

In the unknown parameter $\theta$, the position of the mobile phone is $\mathbf{y} \in \mathbb{R}^d$, $d = 2$ for the 2-D coordinate. Thus, $\theta = [\mathbf{y}, \delta_r]^T \in \mathbb{R}^{(d+1)}$ and $\mathbf{J}_\theta$ is a $(d+1) \times (d+1)$ matrix. To characterize the position estimation accuracy $\sigma_p$, only the first $d \times d$ terms in $\mathbf{J}_\theta$ should be utilized as

$$\begin{aligned} \mathrm{CRLB}_{\mathbf{y}} = \sigma_p &= \sqrt{\mathrm{trace}(\mathrm{E}_r\{(\hat{\mathbf{y}} - \mathbf{y})(\hat{\mathbf{y}} - \mathbf{y})^T\})} \\ &\geq \sqrt{\mathrm{trace}(\mathbf{J}_{\theta|(d\times d)}^{-1})} \end{aligned} \tag{4.35}$$

The elements of Fisher information $\mathbf{J}_\theta$ is defined in (4.33), where $\partial \mathbf{f}(\theta)/\partial\theta$ can be obtained as a $d \times 1$ vector. Using 2-D coordinate, i.e., $d = 2$. And $\partial \mathbf{f}(\theta)/\partial\theta = \begin{bmatrix}(x - x_m)/||\mathbf{y} - \mathbf{x}_m|| & (y - y_m)/||\mathbf{y} - \mathbf{x}_m|| & 1\end{bmatrix}$, if only the accuracy of $(x, y)$ is considered in unknown parameter $\theta$. Assume the ranging variance from all the anchor nodes are the same as $\sigma_r^2$, thus $\sigma = \sigma_r^2$.

For the 2-D coordinate, denote the unit vector from the $m$-th anchor node to target as $\alpha_m = (x - x_m)/r_m$ in $x$ domain; $\beta_m = (y - y_m)/r_m$ in $y$ domain, where $r_m = ||\mathbf{y} - \mathbf{x}_m||$. $\mathbf{J}_{\theta|(2\times2)}$ can be shown as

$$\mathbf{J}_{\theta|(2\times2)} = \begin{bmatrix} \sum_{m=1}^M {\alpha_m}^2/\sigma_r^2 & \sum_{m=1}^M \alpha_m\beta_m/\sigma_r^2 \\ \sum_{m=1}^M \alpha_m\beta_m/\sigma_r^2 & \sum_{m=1}^M {\beta_m}^2/\sigma_r^2 \end{bmatrix} \tag{4.36}$$

### 4.3.3 Anchor Network Coverage

Relying on the anchor network for mobile phone localization, the placement of anchor nodes is very important in achieving high resolution results. To evaluate the effect caused by the geometric layout of anchor nodes, the term of geometric dilution of precision (GDOP) is often used. GDOP can be defined as $GDOP = \sigma_p/\sigma_r$, where $\sigma_p$ and $\sigma_r$ indicates the variance of position and ranging results, respectively. $\sigma_p$ is the variance of localization results obtained in (4.35). The GDOP quantifies the amplification of the ranging error in the position result when passing through the position calculating unit.

For an unbiased estimator, GDOP can be written as

$$\begin{aligned} GDOP &= \sqrt{\mathrm{trace}(\mathbf{J}_{\theta|(2\times2)}^{-1})}/\sigma_r = \sqrt{\mathrm{trace}(\tilde{\mathbf{J}}_{\theta|(2\times2)}^{-1})} \\ &= \sqrt{\mathrm{trace}\left(\begin{bmatrix} \sum_{m=1}^M {\alpha_m}^2 & \sum_{m=1}^M \alpha_m\beta_m \\ \sum_{m=1}^M \alpha_m\beta_m & \sum_{m=1}^M {\beta_m}^2 \end{bmatrix}^{-1}\right)} \end{aligned} \tag{4.37}$$

where $\alpha_m$ and $\beta_m$ are the unit vector in the $x$ and $y$ directions.

(4.37) only depends on the positions of the anchor node $\mathbf{x}_m$ and mobile phone $\mathbf{y}$; it is independent of the ranging noise variance. The difference between GDOP (4.37) and CRLB (4.35) lies in the ranging variance $\sigma_r$.

The GDOP calculated by (4.37) can be used as a metric for evaluating the anchor network coverage. The value of GDOP is determined by the relation between the position of mobile phone ($\mathbf{y}$) and anchor network ($\mathbf{x}_m$). Smaller value of GDOP means good coverage. If the location estimation result ($\hat{\mathbf{y}}$) is available, the corresponding GDOP value of (4.37) can be calculated. If the GDOP value for $\hat{\mathbf{y}}$ is higher than a threshold, then the confidence level of $\hat{\mathbf{y}}$ should be lowered. Such a scheme could be used as a post-position constraint to filter out the localization results. If the ranging variance is available or set as a constant, using CRLB in (4.34) as the metric for evaluating the coverage is equivalent to GDOP. The detailed process is elaborated in Subsection 4.4.3.

## 4.4 Position Refinement

### *4.4.1 NLOS and Error Mitigation*

Most of the existing literature on mobile phone localization focuses on the case where the ranging measurements are known with some slight perturbations, i.e., using zero-mean Gaussian noise to represent the ranging error. This assumption is only effective when all the ranging results are in line-of-sight (LOS) condition with no bias or large error. However, in practice, a significant portion of the ranging results contain outliers.

If the large bias errors of ranging are not accounted for, the localization results could result in significant errors or even diverge. The challenge of the problem arises from the unknown *a priori* information about the bias or NLOS condition. There has been some work in the literature that addresses this case and performs NLOS identification and mitigation. However, these approaches focus on mitigating the NLOS ranging measures by using the channel identification results. One drawback of these kinds of approaches lies their requirement on the additional channel measurement.

For NLOS ranging measurements, (4.22) could be written as a new form of

$$\hat{\mathrm{r}}_m = \mathbf{r}_m + \delta_r + \delta_m + n_m \tag{4.38}$$

where $\delta_m$ is the biased term for $M$ anchors, $m = 1, \cdots, M$. For NLOS conditions, $\delta_m > 0$.

The result of $\{\mathbf{y}, \mathbf{Y}, \delta_r, \delta\}$ obtained from (4.32) contains the position ($\mathbf{y}$) of the mobile phone, and the system bias ($\delta_r$). For most cases, the number of NLOS ranging measurements in one location calculation should be less than $M/2$; otherwise, the location calculation result is problematic. If we ignore the value of $\delta_m$ in (4.32), the obtained relaxed result is still acceptable due to the robustness of the SDP approach. However, the accuracy of the relaxed result could be further improved by using post-position processing. Using the estimated result of $\hat{\mathbf{y}}$, the distance between each pair of anchor nodes could be calculated by $\bar{r}_m = ||\hat{\mathbf{y}} - \mathbf{x}_m||_2$. The difference between $\bar{r}_m$ and the ranging result $\hat{r}_m$ is the estimated value of $\delta_m$ as

$$\hat{\delta}_m = \hat{r}_m - ||\hat{\mathbf{y}} - \mathbf{x}_m||_2 \tag{4.39}$$

where $\hat{\delta}_m$ is the estimated NLOS bias value, which is a byproduct of the SDP result of (4.32). If the value of (4.39) for the $m$-th anchor is larger than a threshold $\eta_{nlos}$, the $m$-th ranging result should be mitigated in the post-position process. Using the NLOS mitigated ranging results to perform position refinement will be illustrated in Subsection 4.4.2. Unlike the conventional NLOS identification approach that relies on channel measurement, using (4.39) for NLOS identification does not require additional information.

### 4.4.2 Steepest Descent Approach

The calculated mobile phone position $\mathbf{y}$ from Section 4.2 is a global optimal solution ensured by the SDP approach. SDP relaxes the initial non-convex problem to convex and achieves global optimal value. Although the SDP approach has been approved such that the relaxation has a tight bound to the original problem, performing a local search near the global optimal value can further improve the accuracy.

With the fixed starting point $\delta_r$ available, additional gains in position refinement can be achieved by relying on the unchanged feature of $\delta_r$. Using the sliding-window approach, denote the iterative weighted value of $\delta_r$ as $\bar{\delta}_r$, as $\bar{\delta}_r := a\bar{\delta}_r + b\delta_r$, where $a = 0.7$, $b = 0.3$. According to (4.22), the position refinement can be achieved by minimizing the term of

$$\begin{aligned} \mathbf{y} : &= \arg\min_{\mathbf{y}}(e(\mathbf{y})) \\ &= \arg\min_{\mathbf{y}} \sum_{m\in\xi_M} \left(||\mathbf{y} - \mathbf{x}_m||_2 - (\hat{\mathbf{r}}_m - \bar{\delta}_r)\right)^2 \end{aligned} \tag{4.40}$$

where $e(\mathbf{y})$ is the sum of distance errors between the mobile phone and all the anchor nodes. $\xi_M$ is the set of all the anchor nodes that are used in the position refinement. If all the ranging measurements are LOS, and $\hat{\delta}_m = 0$, then $\xi_M = \{1, \cdots, M\}$. If there exist some NLOS measurements, then $\xi_M$ is the ranging set with NLOS measures mitigated by using (4.39).

Performing the gradient operation $\nabla$ to the error residues $e(\mathbf{y})$ with respect to the anchor node $m$, and we have

$$\begin{aligned} \nabla e(\mathbf{y}) &= 2 \sum_{m\in\xi_M} \left(||\mathbf{y} - \mathbf{x}_m||_2 - (\hat{\mathbf{r}}_m - \bar{\delta}_r)\right) \frac{\mathbf{y} - \mathbf{x}_m}{||\mathbf{y} - \mathbf{x}_m||_2} \\ &= 2 \sum_{m\in\xi_M} \left(1 - \frac{\hat{\mathbf{r}}_m - \bar{\delta}_r}{||\mathbf{y} - \mathbf{x}_m||_2}\right)(\mathbf{y} - \mathbf{x}_m) \end{aligned} \tag{4.41}$$

where

$$\nabla\left(||\mathbf{y}-\mathbf{x}_m||_2-(\hat{\mathbf{r}}_m-\bar{\delta}_r)\right)=\nabla\left(||\mathbf{y}-\mathbf{x}_m||_2\right)$$
$$=(\mathbf{y}-\mathbf{x}_m)/||\mathbf{y}-\mathbf{x}_m||_2 \tag{4.42}$$

When the gradient function of (4.41) is available, the refined position can be updated by

$$\mathbf{y}:=\mathbf{y}+\alpha\nabla e(\mathbf{y}) \tag{4.43}$$

where $\alpha \in (0, 1]$ is the update step size.

By using the steepest descent approach in (4.43), the objective function of (4.40) can be minimized by performing a local optimization above the global optimized value obtained by SDP. The overall performance of (4.43) can be guaranteed by providing an initial value $\mathbf{y}$ from SDP results.

### *4.4.3 Coverage Constraint*

The obtained mobile phone position $\mathbf{y}$ of (4.43) could still be affected by interference or large bias error. Using some criterion to filter out the error result is the last step before delivering the final result to the user.

The GDOP value of (4.37) can be calculated by inserting $\mathbf{y}$ of (4.43). If the GDOP value is higher than a threshold, then the result $\hat{\mathbf{y}}$ should be mitigated. Specifically, if the calculated position result falls outside of the anchor network coverage, the confidence of the result is low (high GDOP value). The threshold is determined by the minimum acceptable resolution and calculated by experiment. Using the GDOP as the coverage constraint, the problematic localization results could be filtered out. If the ranging variance is available, using CRLB of (4.34) instead of GDOP is equivalent. The filtered result of $\hat{\mathbf{y}}$ is the final calculated position of the mobile phone, and deliver to the user for other location-based-services (LBS).

## 4.5 Numerical Results

In Section 4.2, we derive the pseudorange (PR)-based SDP algorithm to obtain the position with relaxed global optimal value. Due to the unknown system parameter $\delta_r$, the obtained ranging result is not the true distance between an anchor and a phone but a pseudorange. To refine the position results, the steepest descent (SD) approach based on local search and mitigated NLOS ranges is proposed in Section 4.4. For simplicity, we name the pseudorange-based SDP algorithm "SDP-PR"; "SDP-PR-SD" is the SD refined version in additional to "SDP-PR."

To illustrate the effectiveness of the SDP algorithm, we compare our proposed "SDP-PR" with other least-square (LS) approaches, e.g., classic TOA-based LS ("LS-Classic") [16], and revised LS approach for pseudorange measurements ("LS-PR") [95]. Regarding the position refinement, classic LS can also be applied after the semidefinite relaxation instead of our SD approach, denoted as "SDP-PR-LS." Using LS along with SPD has higher computational complexity than the steepest descent (SD) due to the inverse matrix calculation requirement in LS. To compare the performance of the low complexity refinement approach "SDP-PR-SD," we compare it with more complex "SDP-PR-LS" in terms of achievable performance.

We use root mean squared error (RMSE) of the position result as the performance metric, and conduct 1000 Monte Carlo simulations with $1/\sigma^2 = -30 \sim 30$dB, where $\sigma$ is the ranging noise variance. Eight anchor nodes are simulated, and the position matrix in the 2D plane is

$$\mathbf{x}_m = \begin{bmatrix} 0 & 0 & 80 & 80 & 0 & 40 & 80 & 40 \\ 0 & 80 & 0 & 80 & 40 & 0 & 40 & 80 \end{bmatrix}^T \tag{4.44}$$

where $m = 1, \ldots, 8$.

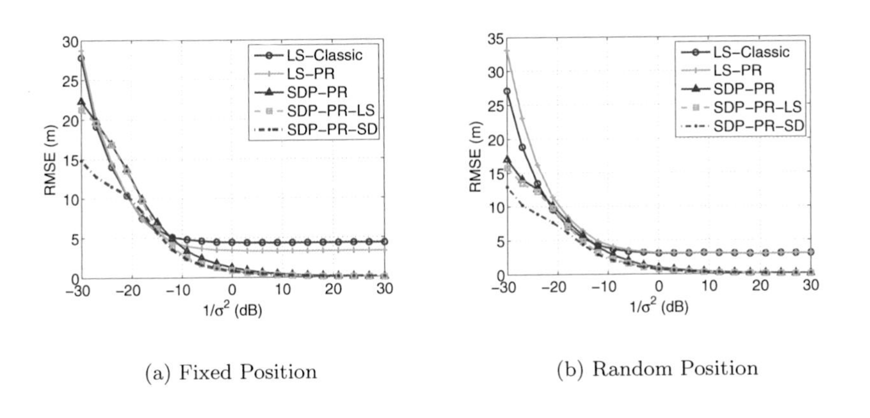


**Figure 4.1: Comparison of LS approaches to SDP and revised SDP algorithms when the mobile phone is in fixed position of [30, 20] (a) and random position with $x$ and $y$ distributed between (20, 80) (b).**

When the mobile phone is placed at $\mathbf{y} = [30, 20]^T$, the RMSE performance comparisons of these five approaches are shown in Fig. 4.1(a). The RMSE results when the mobile phone is randomly placed in a region are shown in Fig. 4.1(b). From Fig. 4.1(a) and Fig. 4.1(b), we observe that SDP-based approaches outperform LS-based approaches in almost all the $1/\sigma^2$ regions. In terms of the postprocessing refinement, using LS and steepest descent after SDP processing can all improve the localization performance. From the results of "SDP-PR-LS" and "SDP-PR-SD," we observe that "SDP-PR-SD" outperforms more complex "SDP-PR-LS," especially when the ranging noise variance is high. These results demonstrate the effectiveness of the proposed SDP approach along with the steepest descent post-processing.

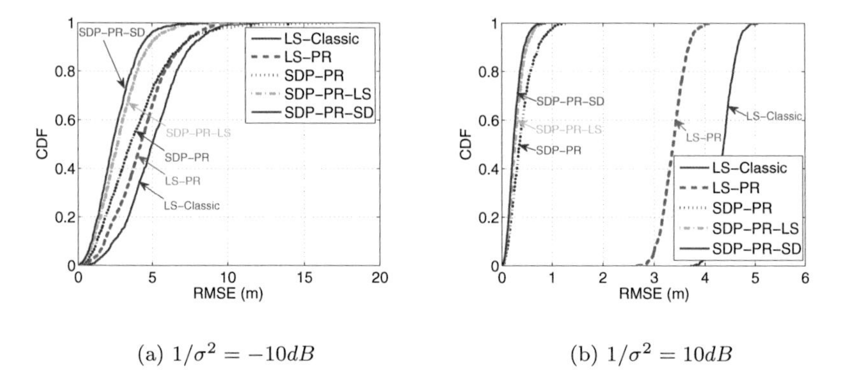


(a) $1/\sigma^2 = -10dB$ (b) $1/\sigma^2 = 10dB$

**Figure 4.2: Cumulative distribution of LS approaches to SDP and revised SDP algorithms when the mobile phone is in [30, 20], $1/\sigma^2 = -10dB$ (a) and $1/\sigma^2 = 10dB$ (b).**

Fig. 4.2(a) and Fig. 4.2(b) show the cumulative distribution function (CDF) of the calculated position RMSE when $1/\sigma^2 = -10$dB and $1/\sigma^2 = 10$dB, respectively. The CDF results show the similar relative performance as in Fig. 4.1(a) and Fig. 4.1(b). SDP approaches outperform LS with larger relative gains when the ranging error decreases. "SDP-PR-SD" performs best among these five approaches. Although the performance gains of "SDP-PR-SD" over "SDP-PR-LS" is relatively small in Fig. 4.2(b), the low complexity property of "SDP-PR-SD" makes it more applicable for real systems.

The CRLB of the position estimation with 8 anchor nodes and $1/\sigma^2 = 10dB$ is shown in Fig. 4.3 when the mobile phone is placed at different positions. This figure can be used to illustrate the theoretical performance bound with respect to the ranging variance and the geometric placement of the anchor nodes. If not using the assumed ranging variance, CRLB is equivalent to the metric of GDOP in (4.37). Moreover, if the estimated mobile phone position is substituted into the CRLB, the obtained value of CRLB or GDOP can be used as the metric to evaluate the coverage. If the position result lies outside of the coverage, then this position result should be mitigated. Using this post-position filter in position refinement of "SDP-PR-SD," the obtained localization accuracy can be further improved.

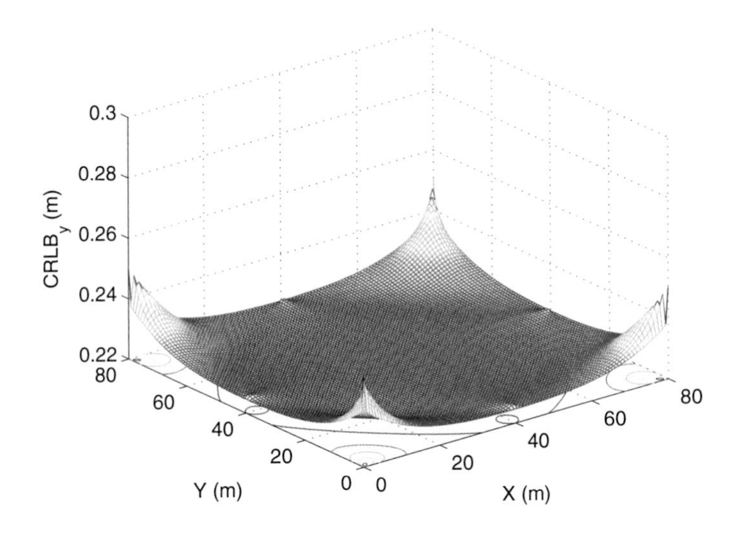


**Figure 4.3: The CRLB of the position estimation with 8 anchor nodes and $1/\sigma^2 = 10dB$ when the mobile phone is placed at different positions.**

## 4.6 Experimental Evaluation

To make the real system work efficiently, a lot of interferences and abnormal conditions should be handled. A good algorithm should be superior in theory and simulation and also superior with robustness in real-world experiments.

### *4.6.1 Experiment Setup for Localization*

To evaluate and compare the performance of different localization algorithms, we deployed the anchor network in a typical office environment as shown in Fig. 4.4. The mobile phone performs ranging and calculates its own position by demodulating the beacon information transmitted from anchor nodes. This environment is polluted with normal voice sound and other acoustic interference, e.g., the sound noise from the fan in a computer. To evaluate the various localization algorithms, we conducted the measurements and drove these localization algorithms with experimental data. The CRLB of the position estimation for the experiment environment is shown in Fig. 4.5 when the ranging accuracy is assumed as 2cm. Fig. 4.5 can be used to illustrate the theoretical performance lower bound with different ranging accuracies. Without the assumed ranging variance, CRLB is equivalent to the metric of GDOP in (4.37). Moreover, the obtained value of CRLB or GDOP can be used as the metric to evaluate the coverage and perform post-position filtering.

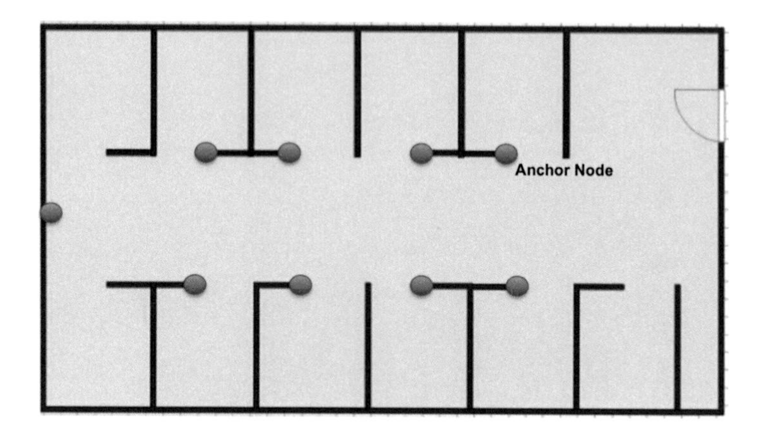


Figure 4.4: The experiment environment.

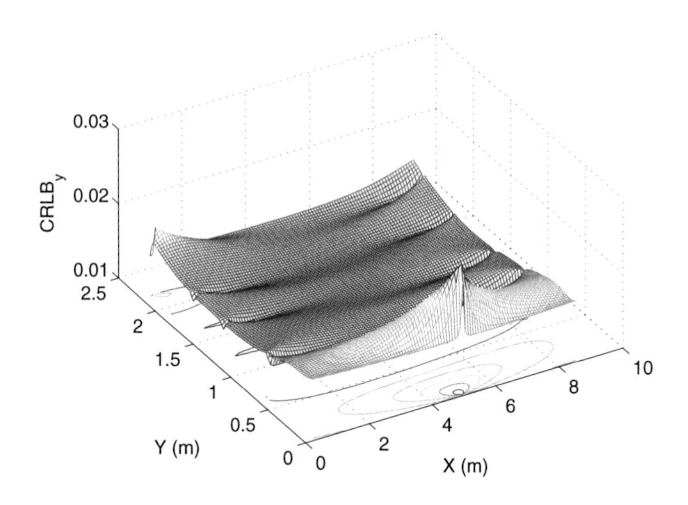


Figure 4.5: The CRLB of the position estimation for the experiment environment when the ranging accuracy is assumed as 2cm.

## 4.6.2 Localization Results

**Performance Comparison.** Similar to Section 4.5, we use the CDF results to compare the performance of different algorithms. The algorithms compared are still the same as in Section 4.5, i.e., "LS-Classic," "LS-PR," "SDP-PR," "SDP-PR-LS," and "SDP-PR-SD." Unlike Section 4.5, the RMSE results with a variable of $1/\sigma^2$ are not available in experiments, due to the uncontrollability of the ranging variance.

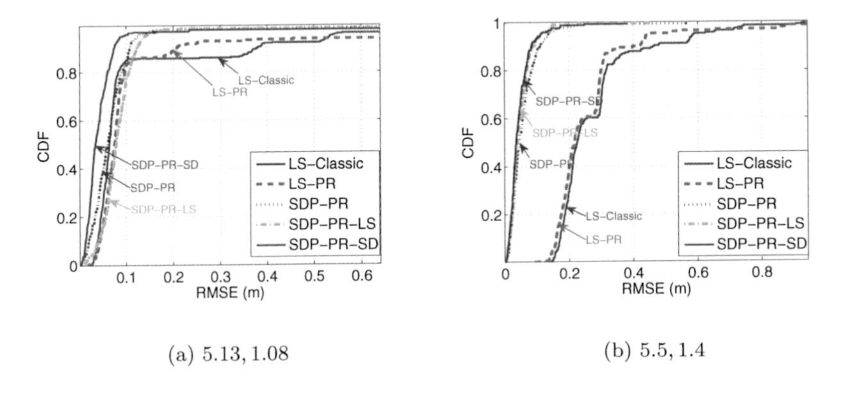


**Figure 4.6: Cumulative distribution of LS approaches to SDP and revised SDP algorithms when the mobile phone is in $[5.13, 1.08]$ (a) and $[5.5, 1.4]$ (b) by using experimental data.**

**Table 4.1: The localization accuracy of different algorithms with** 80% **probability when the mobile phone is placed near Pos1** $[5.13, 1.08]m$ **and Pos2** $[5.5, 1.4]m$

| | LS-Classic | LS-PR | SDP-PR | SDP-PR-LS | SDP-PR-SD |
|---|---|---|---|---|---|
| Pos1 | 0.083 | 0.097 | 0.09 | 0.105 | 0.06 |
| Pos2 | 0.32 | 0.30 | 0.082 | 0.069 | 0.065 |

Fig. 4.6(a) and Fig. 4.6(b) show the CDF of the position error when the mobile phone is placed near $[5.13, 1.08]m$ and $[5.5, 1.4]m$, respectively. The SDP-based approaches perform better than the LS-based approaches in these two cases. By performing position refinement after the SDP results, "SDP-PR-LS" and "SDP-PR-SD" are superior than other approaches. Using the steepest descent (SD) as post-processing is less complex than the LS approach, and "SDP-PR-SD" even outperforms "SDP-PR-LS" in most situations with different performance gains, e.g., the available gain in Fig. 4.6(a) is larger than in Fig. 4.6(b).

For more detailed comparison, we created Fig. 4.7 by listing the localization accuracy that is achieved by 80% measurements. From Fig. 4.7, we know that the localization accuracy of "SDP-PR-SD" is near 6cm, i.e., 80% of position results within 6cm of error. Achieving such a high-accuracy position is low-cost and only relies on the normal mobile phone at the user's side. With such high precision at low cost, we expect our localization algorithm will enable a large spectrum of indoor location-based services in museums, stores, libraries, and hospitals, just to name a few.

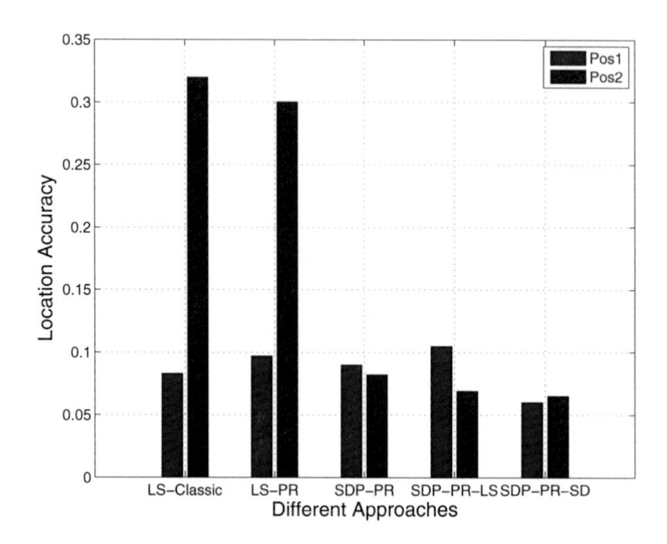


**Figure 4.7: The localization accuracy of different algorithms with** 80% **probability when the mobile phone is placed near Pos1** $[5.13, 1.08]m$ **and Pos2** $[5.5, 1.4]m$.

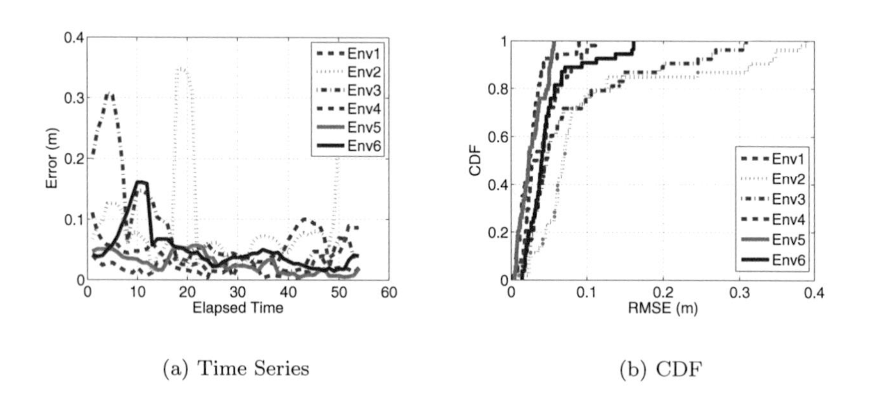


(a) Time Series

(b) CDF

**Figure 4.8: The time series (a) and CDF (b) of localization error at different spots.**

**Performance Comparison at Different Spots**. To demonstrate the overall localization performance at different spots, we calculated the localization accuracy of "SDP-PR-SD" when its user stands still at different spots. In this case study, to support quantitative analysis, we used the time series (Fig. 4.8(a)) and CDF (Fig. 4.8(b)) of the localization error to illustrate the performance. If using 80% probability, the localization error for all the spots is in the range between $4cm$ and $10cm$. These results are very accurate for indoor mobile phone-based localization. Localization traces of three moving subjects are illustrated in Fig. 4.9. The moving traces demonstrate the localization accuracy and real-time feature of our system.

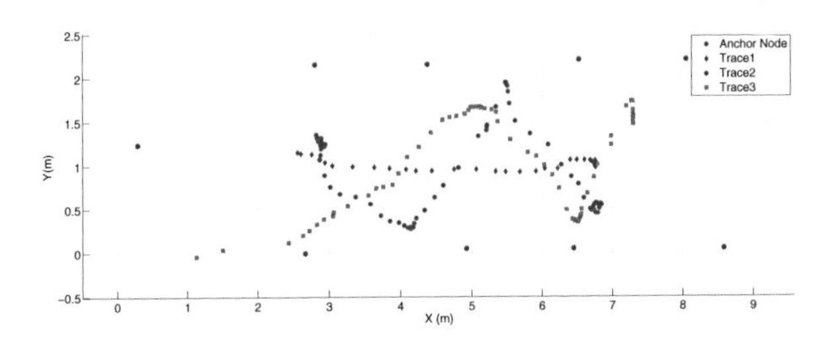


**Figure 4.9: Three different moving traces.**

## 4.7 Conclusion

To address the challenges of utilizing the audible-band acoustic signal, i.e., small coverage and NLOS sensitivity, we propose semidefinite programming to ensure global optimal position values, along with steepest descent refinement and NLOS mitigation. The metric of GDOP is introduced to quantify the coverage and evaluate the quality of estimated position value of a mobile phone. By comparing it with other classic localization algorithms and post-processing techniques, our proposed "SDP-PR-SD" method achieves best performance in different simulation and experiment scenarios.

# Chapter 5

# Guoguo: Enabling Fine-Grained Smartphone Localization

**Kaikai Liu**

*Assistant Professor, San Jose State University*

**Xinxin Liu**

*Privacy Engineer, Google Inc*

**Xiaolin Li**

*Associate Professor, University of Florida*

## CONTENTS

## 5.1 Introduction

As one of the two key components of a mobile context (time and location), localization has been the subject of extensive works ranging from algorithms, models, and supporting technologies, to systems and applications. Current coarse-grained (room-level or meter-level) localization on a smartphone has enabled a lot of mobile services, such as location-based services, maps, and navigation systems. However, these services are severely limited when applied in more pervasive indoor environments due to low resolution. Indoor users can hardly navigate like those using outdoor GPS services. The major difference is as follows: meter-level (e.g. GPS with five-meter accuracy) localization is sufficient to navigate a car (meter-level footprint) on a street (several-meter footprint); but it is far from sufficient to navigate a user (foot-level footprint) in a library (with half-meter-wide isles and inch-level books) or a shopping mall (with inch-level items). Smartphone-based accurate "indoor GPS" or IPS (indoor positioning system) have been long awaited to improve indoor mobile services and enable new services. Despite significant efforts on indoor localization in both academia and industry in the past two decades [6, 8, 10, 15, 85, 113], highly accurate and practical smartphone-based indoor localization remains an open problem. Some accurate localization solutions cannot be readily converted to smartphone-based ones due to various constraints.

The technology that enables centimeter- or decimeter-level location accuracy and integrates with context-aware mobile services has the potential to revolutionize users' mobile experiences. Using ranging, angle, displacement, or fingerprinting-based sensing is required for accurate indoor localization. However, none of the existing solutions could achieve the desired performance at low cost and low complexity. Time-of-arrival (TOA)-based ranging approaches, e.g., special devices using ultra-wideband or ultrasound signals, are more reliable and accurate. However, the increased complexity and additional devices make them not practically useful for conventional users. Other low-complexity approaches rely on existing sensors in smartphones to infer current location, e.g., WiFi, fingerprinting, compasses, and accelerometers. However, their low accuracy and prerequisites like site survey or pairing limit their applications.

Recent research on leveraging ubiquitous microphone sensors in a smartphone introduces a convenient approach to ranging by using the audible band acoustic signal (less than 20kHz). A microphone sensor is inexpensive and has potential for highly accurate ranging due to the low transmission speed of acoustic signals. However, the limited bandwidth of a microphone, strong attenuation of aerial acoustic signals, as well as various interferences in the audible band, pose significant challenges in using acoustic signals for indoor localization. Although using acoustic signal to perform ranging-based localization suffers from various issues, such as short operation distance, low update rate, and sound pollution, the potential of centimeter-level localization accuracy motivates us to design better solutions to overcome its drawbacks.

In this chapter, we propose an indoor localization system called "Guoguo,"[1] and further improve its performance in terms of coverage, accuracy, update rate, and sound pollution [57]. We make its acoustic beacons imperceptible to humans and improve detection sensitivity for better location coverage. Rather than simply using "Beep" signals to enable passive sensing for multiple smartphone users, we designed transmission waveform, wide-band modulation, and one-way synchronization and ranging schemes. We designed a transmission scheme of the acoustic beacon that follows the high-density pseudo-codes to enable anchor node identification without radio assistance on a smartphone. We propose a *symbol-interleaved* beacon structure to overcome the drawback of the low transmission speed of acoustic signal and improve the *location update rate*. To improve the accuracy of ranging, we propose a fine-grained adaptive time-of-arrival (TOA) estimation approach that exploits the details of the beacon signal and perform non-line-of-sight (NLOS) identification and mitigation. By combining all these techniques together, we implement the prototype system with anchor nodes and a localization processing app in a smartphone, and make it work in a realistic environment.

The rest of the chapter is organized as follows. Section 10.8 summarizes the related work. Section 5.2 introduces the design consideration and system

[1] "Guoguo" means katydid/bush-cricket in Mandarin Chinese. Guoguo, famous for its beautiful sound, has represented a pet culture in ancient China for thousands of years

architecture. Section 5.3 presents the acoustic beacon and anchor network design. Section 5.4 presents the ranging and localization approaches by using smartphone. Section 5.6 presents the experimental evaluation of our proposed approach. Section 5.7 concludes this chapter.

## 5.2 System Design

### 5.2.1 Motivation

Various existing localization solutions rely on different assumptions and are dedicated to specific applications. For most existing solutions using only built-in sensors in a smartphone, rough position information can be obtained for building- or room-level navigation. For indoor mobile services like shelf-to-shelf navigation, location-aware and context-aware information recommendation, and virtual-reality interaction, fined-grained indoor localization with foot-level accuracy is critical.

In terms of system complexity, the users' side is more stringent, especially in hardware requirements. It is very hard to persuade consumers to purchase special devices for indoor localization. Even with special devices, the integration with the mobile services provided by smartphones is also a difficult problem. To enable centimeter- or decimeter-level localization accuracy, and only require a smartphone on the user side, the investment in indoor infrastructure is necessary and has potential for other interesting applications, e.g., surveillance and monitoring. For normal retail stores, several hundred dollars' investment in indoor infrastructure to enable indoor "smart" shopping could attract more consumers to enjoy mobile services, and in turn boost business.

With these two points in mind, we focus on developing a fine-grained smartphone-based indoor localization system leveraging an indoor low-complexity anchor network.

### 5.2.2 Design Challenges and Considerations

To meet the localization accuracy, TOA-based ranging is more appropriate [80]. Limited by the smartphone, using acoustic signal for TOA-based ranging is an appropriate solution. However, making the acoustic signal for ranging and localization in real situations has a lot of challenges.

First, the coverage of the anchor network should be sufficient for a room or a retail store with less than 10 anchor nodes placed to minimize the infrastructure cost. The possible solution is to increasing the transmit power or detection sensitivity. However, increase the transmit power could cause sound pollution and high energy cost. [62, 81] all suffer from sound pollution caused by the audible signal. H. Liu et al. [52] achieves less pollution, but only works within 3 meters. Maintaining a low transmit power and keeping the acoustic beacon unnoticeable are the two prerequisites in the anchor network design.

Thus, improving the detection sensitivity in the smartphone by using an advanced signal processing technique is the feasible solution to ensure sufficient coverage.

Second, sufficient update rate is required to track users' movement. If the localization update rate is too sluggish, no user could be patient enough to await their location results. There are also not enough time margins for mobile services. The existing acoustic ranging solution proposed in [52] shows 7.8s delay in one location calculation; authors in [74] perform localization for desktop and calculate the running time for 5 nodes in the range of 3.7s to 285s. Possible solutions for improving the location update rate include better design of the anchor network protocol, real time implementation of the algorithm, and eliminating the use of WiFi assistance (it takes about 2s for one scan).

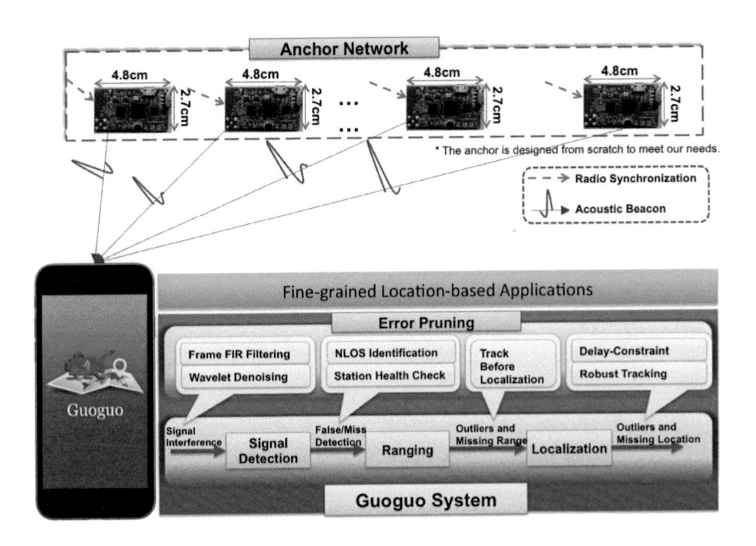


**Figure 5.1: Conceptual architecture of the Guoguo System.**

Third, the system should support multiusers simultaneously. However, solutions used in [81, 52, 74] utilize the two-way approach for ranging. Two-way ranging eliminates the synchronization requirement, but suffers from the limitation of only allowing one user to access it at one time. Sound source localization [108] is another way to eliminate the synchronization by using active transmit mode for users; however, the random access time of the user making the sensor network could only handle limited users at the same time. One-way passive ranging should be utilized to maximize the multiuser capacity, which could support unlimited number of users in theory.

### *5.2.3 System Architecture*

In order to meet the requirements mentioned in Section 5.2.1 and 5.2.2, we propose to the develop an anchor network with better coverage and make the beacon unnoticeable. With several low-complexity anchor nodes mounted on indoor places, the smartphone can localize itself by receiving more than three beacon signals from the anchor nodes. The simplified architecture of our proposed localization system is shown in Fig. 5.1. On the receiver side, we implement advanced signal processing and a ranging algorithm in a smartphone to achieve accurate localization in passive mode. The cloud server provides mobile services for potentially massive users in real-time response mode, e.g., localization, navigation, and location-based services, minimizing computation and storage cost in a smartphone. We adopt the NoSQL database to handle massive user concurrent access and big data services in the future, e.g., business analytics and intelligence, mobile social network dynamics and evolutions.

Specifically, the anchor network periodically transmits modulated acoustic beacon signals based on the token of the controller. By detecting and extracting the desired information bits embedded in beacon signals, the smartphone demodulates the symbols and calculates the relative TOA. The algorithm in the smartphone eliminates erroneous measurements using statistical pruning methods, and accesses the anchor position by matching the pseudocodes. Finally, the smartphone can be aware of its fine-grained position by accessing the localization results.

## 5.3 Anchor Network Design

The key notations are listed in Table 5.1.

**Table 5.1: Notation**

| Symbol | Description |
|---|---|
| $M$ | Number of anchor nodes ($m = 1, \ldots, M$) |
| $L$ | Pseudocode length |
| $\mathbf{p_m}$ | Pseudocode sequence for $m$-th anchor node |
| $F_s$ | Sampling rate of smartphone |
| $T_s$ | Symbol duration of acoustic beacon |
| $v_s$ | Speed of the acoustic signal |

### 5.3.1 Design Criterion

The timing accuracy of existing radios, e.g., Bluetooth or Wi-Fi, on a smartphone, is not accurate enough for TOA-based localization. The mobile OS (Android/iOS) introduces additional random delay in radio processing. Moreover, using Wi-Fi for localization would impact the Internet connection and battery life. We eliminate the use of radio signal in a smartphone, and only require a microphone to participate in the passive acoustic beacon sensing. Such a low hardware requirement on the user's side can help us simply extend our system to other devices that contain a microphone and computation resources rather than limiting ourselves to the smartphone platform. All anchor nodes are synchronized by Zigbee radio and transmit the acoustic beacon signal to broadcast its unique location. We propose approaches to jointly solve the one-way anchor–smartphone synchronization during the location estimation process. Compared with systems using active ranging or radio-assisted ranging, e.g., Cricket [85] and Beep [62], our solution features the following salient advantages: no need for a special radio signal (e.g., ZigBee) that is not available on a smartphone; no need of special devices for ranging assistance; one-way passive ranging to support massive numbers of users.

### 5.3.2 Anchor Node Hardware Design

Targeting fine-grained indoor location-based services, e.g., retail store, museum, office, and classroom, several hundred dollars investment in the anchor infrastructure to enable centimeter-level location-related services could be profitable. To minimize the infrastructure cost and promote the real system deployment, we designed our own hardware for the anchor node with low-complexity and additional features, e.g., solar powered, battery charging, pluggable sensor board, remote program upload, wireless speaker/microphone. The comparison of our "Guoguo" hardware vs. two quarters is shown in Fig. 5.2. The total BOM price for the anchor node is less than $10. With one anchor

**Figure 5.2: Guoguo anchor node vs. quarter.**

covering every 10–20 meters in an indoor space (e.g. museum, shopping center), it is feasible and cost effective to deploy such systems in many indoor environments. For example, with marginal investment in the indoor infrastructure, a retail store can enable "smart" shopping experiences for smartphone users with fine-grained shelf-to-shelf navigation, location-aware recommendation and advising, and physical item-searching and navigation.

### 5.3.3 Transmitter Waveform Design and Modulation

Using audible-band acoustic signal as a beacon for ranging and communication, we must contend with a variety of noises to ensure its accurate ranging due to the highly populated frequency band below 20kHz. The ranging accuracy directly depends on the signal-to-noise ratio (SNR) and effective bandwidth. However, higher transmitter signal bandwidth and power will generate noise that disturbs users. The standard microphone in a mobile phone can only support bandwidth of 200Hz $\sim$ 20kHz. To minimize the sound while perform ranging, we choose 15kHz $\sim$ 20kHz as the operating band. The reason is that our ear is less sensitive to the high frequency signal, while the microphone in a smartphone could still receive signal in this boundary band. With well-controlled transmitter signal power, the acoustic beacon could be unnoticeable to humans while still detectable for smartphones. We use spread-spectrum and ultra low-duty-cycle pulse sharpening techniques to ensure the proper operation of the acoustic beacon under realistic environments, as well as improving the ranging accuracy. Unlike the less-sophisticated acoustic signal used in Cricket and Beep, the beacon signal used in *Guoguo* is wide-band modulated with short duration in the time-domain and unnoticeable to humans.

The task of the anchor nodes that transmit modulated acoustic sharp pulse to the mobile receiver is to realize synchronization and ranging, as well as to inform its unique pseudocode $\mathbf{p} = [p_{m,i}]$, $m = 1, \ldots, M$, $i = 1, \ldots, L$, where $M$ is the number of the anchor nodes; $L$ is the pseudocode length. The pseudocode $\mathbf{p}_{m,i} = 0, 1$ is also the information bits carried by the beacon signal. For the symbol waveform, we chose to use the second derivative of the Gaussian (Doublet) Pulse [104] and multiply it to the carrier wave. The waveform could be written as

$$g(t) = \frac{A}{\tau}\left[1 - 4\pi\left(\frac{t}{\tau}\right)^2\right] \exp\left[-2\pi\left(\frac{t}{\tau}\right)^2\right] \cos(2\pi f_c t) \tag{5.1}$$

where $\tau$ is the pulse width parameter, $f_c$ is the carrier wave frequency. We truncate the Doublet pulse by their $3\tau$ to approach the real-time condition. The short duration of the Doublet pulse in the time-domain could contribute to smaller symbol duration, higher ranging rate, and better location update rate.

The center frequency $f_c$ of the modulated pulse is controlled by the on-chip timer and working at 18kHz. In addition to $f_c$, $\tau$ is tuned to ensure the effective bandwidth of (5.1) lies in between 15kHz $\sim$ 20kHz. The multiplication with the carrier wave in (5.1) is also a spread spectrum process that extends the initial narrow band $f_c$ to wide band signal $g(t)$ by using the Gaussian Doublet Pulse sharping. The use of ultra low-duty cycle pulse (5.1) has the feature of higher data rate, higher location refresh rate, better multipath resolution, lower energy consumption, and smaller sound pollution. To balance between the system complexity and the sophisticated modulation scheme, we chose to perform 2-PAM modulation with symbol duration as $T_s$; i.e., transmit sharp pulse $g(t)$ represents symbol "1," no pulse for symbol "0."

### 5.3.4 High Density Pseudocode Sequence

One design challenge faced in selecting the pseudocode is to utilize the proper pseudo-codes with enough code distance redundancy to separate different anchor nodes, e.g., utilize orthogonal codes. In addition to communication, every symbol "1" in $\mathbf{p}$ will contribute to one ranging measurement, thus the high density bit "1" could improve the location update rate.

The conventional communication process relies on the preamble part of a frame to perform synchronization, then follows the data bits. For our designed anchor network, the beacon signal should only contain the pseudocodes, and transmits cyclically to save round time for higher location update rate. The stringent requirement on efficiency making the removal of unnecessary parts of bits is especially important for Guoguo due to the low transmission speed of acoustic signal. However, most orthogonal pseudo-codes are not cyclic orthogonal; it will lose orthogonality due to the cyclical transmission. Using Walsh-Hadamard codes, for example, only $L+1$ codes with length $L$ are orthogonal to each other in all phases among Hadamard Matrix $H(2^L, L)$. Moreover, the balanced "1" and "0" in these pseudo-codes does not comply with our requirement on high "1" density sequence.

In order to meet the special requirement of pseudocode for our anchor network, we select three distinct maximum sequences as $[ms_1, ms_2, ms_3]$ with length of $2^L$, while $L$ is the pseudocode length. To increase the "1" density, we perform the plus and decision process to combine these three m-sequences as a new sequence $ms'$, as

$$ms' = [ms_1 + ms_2 + ms_3] \gtrless_0^1 0.5 \tag{5.2}$$

where the length of $ms'$ is $2^L$. The reason that we choose three maximum sequence is to balance the "1" density and the number of available new sequences. Reshape $ms'$ into a $(L', L)$ matrix as $\mathbf{m} = [m_{a,b}]$, with $L' = 2^{(L-1)}/L$, $a = \{1, \ldots, L'\}$, $b = [1, \ldots, L]$. $L$ is the code length, and total $L'$ sequences in $\mathbf{m}_{a,b}$. To maintain the minimum code distance $d_{tol}$ between each sequence, we only select sequences in $\mathbf{m}$ that satisfy the conditions in any phases such that

$$p_{m,i} = \arg_{\hat{a}}\{p_{\hat{a},i} | \min ||m_{a,i} - p_{k,i+\triangle_i}||_1 \geq d_{tol}\}, \quad (5.3)$$
$$\forall k = \{1, \ldots, m-1\}, \forall \triangle_i = \{1, \ldots, L\}$$
$$\hat{a} \in \{1, \ldots, 2^{(L-1)}/L\}, m = m + 1$$

where $|| \cdot ||_1$ calculates the code distance between $m_{a,i}$ and the selected sequences $p_{k,i+\triangle_i}$, while $i + \triangle_i$ performs cyclic phase change for the sequence. (5.3) filters sequences in $\mathbf{m}$ that meet minimum distance $d_{tol}$ among selected sequence sets and adds into $p_{m,i}$. When $L = 16$, among a total of $2^{(L-1)}/L = 2048$ sequences, only 12 sequences satisfy (5.3). These sequences can be assigned to $M$ anchors, where $M = 9$ for a micro-cell in our system.

### 5.3.5 Anchor Network Synchronization

There are two kind of beacons transmitted by the anchor nodes in Guoguo. The first kind of beacon is a radio beacon, which synchronizes all the anchor nodes to the controller's clock based on message passing; the second kind of beacon is transmitted by the anchor nodes from the acoustic speaker, which provide ranging and synchronization information to the smartphone. In this sub-section, we mainly focus on the radio beacon, which synchronizes the whole anchor network.

**Anchor Network Synchronization**. To enable the trilateration in a smartphone based on the TOA acoustic ranging results, all these TOA results should be based on the same timing-pace, i.e., synchronized. The accuracy of the synchronization between anchors directly affects the overall ranging accuracy. For example, the transmission speed of the acoustic signal is near 340m/s; if the synchronization error is around 10 micro seconds, the resulting ranging error is at least 3.4 meters. Thus, sub-micro-second-level synchronization accuracy is the prerequisite of the Guoguo system.

The anchor network structure features several anchor nodes and one controller (which could also be one of the anchor nodes). The controller provides basic timing via radio beacon passing for the whole network, and all the anchor nodes are synchronized to the controller by receiving the radio beacon passively and periodically. Such a scheme can guarantee that the transmitted signals from anchor nodes are synchronized to a common timing source. The controller beacon is realized by using the existing radio chips, e.g., Zigbee radio. The data domain of the radio beacon contains the anchor token, and current beacon index. The anchor node performs its own processing when the anchor token is received, i.e., estimate the received beacon period, transmit its own acoustic beacon according to the estimated timing pace. Denote the controller beacon interval is $T_p$, every anchor node executes the similar processing, the delay can be viewed as fixed and making the acoustic beacon timing equals to $T_p$.

**Acoustic Beacon and Multiplexing.** Due to the reason that there is no commercial acoustic communication physical layer in mobile devices, we need to design our own processing to realize the communication capability. The ranging and synchronization capabilities of the anchor–smartphone pair are all based on the acoustic communication channel, while a good beacon structure can enable multiplex of anchors and improve the overall throughput and data rate.

To enable sufficient signal-to-noise ratio when detecting the acoustic beacon signal, we designed such that the smartphone only detects one acoustic beacon at one frame duration. The guard time between each beacon is $T_g$; the frame duration is $T_f = LT_s + T_g$. One beacon period of anchor can be written as $T_p := T_f = LT_s + T_g$. Multiple anchors should be synchronized and transmit their own acoustic beacon sequence (length $L$) using the time division multiple access (TDMA) approach to avoid the collision between each beacon. The total round period $T_r$ equals to $T_r = MT_p = M(LT_s + T_g)$. The mobile phone can differentiate all the anchor nodes after $T_r$, thus the synchronization time between the anchor network and smartphone is $T_{syn} = M(LT_s + T_g)$. The localization process after synchronization also needs $M$ beacons to obtain ranging information from all the anchors, i.e., update rate is $T_{up} = T_{syn} = M(LT_s + T_g)$. When $M = 9$, $L = 16$, $T_s = 0.0781s$, assume $T_g = T_s$, then $T_{up} = T_{syn} = 11.9493s$. Such a long time of synchronization and position update is caused by the low transmission speed of acoustic signal, and it is not fast enough for tracking the movement of humans. One possible way to increase the update rate is to lower the symbol duration $T_s$, with faster symbol transmission rate. However, symbol duration is restricted by the delay spread or coherence bandwidth of the channel, and could not be further reduced without sacrificing the multipath resistance. To improve the speed of localization without relying on the compression of the symbol interval, a new beacon structure should be developed.

### 5.3.6 Symbol-Interleaved Beacon Structure

One possible solution is interleaving the $L$ length symbols into different beacon periods, and sending the received adjacent symbols from different anchors. In this way, the round time for one location calculation can be significantly reduced; we call this *symbol-interleaved* acoustic beacon structure. Unlike the conventional TDMA that transmits the whole frame within a beacon period, i.e., $T_p := T_f = LT_s + T_g$, we divide the whole frame into symbols and transmit one symbol in each beacon period without $T_g$; the beacon period $T_p$ can be decreased to $T_p := T_s$.

To enable code matching under reduced frame length, the receiver maintains $L$ length pipeline, and performs code matching in an iterative way. When the symbols in the pipeline are matched with one sequence in $\mathbf{p}$, the anchor node can be identified. Using *symbol-interleaved* beacon structure, the initial synchronization time $T_{syn}$ cannot be lowered, but the following *update time interval* can be reduced to $T_{up} = MT_s = 0.7029s$. The refresh of the location data for every 0.7029 seconds is sufficient for tracking slow-moving humans.

The transmitted beacon sequence from the anchor network in the $j$-th period can be written as $p_{c(j)}(a(j), b(j))$, where $a(j) \equiv [j \mod M]$ is the index of the anchor node; $b(j) \equiv \lfloor j/L \rfloor$ is the index of the pseudocode in one frame; $c(j) \equiv \lfloor j/(ML) \rfloor$ is one round measurement. The transmitted beacon from the anchor network can be modeled as

$$g_t(t) = \sqrt{\varepsilon} \sum_{j=0}^{N_s-1} p_{c(j)}(a(j), b(j)) \cdot g(t - jT_s) \tag{5.4}$$

$g(t)$ is the acoustic sharp pulse designed in (5.1); $\varepsilon$ is the signal energy. The number of total transmitted symbols is $N_s$.

## 5.4 Smartphone Localization

### 5.4.1 Design Workflow

In the smartphone, the *Signal Detection and Demodulation* module performs audio signal recording and filtering, detecting and extracting the embedded beacon signal. The detection result is the digital symbols. The *Code Matching* module matches the demodulated digital symbols to determine the anchor ID and its predefined location. The *TOA Estimation* module obtains pseudo-ranges by measuring the arrival time of the signal. The *Relative Distance* module accumulates all the distance from different anchor nodes until a sufficient number of measurements is available for localization. The *Localization* module performs location calculation by using the measured pseudo-distance pairs and available anchor positions. The technical details of all these modules are elaborated in the following subsections.

### 5.4.2 Symbol Detection and Demodulation

In the receiver side, i.e., the smartphone, we need to detect the signal and demodulate the information bits in the received signal $r(k)$. The received signal constitutes $\xi_j$ multi-paths, and these multi-paths can be utilized to extract the symbol. Thus, the detection problem can be written so as to detect the signal that is present or not in the $j$-th symbol. The received signal waveform in the $j$-th symbol period could be written as

$$r_j(k) = \sqrt{\varepsilon} \sum_{l=0}^{\xi_j-1} A_j^l g_j(k - k_j^l) + n_j(k) \tag{5.5}$$

Assuming the sampling rate is $F_s$, for symbol duration $T_s$, the total sampling point in one symbol is $M_o = T_s \times F_s$. For the low-duty-cycle pulse used in

(5.1) as the symbol waveform, the actual signal length is shorter than the total symbol length. Assuming the multi-path delay spread coefficient is $\alpha$, the average sampling points for the signal region is $M_p = \alpha \times (3\tau) \times F_s$. Thus, the two conditions of the hypothesis for detecting the signal can be written as

$$\begin{cases} H_0: & \mathbf{r}_{j,k} = \mathbf{n}_{j,k} \quad k = 1 \cdots M_o \\ H_1: & \begin{cases} \mathbf{r}_{j,k} = \sqrt{\varepsilon}\mathbf{s}_{j,k} + \mathbf{n}_{j,k} & k = k_j^0 \cdots k_j^0 + M_p - 1 \\ \mathbf{r}_{j,k} = \mathbf{n}_{j,k} & k = 1 \cdots k_j^0 - 1, k_j^0 + M_p \cdots M_o \end{cases} \end{cases} \tag{5.6}$$

where $\mathbf{n}_{j,k}$ is the matrix form of the noise $n_j(k)$ and $\mathbf{s}_{j,k}$ is the $k$-th sampling point for the signal in the $j$-th symbol. The symbol synchronization process is to detect the signal region in the noise background, i.e., detect $H_1$ condition out of $H_0$, while the TOA estimation is to detect the first path of the signal and its delay $k_j^0$.

To detect the signal region ($H_1$ condition), the decision vector $\mathbf{z}_j$ can be obtained by using the generalized likelihood ratio test (GLRT) [42]. The decision vector could be derived as the form of $\mathbf{z}_{j,k} > \eta_{syn}$, with the $j$-th symbol declared present if the inequity condition is satisfied and returns an estimated value of the signal region $k_p$. The threshold $\eta_{syn}$ is chosen to maintain a constant false alarm rate (CFAR) [42], and written as the form of $\beta\sigma$ where $\sigma$ is the noise variance, and $\beta$ is calculated in the experiment by using the given false alarm rate. Then $\hat{p}_{c(j)}(a(j), b(j))$ in the receiver will be set to "1," otherwise $\hat{p}_{c(j)}(a(j), b(j)) = 0$, where $c \equiv \lfloor j/(ML) \rfloor$. $\hat{p}_{c(j)}(a(j), b(j))$ is one estimated version of $p(a(j), b(j))$, and $c$ is one round measurement.

### 5.4.3 TOA Estimation

After detecting the symbols in Section 5.4.2, more detailed time-of-arrival (TOA) estimation should be performed to estimate the first path sample $k_j^0$ in the whole symbol duration. The TOA estimation provides ranging information that is needed for localization, and its accuracy directly affects the overall position resolution. The TOA estimation problem can be written so as to detect $k_j^0$ in the $j$-th symbol of (5.5) as

$$\hat{k}_j^0 = \min_k (k | r_j(k) > \eta_{toa}), k \in [k_p - J_p, K_p + M_o] \tag{5.7}$$

where $k_p$ is the rough signal region obtained during signal detection; $J_p$ is the step length used for jump-back-and-search-forward (JBSF) [22]; $\eta_{toa}$ is the TOA estimation threshold; $\hat{k}_j^0$ is the estimated TOA path of $k_j^0$. The TOA estimation threshold $\eta_{toa}$ can be dynamically adapted to balance the false alarm and misdetection probability. Based on our previous work [53], we can write the *TOA detection probability* as

$$P_d^{toa} \approx \frac{P_d(z_k)}{N_k P_{fa}(z_k)}(N_k P_{fa}(z_k) - C_{N_k}^2 (P_{fa}(z_k))^2) \\ = P_d(z_k)[1 - \frac{1}{2}(N_k - 1)P_{fa}(z_k)] \tag{5.8}$$

where $P_d(z_k)$ and $P_{fa}(z_k)$ are the detection probability and false-alarm probability for the single sampling point. From (5.8), we know that increasing $P_d(z_k)$ or decreasing $P_{fa}(z_k)$ of the single sampling point can contribute a better $P_d^{toa}$. Small $N_k$ also helps to improve the performance such that less sampling points are detected.

Since $P_d(z_k)$ and $P_{fa}(z_k)$ is a function of $\hat{\eta}_{toa}$, $P_d^{toa}$ can be written as $P_d^{toa}(\hat{\eta}_{toa})$. The maximum value of $P_d^{toa}$ can be achieved when $\hat{\eta}_{toa} = \hat{\eta}_{toa}^{oth}$, as

$$\hat{\eta}_{toa}^{oth} = \arg\max_{\hat{\eta}_{toa}}(P_d^{toa}(\hat{\eta}_{toa})) \tag{5.9}$$

Then, $\hat{\eta}_{toa}^{oth}$ can be written as

$$\hat{\eta}_{toa}^{oth} = \frac{1}{2}\sqrt{\varepsilon}\hat{A} + \frac{\sqrt{\pi}}{2\sqrt{2}}(1 - \frac{1}{N_k - 1})\hat{\sigma} \tag{5.10}$$

where $\sqrt{\varepsilon}\hat{A}$ and $\hat{\sigma}$ are the estimated signal and noise value. By using (5.10) in TOA estimation, the optimized TOA performance can be achieved under the criterion of maximum TOA detection probability, as shown in (5.9).

With $\hat{k}_j^0$ available, the distance can be obtained by multiplying the speed of acoustic signal $c$ as $\hat{r}_{m,j} = c \times \hat{k}_j^0 / F_s$. The practical formula of sound speed in air can be written as $c = 20\sqrt{\vartheta + 273.15}$m/s, where $\vartheta$ is the temperature in the air and can be measured by the anchor nodes.

### 5.4.4 Acoustic Beacon Synchronization and Code Matching

Performing code matching between predefined pseudo-codes $p_{m,i}$ and the estimated information bits $\hat{p}_{c(j)}(a(j), b(j))$, the $m$-th anchor node can be identified if these two sequences match. When symbols with total $M \times L$ length have been received, the code matching process can be utilized to synchronize between the anchor node and the smartphone, by

$$[\Delta_a, \Delta_b] = \\ \arg\min_{\Delta_a, \Delta_b} ||\hat{p}_{c(j)}(a(\hat{j}) + \Delta_a, b(\hat{j}) + \Delta_b) - p_{m,i}||_1 < d_{tol} \tag{5.11}$$

where the beacon period index $j \in [j_0, j_0 + ML]$, $j_0 = c(j) \times (ML)$ is the starting index of the symbol sequence, $ML$ is the total length of symbols that are used to perform (5.11), and $\hat{j} = j - j_0$, $c(j)$ can be used to illustrate the index number of code matching. $a(j) + \Delta_a$ and $b(j) + \Delta_b$ is the cyclic shifting in $\hat{p}_{c(j)}(\cdot)$. $d_{tol}$ is the detection threshold illustrating the tolerance for bit error.

With the offsets $\Delta_a$ and $\Delta_b$ available for the anchor node index and pseudocode sequence index, the mobile phone can aware the anchor node index $(\hat{m})$ and sequence index $(\hat{i})$ of the current received symbol $j$ as

$$\begin{aligned} \hat{m} &= [(j - j_0) \mod M] + \Delta_a \\ \hat{i} &= [(j - j_0) \mod L] + \Delta_b \end{aligned} \tag{5.12}$$

where $j_0 = \lfloor j/ML \rfloor \times (ML)$.

### 5.4.5 Distance Update

After the synchronization in Section 5.4.4, and obtaining $\hat{m}$ and $\hat{i}$ in (5.12), then every $M$ symbol can obtain one distance measurement group with the same pseudo sequence index $\hat{i}$. Such a group of measurements is the minimum tuple of ranging for one position update, and every measurement in such a group is from different anchor nodes. For the $j$-th symbol, the index of the group is $j_g = \lfloor j/M \rfloor$, with each element representing the TOA value $\hat{r}_{m,j}$. Denote the TOA estimation matrix as $\mathbf{r} = r_{m,j_g}$, $m = [1, M]$. For notation convenience, one group of measurements from all anchor nodes can be represented as $\mathbf{r}_g = r_{m,j_g}$, where $j_g$ is a fixed value, and $\mathbf{r}_g$ is a measurement vector.

Due to some symbol-miss in real situations and none distance measure during "0" symbols, the TOA value is not fully available for all $M$ anchor nodes in one round time, i.e., the obtained ranging value $\mathbf{r}_g$ is a sparse vector. To improve the reliability of ranging results, we perform robust regression for the TOA estimation matrix $\mathbf{r}$ with an appropriate length of $W$, and obtain a new version of $\mathbf{r}_g$. The rationale for the robust regression is to use the outliers-filtered historical data to estimate current sparse measurement. Since the different TOA value corresponds to its own symbol period, the latest $W$ column of the matrix $\mathbf{r}$, $\mathbf{r}_g = [r_g(m)]$ can be calibrated by

$$r_g(m) = \frac{1}{N_d(m)} \sum_{\Delta_g=0}^{W-1} \left[ r_{m,(j_g-\Delta_g)} + \Delta_g \times N_k \right] \tag{5.13}$$

where $r_{m,(j_g-\Delta_g)}$ only counts when it contains TOA measurement, $N_d(m)$ is the total number of effective measurements available in the $W$ length matrix $\mathbf{r}$, and $m$ represents the $m$-th row. Thus, vector $r_g(m)$ can be used as the current distance measurement, with each TOA value indicating the pseudo-range from the $m$-th anchor to the mobile device.

The measured pseudo-ranges between the anchor nodes and the mobile phone are $\hat{r}_m = \mathcal{C} \times r_g(m)$, where $\mathcal{C}$ is the speed of the aerial acoustic signal.

Denoting the true distances as $r_m$, $m = [1, \ldots, M]$, the unknown starting point as $\delta_r$, $\hat{r}_m$ can be written as

$$\hat{r}_m = r_m + \delta_r + n_m \tag{5.14}$$

where $n_m$ is the distance measurement noise for $\hat{r}_m$.

### 5.4.6 Location Estimation

With the measured distance from $M$ anchor nodes available, multilateration can be performed to localize the smartphone. Assuming the real position of the smartphone is $\mathbf{p} = (x, y)$, the position of the anchor node is $\mathbf{p}_m = (x_m, y_m)$, then the real distance from the smartphone to the anchor node $m$ can be written as $r_m = \sqrt{(x - x_m)^2 + (y - y_m)^2}$. We define the vector of unknown parameter as $\theta = [x \quad y \quad \delta_r]^T$. The localization purpose is to obtain the estimated value of $\theta$ from observations, where $[\hat{x}, \hat{y}]$ in $\hat{\theta}$ is the estimated position.

However, (5.14) contains the nonlinear term. Rather than estimate $\theta$ directly, we select one of the $M$ measurements as the reference $f$, and use this reference to cancel out the nonlinear term. The selection of this reference could be random or based on the code-matching quality (the mismatch residues). To better illustrate this process, we can square both sides of (5.14) and subtract the reference measurement $f$ as

$$\begin{aligned} &[(x_m^2 + y_m^2 - \hat{r}_m^2) - (x_f^2 + y_f^2 - \hat{r}_f^2)] + 2\hat{r}_m\widetilde{n}_m - 2\delta_r n_m \\ &= 2(x_m - x_f)x + 2(y_m - y_f)y - 2\delta_r(\hat{r}_m - \hat{r}_f) \end{aligned} \tag{5.15}$$

where $\widetilde{n}_m$ is the differential noise term of $n_m - n_f$. Equation (5.15) can be expressed in matrix form as

$$\mathbf{A}\theta = \nu + \mathbf{p_n} \tag{5.16}$$

where $\mathbf{A} = \begin{bmatrix} x_m - x_f & y_m - y_f & \hat{r}_f - \hat{r}_m \\ \cdots & & \\ \cdots & & \cdots \end{bmatrix}$, $\theta = \begin{bmatrix} x \\ y \\ \delta_r \end{bmatrix}$ and $\nu = \frac{1}{2}[(x_m^2 + y_m^2 - \hat{r}_m^2) - (x_f^2 + y_f^2 - \hat{r}_f^2)]_{(M-1)\times 3}$, $m = 1, \ldots, M$ when $m \neq f$. $\hat{r}_m$ is the measured distance obtained from (5.14). $\mathbf{p_n} = 2\hat{r}_m\widetilde{n}_m - 2\delta_r n_m$ is the noise term with its first moment as $\mathrm{E}(\mathbf{p_n}) = 0$ and the covariance matrix of vector $\mathbf{p_n}$ as $\widetilde{\sigma} = \mathrm{Cov}(\mathbf{p_n})$. The diagonal elements of the covariance matrix are $2\hat{r}_m(\sigma_m^2 + \sigma_f^2) + 2\delta_r\sigma_m^2$, with other elements in the matrix as $2\hat{r}_m\sigma_f^2$.

Then (5.16) can be formulated as a least-square (LS) problem, and we can obtain the solution as

$$\hat{\theta} = (\mathbf{A}^T\widetilde{\sigma}^{-1}\mathbf{A})^{-1}\mathbf{A}^T\widetilde{\sigma}^{-1}\nu \tag{5.17}$$

Initially, the value of the covariance matrix $\widetilde{\sigma}$ is unknown, and we could initialize $\widetilde{\sigma}$ with all "1s" in its diagonal and "0s" for other elements. For the following snapshots, $\widetilde{\sigma}$ could be estimated, and we can substitute $\widetilde{\sigma}$ into (5.17) to improve estimation accuracy.

## 5.5 Error Pruning Techniques

Error pruning techniques are essential for the correct operation of the indoor localization system, especially in sound interference. The source of potential errors could be the noise and interference, blockage, indoor non-light-of-sight (NLOS) transmission, anchor network failure.

### 5.5.1 *Signal Level Resistance*

To deal with the signal noise issues, we perform spread spectrum in transmission and correlation processing in the receiver to obtain noise resistance. We use the Gaussian second derivative pulse as the waveform, which has the properties of high multipath and timing resolution, low energy consumption, robustness to noise, and low spectrum emission.

Compared with noise, harsh indoor environments with blockage and high-density multi-paths are more challenge. The blocked localization beacon could either be attenuated or introduce additional delay that prolongs the obtained ranging distance; we call this NLOS bias effects. Using these problematic ranging results could cause over-fitting in localization calculation, or even make the final results diverge. Thus, how to identify and mitigate these prolonged or problematic ranging measurement is crucial in localization.

One of the advantages of using acoustic signal for ranging lies in its high timing resolution and the full access of the acoustic channel information. Extracting features from estimated channel condition, the goodness of transmission could be evaluated. For example, if the delay spread of the estimated channel is significantly larger than normal conditions, the ranging result from this channel has a high probability of blockage, i.e., NLOS condition. By lowering the weighting efficiency of these ranging measurements during localization, overall location result could be more stable. We use the combined metrics of RMS delay spread, Kurtosis, and mean excess delay to identify the NLOS channel condition, and assign lower weight for the measurements from the NLOS channel during localization processing.

### 5.5.2 *Track before Localization*

TOA ranging results $\mathbf{k}_m$ from $M$ anchor nodes are the prerequisite for ranging-based localization. In this section, we analyze the sources of ranging errors and design new approaches to optimize the TOA ranging results $\hat{\mathbf{r}}_m$.

The obtained pseudorange $\mathbf{r}_m$ can be written as

$$\hat{\mathbf{r}}_m = \delta_r + \left[\mathbf{v}_m + \frac{v_s(f_{loc} - f_m)}{F_s} + v_s\right]\mathbf{t}_m + \mathbf{a}_m\mathbf{t}_m^2 \tag{5.18}$$

where $\delta_r$ is the unknown delay that maps the pseudorange to the real distance; $\mathbf{t}_m$ is the TOA vector for the $m$-th anchor; $(f_{loc} - f_m)$ is the clock drift between the smartphone $f_{loc}$ and anchor node $f_m$; $\mathbf{v}_m$ and $\mathbf{a}_m$ are the relative moving speed and acceleration of the smartphone relative to the $m$-th anchor. Denote $\mathbf{r}_m = \mathbf{t}_m v_s$, as the real distance from the $m$-th anchor to the smartphone when the acoustic beacon is transmitted. If ignoring the movement of the user and clock drift, i.e., $\mathbf{v}_m, \mathbf{a}_m \ll v_s$, $\mathbf{t}_m$ and $(f_{loc} - f_m)$ are all small, then the slope of $\hat{\mathbf{r}}_m$ is ignored and $\hat{\mathbf{r}}_m \approx \mathbf{r}_m$.

To study the time series of the pseudorange $\hat{\mathbf{r}}_m$ in an experiment, Fig. 5.3 shows the results from four different anchor nodes when a user is standing still and moving. When the user is standing still, the time series of $\hat{\mathbf{r}}_m$ is linear as shown in Fig. 5.3(a), where the slope is $v_s(f_{loc} - f_m)/F_s$. $\hat{\mathbf{r}}_m$ increases over time due to the clock difference ($f_{loc} > f_m$). From Fig. 5.3(a), we can calculate the clock difference $f_{loc} - f_m = 1.4$Hz, and the equivalent moving speed of the drift is 0.01m/s.

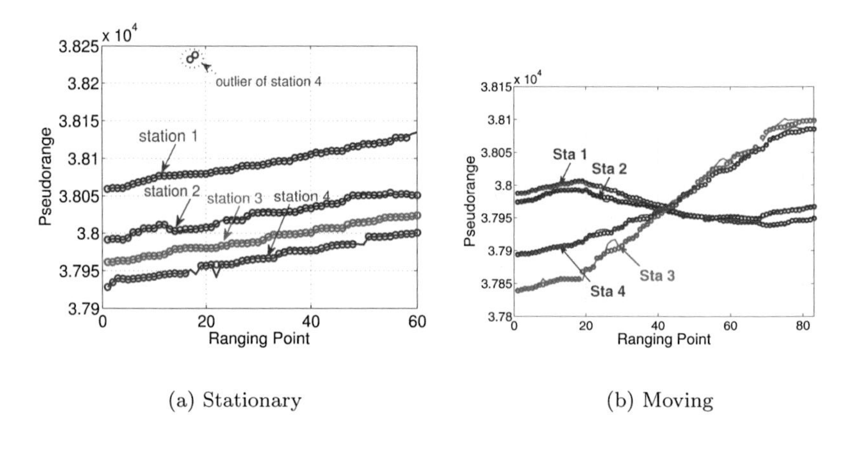


(a) Stationary

(b) Moving

**Figure 5.3: Ranging results for (a) stationary and (b) moving users.**

Fig. 5.3(b) shows the pseudorange results when a user is moving. The time series of $\hat{\mathbf{r}}_m$ becomes a curve with the first derivative as the effective velocity ($\mathbf{v}_m$) to the anchor node $m$, whereas the second derivative is the acceleration $\mathbf{a}_m$.

Tracking $\hat{\mathbf{r}}_m$ before localization could provide more robust and smoothed ranging results for the location calculation step. From Fig. 5.3(a), we also know that some of the ranging points are missing or become outliers due to detection errors. Conventional tracking approaches, e.g., a Kalman filter, could not provide satisfactory results. The real measurement with missing data and outliers does not follow Gaussian distribution and could make the Kalman filter diverge.

Our proposed solution is to estimate the *health status* of each station by monitoring $\hat{\mathbf{r}}_m$, and perform smoothing and tracking for the healthy stations.

If the measurement from one station has a high probability of error, it is better to mitigate the ranging measurements from this station in this round. For the missing data in healthy stations, a spline interpolation process is utilized. After the outlier mitigation and missing data interpolation, a modified robust Kalman filter is applied for further smoothing the ranging observations [84].

### 5.5.3 Location Tracking

The location of $\mathbf{y}$ is estimated via batch processing in DC-SDP. The high correlation feature of adjacent $\mathbf{y}$ could be utilized for smoothing and tracking the final location results. A 2D robust Kalman filter [84] is applied for its outlier robustness.

The state of the smartphone could be stationary, moving, or in the transition (from stationary to moving or the reverse). Without considering the status of $\mathbf{y}$, the Kalman filter can achieve limited performance because different coefficients of the Kalman filter should be applied for different states.

To make the tracking algorithm adaptive to the state of the target (smartphone), we perform stationary detection concurrently with the tracking process. From Fig. 5.3(a) and Fig. 5.3(b), when the target is stationary, the trend of the pseudorange $\hat{\mathbf{r}}_m$ increases over time with constant slope of $(v_s(f_{loc} - f_m)/F_s)$. The clock of the anchor nodes is synchronized through Zigbee radio, i.e., $f_m = f_{const}, m = 1, \ldots, M$, thus the slope of $\hat{\mathbf{r}}_m$ has the same value for each station. By performing common slope detection, the state of the target is identified, and distinct coefficients of the tracking filter could be tuned for best performance in each state.

## 5.6 Performance Evaluation

In this section, we perform system performance evaluation by using Apple's iPhone 4S and iPod Touch 5 (iTouch5) without any modification of the hardware or jailbreak of the operating system.

The 3dB pulse width of the transmitted signal as in (5.1) is chosen with $T_w = 1.5ms$ to meet the bandwidth constraint. A sharper pulse results in better multipath robustness and time-domain resolution, but with increased bandwidth occupancy. Restricted by the bandwidth of a microphone, the effective pulse energy and operating distance would be decreased when more frequency components are outside the receiver band. The symbol duration is chosen as $T_s = 0.0781s$, resulting in the pulse duty cycle of $R = 1.92\%$. Shorter symbol duration leads to a higher data rate and location refreshing rate, but restricted by the multipath environment. The choice of these parameters is a tradeoff between the achievable resolution and maximum operating distance.

### 5.6.1 Sound Pressure Level Measurement

The Guoguo System uses audible-band signal as a beacon. However, users might be concerned about the noise effect of the acoustic signals in indoor environments.

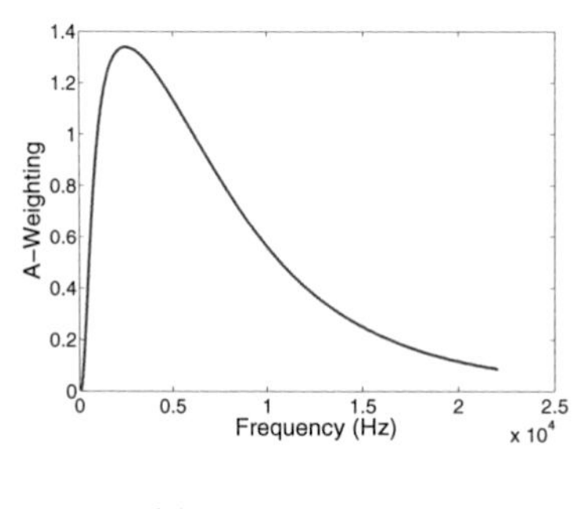


(a) A-weighting curve

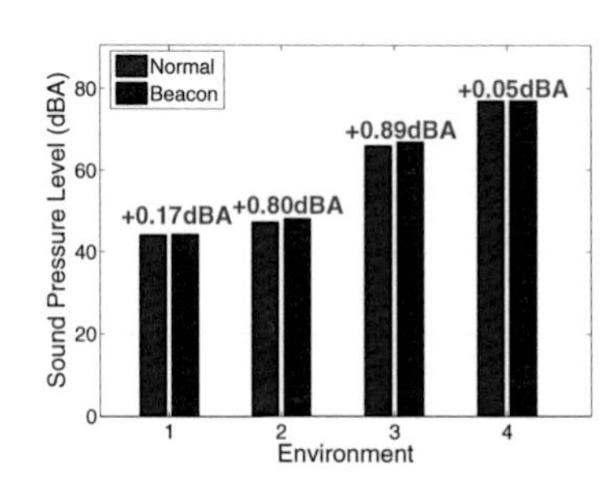


(b) Sound Pressure Level

**Figure 5.4: (a) The perception level of human ears; (b) comparison of sound pressure level in four scenarios.**

To make our transmitted acoustic signal unperceptible to users, we designed the waveform of the beacon signal with appropriate transmission power. To quantify the results, from the perception curve of human ears (*A-weighting* coefficients), we derive the *sound pressure level* (SPL) value as a metric of sound level. We place four anchor nodes very close to a smartphone (the most noisy case), and measure the difference of SPL. The normal sound background case is named "Normal"; the environment that filled with our beacon signal is called "Beacon". The SPL values for these two conditions in four different environments are shown in Fig. 5.4(b). The maximum difference of the SPL caused by the "Beacon" is 0.85dBA, which is below the perception level of the human ear. Therefore, we can conclude that our transmitted acoustic beacon signal is completely ignorable and *disturbance-free* even in extreme cases.

### 5.6.2 Maximum Operation Distance

Limited by the available maximum distance in a room, we move the evaluation of the *maximum operation distance* in an aisle environment from 1.8 $\sim$ 24.4m under two configurations: no sound interference (SI) and with sound interference (SI), where the SI cased is simulated by playing video sound near the mobile device. The quantified *signal level* is shown in Fig. 5.5, where higher signal level means stronger SI.

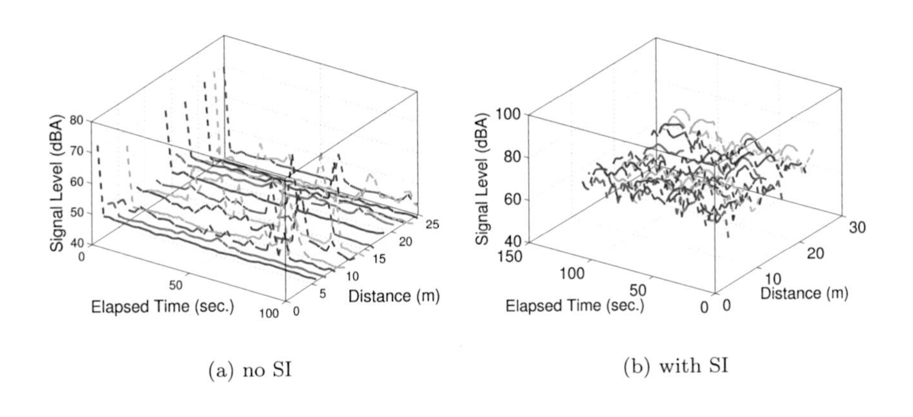


**Figure 5.5:** ***Signal level*** **(dBA) in an aisle.**

The *ranging accuracy* in Fig. 6.8 shows the *maximum operation distance* for our proposed scheme is around 15 $\sim$ 20m. Even with SI, the ranging accuracy at the *maximum operation distance* between 15 $\sim$ 20m is still satisfactory.

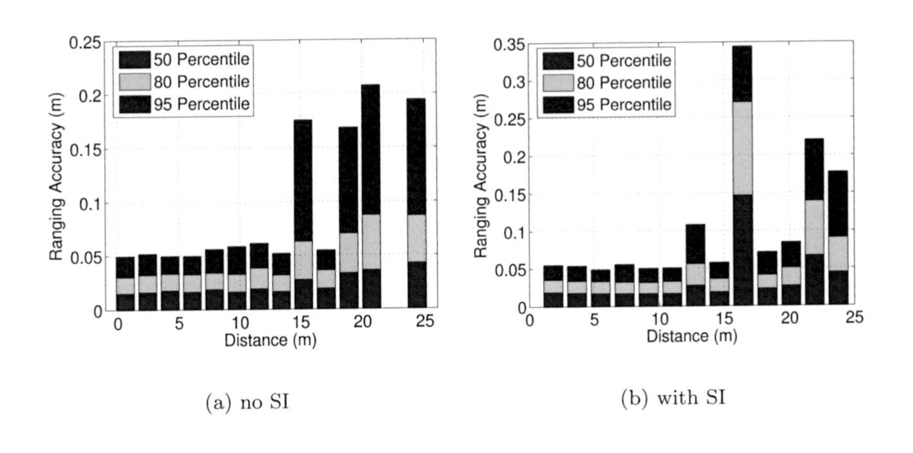


**Figure 5.6:** ***Ranging accuracy*** **in an aisle.**

### *5.6.3 Localization in Office Environments*

**Experiment Setup**. We deploy 9 anchor nodes in a typical office environment to evaluate the performance of Guoguo, as shown in Fig. 5.7.

A total of 12 cases have been tested: 6 of them were tested in a quiet environment for theoretical performance; the other 6 cases were tested under

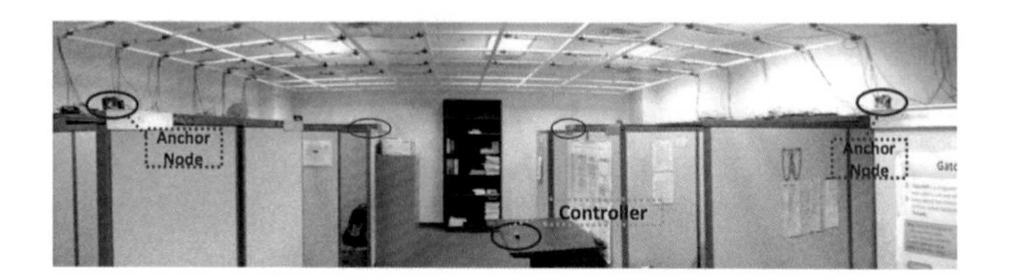


**Figure 5.7: Experiment setup in an office environment.**

SI. The experiment configuration is summarized in Table 5.2. The last term is the effective number of accessed anchor nodes ($N_{eff}$) during localization calculation. The number of $N_{eff}$ is less than the total number of deployed anchors. This is often caused by interference, blockage, and NLOS during the localization process.

**Table 5.2: Experiment configuration in an office environment (some anchors are not shown in the photo)**

| ID | Use Cases | Length (s) | Background | $N_{eff}$ |
|---|---|---|---|---|
| 1 | Env1 | 60.91 | Quiet | 6.86 |
| 2 | Env2 | 60.97 | Quiet | 6.15 |
| 3 | Env3 | 60.61 | Quiet | 8 |
| 4 | Env4 | 60.64 | Quiet | 7.77 |
| 5 | Env5 | 60.53 | Quiet | 8.24 |
| 6 | Env6 | 60.64 | Quiet | 8.72 |
| 7 | P4snoise1 | 522.97 | Sound | 5.03 |
| 8 | P4snoise2 | 300.48 | Sound | 6.11 |
| 9 | P4snoise3 | 346.35 | Sound | 5.99 |
| 10 | Touchnoise1 | 398.79 | Sound | 5.97 |
| 11 | Touchnoise2 | 461.03 | Sound | 6.13 |
| 12 | Touchnoise3 | 582.14 | Sound | 6.07 |

**Localization Accuracy in Quiet Environment.** In this case study, to support quantitative analysis, we localize a smartphone when its user stands still at different spots. The location error (LE) is evaluated via Root Mean Square Error (RMSE) between the ground truth and the calculated location value. The cumulative distribution function (CDF) of the localization error is shown in Fig. 5.8(a). If using 80% probability, the localization error for all the cases is in the range between $4cm$ and $8cm$. These localization results are very accurate and sufficient for fine-grained indoor location-based services.

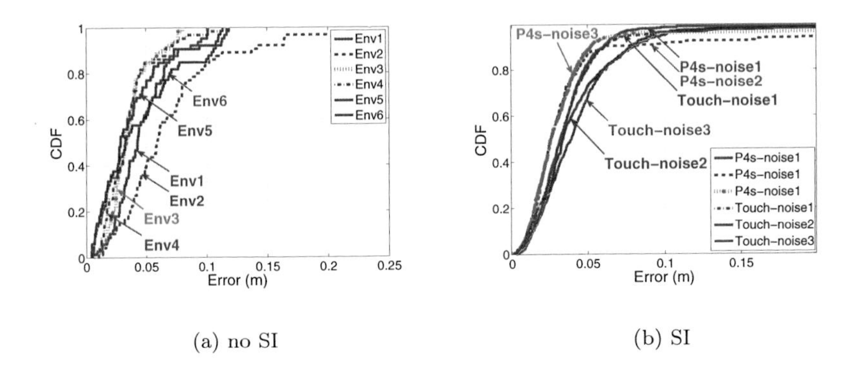


(a) no SI

(b) SI

**Figure 5.8: The localization accuracy in an office environment under (a) no SI; (b) with SI.**

**Localization Accuracy under SI**. To evaluate the localization performance in a realistic environment, we add artificial sound to evaluate the robustness of our proposed system. The SPL level of the added sound is around $40 \sim 90$dBA. The measured sound pressure level (dBA) is calculated from the received acoustic signal at a smartphone during localization.

Fig. 5.8(b) shows the CDF of the localization error (LE) in an office environment. In these six cases with background sound, the final localization accuracy is still within $10cm$, with less than 10% of results slightly affected. From another point of view, the number of effective accessed anchors under background sound is slightly lower than the normal cases from the $N_{eff}$ value in Table 5.2 because some of the acoustic beacons have been mitigated. Since we use more anchors than the minimum requirement (three anchors), these additional anchors can improve our system's robustness under a realistic environment.

**Localization Metrics**. Other metrics that are used to evaluate the system performance are the localization *Miss Rate*, and location *Average Update Time*.

The localization *Miss Rate* evaluates the quality of obtained location values. The definition of *Miss Rate* is $N_{loc}/N_{pos}$, where $N_{loc}$ is the number of obtained location results; $N_{pos}$ is the number of refined location results after the post-processing module. Lower *Miss Rate* means better localization results. The *Miss Rates* for all the cases in Fig. 5.9(a) show very small value, i.e., the localization results are of very good quality.

Another important metric is the *Average Update Time*, representing the refreshing rate of the localization process. If the refreshing rate is too slow, it is hard to keep up with the moving traces of subjects. Due to the low transmission speed of acoustic signal, minimizing *Average Update Time* is nontrivial. From Fig. 5.9(b), we observe that the *Average Update Time* for

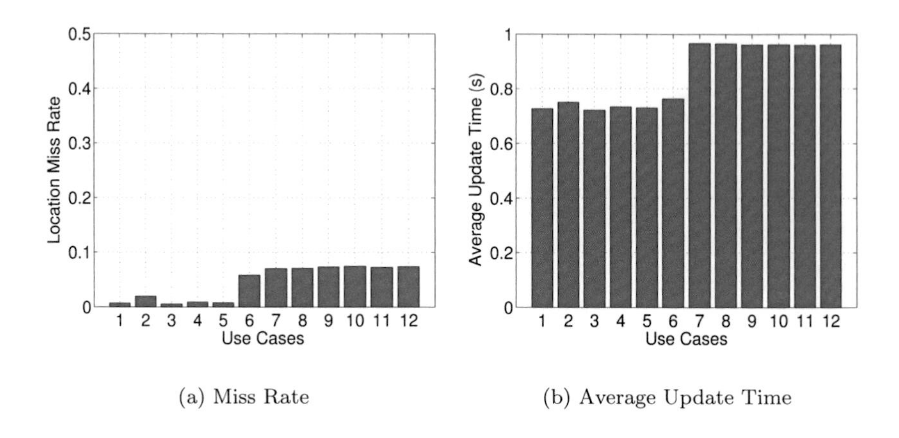


(a) Miss Rate

(b) Average Update Time

**Figure 5.9: Measured performance metrics of ranging rate (a), miss rate (b), and average update time (c) for different measurements in an office environment.**

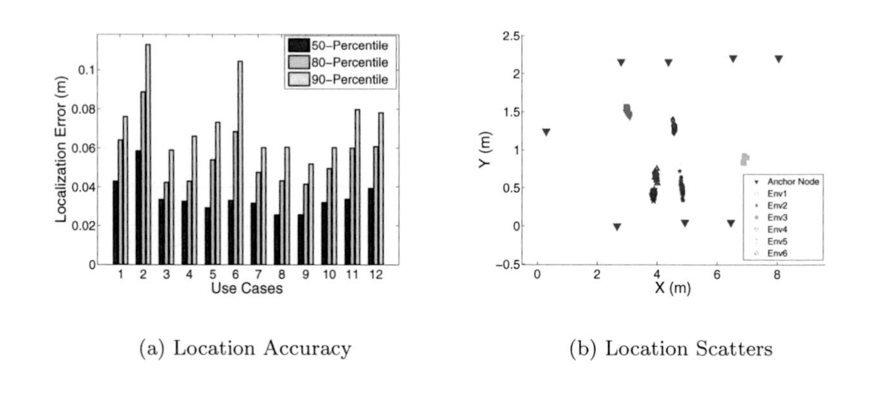


(a) Location Accuracy

(b) Location Scatters

**Figure 5.10: 1) Localization accuracy in different cases; 2) obtained location results as scatters.**

Guoguo is less than one second, sufficient for capturing the traces of moving subjects at modest speed. This rate is significantly faster than other existing approaches that also use acoustic signal for localization.

**Overall Localization Accuracy**. Using 50-percentile, 80-percentile, and 90-percentile probability to evaluate the localization accuracy, the results of all 12 cases are shown in Fig. 5.10(a). The achieved localization accuracy is in the centimeter-level even for the cases with added interference. Such results could push the current coarse-level LBS into fine-grained level in an effective way. The scatters of the localization results for users in 6 different locations

**Figure 5.11: The experiment setup in a classroom environment.**

in an office environment are shown in Fig. 5.10(b). The error surface of the scatter is very small and demonstrates the highly accurate location result.

### 5.6.4 Localization in Classroom Environments

**Experiment Setup.** Similar to Subsection 5.6.3, we deploy 9 anchor nodes in a multimedia classroom environment to evaluate the performance of Guoguo, as shown in Fig. 5.11, with maximum distance larger than 15 meters.

The experiment configuration in this classroom environment is summarized in Table 5.3.

**Table 5.3: Experiment configuration in a classroom**

| ID | Use Cases | Length (s) | Background | $N_{eff}$ |
|---|---|---|---|---|
| 1 | Env1 | 590.43 | Quite | 6.98 |
| 2 | Env2 | 408.70 | Quite | 6.80 |
| 3 | Env3 | 459.76 | Quite | 7.0 |
| 4 | Env4 | 676.39 | Quite | 6.99 |
| 5 | Env5 | 270.49 | Quite | 5.9 |
| 6 | Env6 | 599.47 | Quite | 7.0 |
| 7 | P4snoise1 | 279.38 | Sound | 6.94 |
| 8 | P4snoise2 | 641.74 | Sound | 7.96 |
| 9 | P4snoise3 | 240.40 | Sound | 7.72 |
| 10 | Touchnoise1 | 599 | Sound | 7.0 |
| 11 | Touchnoise2 | 88.79 | Sound | 5.38 |
| 12 | Touchnoise3 | 347.5 | Sound | 5.0 |

**Localization Accuracy under Quiet Environment.** The scatters of the localization results for users in different locations in a classroom environment are shown in Fig. 5.12(a). Similar to Fig. 5.10(b), the error surface of achieved location results is very small. The CDF of the localization results for the measurements in Fig. 5.12(a) is shown in Fig. 5.12(b). For most of the

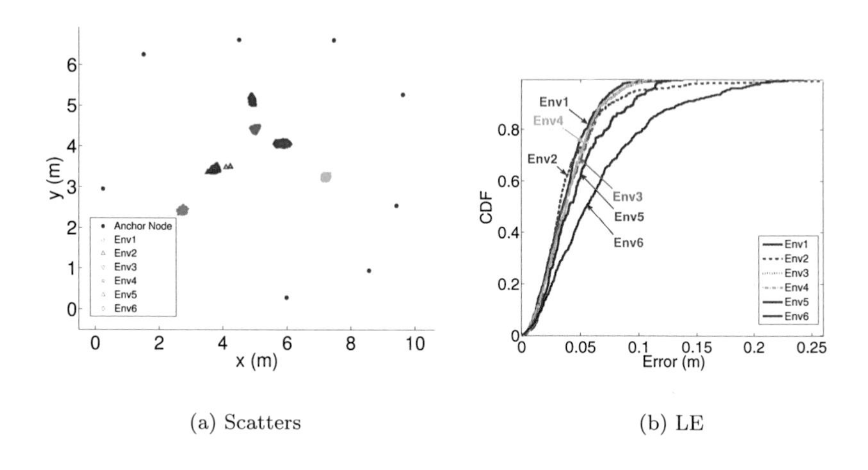


(a) Scatters (b) LE

**Figure 5.12: The scatters (a) and CDF (b) of the localization results in a classroom environment.**

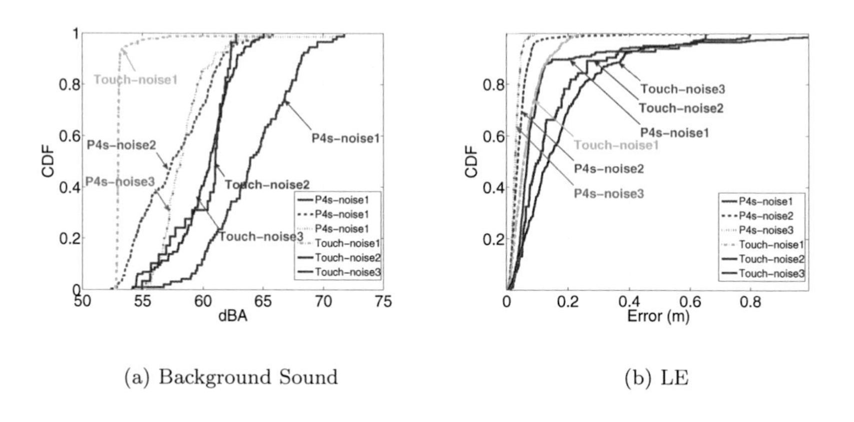


(a) Background Sound (b) LE

**Figure 5.13: The CDF of the background sound (a) and localization error (b) in a classroom environment.**

cases, e.g., using 80% probability, the localization accuracy is very high and in the range from $5cm$ to $10cm$.

**Localization Accuracy under Background Sound**. To evaluate the localization performance under interference, we added artificial sound in the classroom environment by playing the lecture videos. The CDF of the added different background sound is shown in Fig. 5.13(a). The CDF of the localization results under background sound in a classroom environment is shown in Fig. 5.13(b). One interesting observation from Fig. 5.13(b) is that the iPhone4s achieves much better performance than the iTouch5. The reason could be the built-in multi-microphones and noise-canceling mechanisms in iPhone4s, while the iTouch5 is only equipped with one microphone and no noise-canceling hardware.

**Localization Metrics**. The *Ranging Rate* for the measurements in a classroom environment is shown in Fig. 5.14(a), while two cases of iTouch5 suffer from low *Ranging Rate* under background sound. The *Miss Rates* in Fig. 5.9(a) show very small value. The *Average Update Time* for the environment of a classroom is near 0.8 seconds and sufficient for mobility cases.

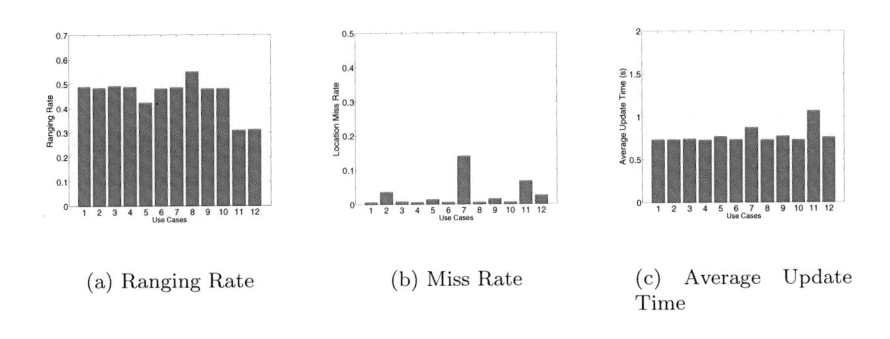


(a) Ranging Rate (b) Miss Rate (c) Average Update Time

**Figure 5.14: Measured performance metrics of ranging rate (a), miss rate (b), and average update time (c) for different measurements in a classroom environment.**

**Overall Localization Accuracy**. The 50-percentile, 80-percentile, and 90-percentile localization accuracy results are summarized in Fig. 5.10(a). Excepts the iTouch5 under background sound, other cases achieve localization accuracy under $10cm$ for most cases. The localization results of iPod Touch5 are slightly worse than iPhone4's, but still sufficient for indoor location-based services.

### 5.6.5 *Impact of Anchor Numbers and Locations*

To evaluate the impact of anchor numbers and location distributions to the location results, we utilize the ranging results from 8 anchors and disable some of the anchors in location estimation. The ranging accuracy of all these 8 anchors are shown in Fig. 5.16(a). There are multiple choices when selecting different numbers of anchors, and the resulting localization accuracy is also different, which depends on the geometric dilution of precision (GDOP) of the anchor network. By selecting all the combinations exclusively, we could

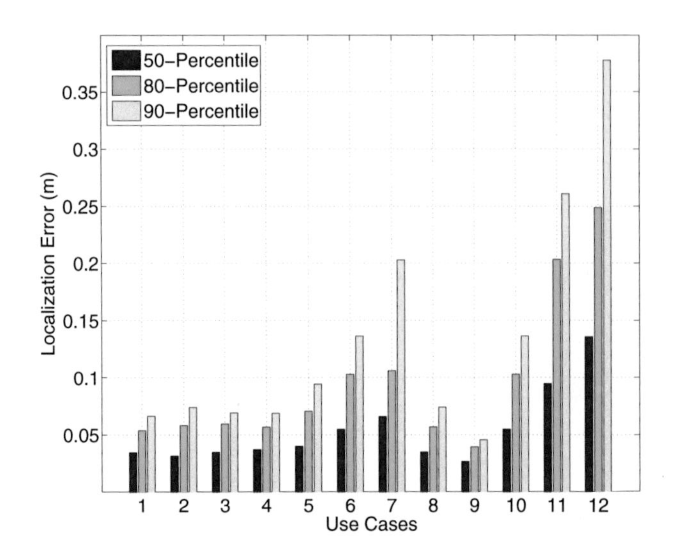


**Figure 5.15: Localization error for different measurements in a classroom environment.**

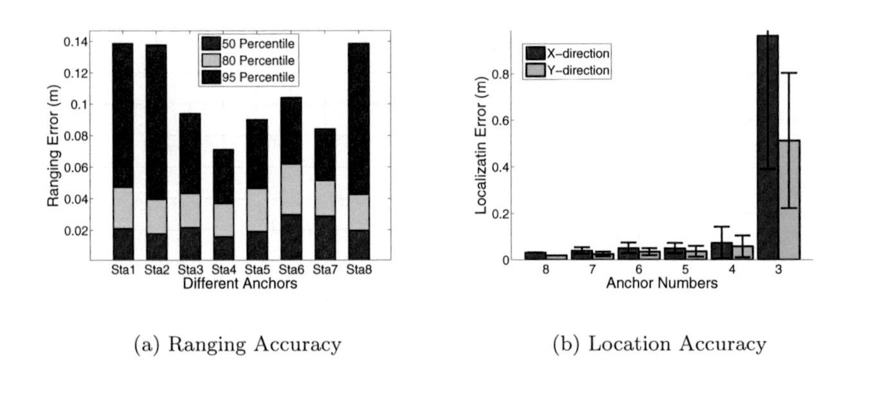


(a) Ranging Accuracy

(b) Location Accuracy

**Figure 5.16: (a) Ranging accuracy for 8 anchor nodes; (b) localization accuracy with respect to anchor numbers.**

calculate the average localization accuracy (using the 80 percentile as an example) and the variance. The result is shown in Fig. 5.16(b), where the error bar shows the value of variance. With the decrease of the anchor numbers, the accuracy also decreases with larger variance. When the anchor number is small, the locations of the available anchor nodes become more important (a biased-placed anchor network could increase the location error significantly).

In Fig. 5.16(b), we show that the location error increases significantly when there are only three anchors. We need at least three measurements to solve the three unknown parameters in (5.17), i.e., location ($x$ and $y$) and unknown delay ($\delta_r$). The large error is understandable since the LS approach needs redundancy to tolerate the ranging errors.

## 5.6.6 Moving Traces

We used stationary localization cases to evaluate and quantify the accuracy of the Guoguo system. To improve the dynamic performance when tracking a moving user, we proposed error-pruning techniques, e.g., signal level resistance, track before localization, and location tracking, to improve the robustness and accuracy. To compare the contributed performance improvement, we calculate the moving trace of a user by performing simultaneousness real-time computation using two different approaches: with and without error pruning. To keep the moving trace intact and easy to compare, we perform moving in a v-shape with modest speed (peak speed around 50cm/s). The calculated ranging results (pseudorange) and final location traces are illustrated in Fig. 5.17 and Fig. 5.18. From Fig. 5.17, we know that the obtained pseudorange is significantly better after applying the error pruning techniques. The location trace in Fig. 5.18(b) also shows clear and accurate results compared with the scattered trace in Fig. 5.18(a) without error pruning.

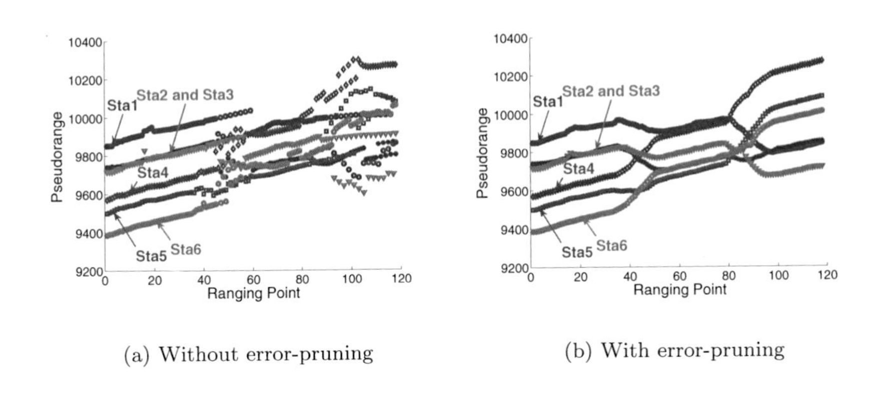


(a) Without error-pruning (b) With error-pruning

**Figure 5.17: Pseudorange results from multiple anchors of a moving user: a) without error-pruning; b) with error-pruning.**

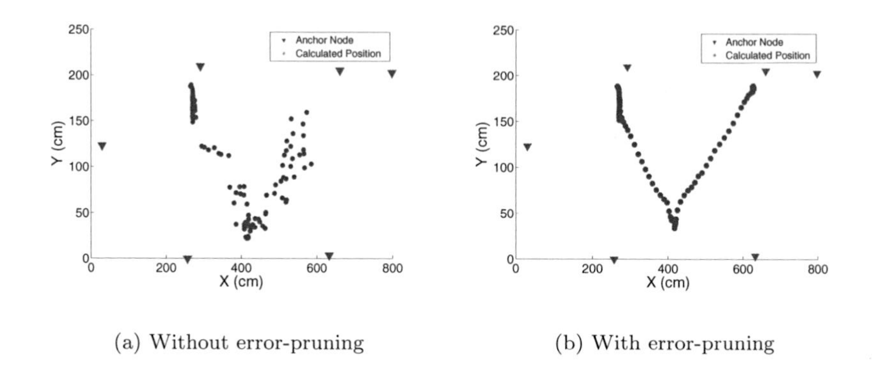


(a) Without error-pruning (b) With error-pruning

**Figure 5.18: Location traces of a moving user: a) without error-pruning; b) with error-pruning.**

## 5.6.7 System Evaluation

For the app, we implemented the signal detection and ranging algorithm in a smartphone leveraging the iOS Grand Central Dispatch (GCD) and vDSP Accelerate framework for better responsiveness and efficiency. To save the battery life of the smartphone, we offload the computation intensive localization and tracking to the server. Leveraging the open-source Redis NoSQL database, we designed a pub/sub framework [92]. Such configuration balances the communication and computation cost. Fig. 5.19 compares our app "Guoguo" to "Chrome" and "Temple Run2" in terms of the energy consumption rate, CPU activity, and network activity on the iOS platform. Guoguo features a stable CPU utilization around 28% (Apple iPhone4S), which is lower than the popular game Temple Run 2, and much lower than Google's iOS version web browser Chrome. The delay caused by the network communication and computation is in the micro-second level, which is ignorable compared with the location update rate.

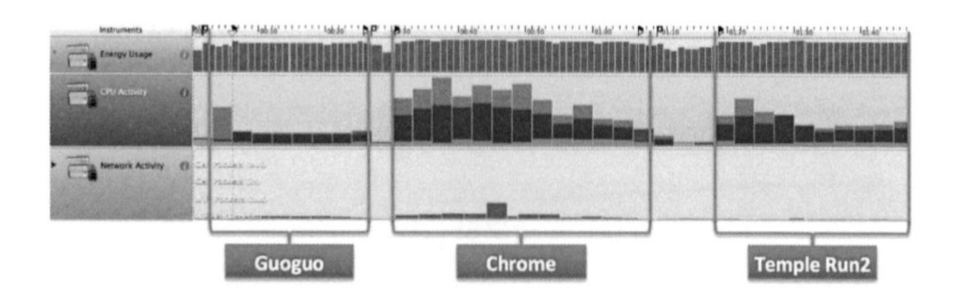


**Figure 5.19: Energy consumption rate, CPU activity, and network activity generated by Xcode Instrument.**

## 5.7 Conclusion

We proposed the Guoguo algorithm and ecosystem to realize the smartphone-based fined-grained indoor localization. For the first time, we can locate a smartphone user at the centimeter-level, which has significant implications for potential indoor location services and applications compared with existing meter-level localization solutions. To address the challenges of utilizing the audible-band acoustic signal in smartphone localization, i.e., strong attenuation, interference-rich, high sound disturbance, and difficulty in synchronization, we proposed comprehensive schemes to improve the localization accuracy and extend coverage without sound disturbance. Significant improvements were achieved in terms of accuracy, cost, and scalability, compared with other existing approaches. Experimental results demonstrated that the achieved average localization accuracy is about $6 \sim 15$cm in typical indoor environments. Guoguo represents a leap of progress in smartphone-based indoor localization, opening enormous new opportunities for indoor location-based services, positioning and navigation systems, and other commercial, educational, or entertainment applications.

# Chapter 6

# Enhancing Location Accuracy and Robustness via Opportunistic Sensing

**Kaikai Liu**

*Assistant Professor, San Jose State University*

**Xiaolin Li**

*Associate Professor, University of Florida*

## CONTENTS

## 6.1 Introduction

Ubiquitous smartphone and location information are enabling new features of location-based services (LBS) around local navigation, retail recommendation, proximity social networking, and location-aware advertising. Recently, the focus is also shifting geographically from outdoor to indoor, where we spend the most money, meet friends, work, and do business.

The indoor location market will be more enormous than the outdoor, since we spend more than 80% of our time indoors in our daily activities, e.g., working, shopping, eating, at the office, at home. Technologically, outdoor localization techniques cannot be directly moved indoors. Satellite-based localization, e.g., GPS, has been one of the most important technological advances of the last half century. No matter how good the systems get for outdoors, their accuracy, coverage, and quality deteriorate significantly in small-scale indoor places. Emerging techniques using existing infrastructure, e.g., WiFi access point, cellular tower, could only achieve limited accuracy, or need extensive war-driving and calibration [6, 8, 15]. Other accurate approaches rely on the deployment of additional infrastructure [48, 57, 81, 85], e.g., dense anchor nodes. These approaches have a high requirement for the minimum anchor number, e.g., at least three anchors for 2-D trilateration. Accessing multiple Wi-Fi access points at the same channel simultaneously, or getting beacon signals from at least 3 deployed anchor nodes, are harsh prerequisites for indoor localization that are hard to meet in real environments. The highly dynamic and mobile setting, where humans are essentially moving, presents further challenges for existing solutions either using existing infrastructure or self-deployed anchor networks. There exist inherent tradeoffs between the localization accuracy and the deployment complexity. Existing low-accurate or high-complexity indoor localization solutions in a mobile phone call for significant innovations in balancing the accuracy and complexity.

In this chapter, we propose a highly accurate and scalable mobile phone localization system via opportunistic anchor access. Unlike [57] which requires a minimum of *three* anchors, our designed approach works under different "anchor" coverage and scales from only one "anchor." For applications with a limited budget for anchor deployment, deploying only one anchor node could achieve significantly better accuracy than non-anchor based approaches, and greatly lower the deployment cost. When the number of deployed anchors increases, our proposed algorithms can adapt to highly accurate results contributed by the additional anchors. With this flexibility, location service operators could select configurations that suit various resolution requirements. Transforming the high level system design goal into a practical working system poses significant challenges: (1) How can we improve the location accuracy significantly even with only one or two anchor nodes? (2) Will our system adapt to higher accuracy with better anchor accessibility? This paper addresses these challenges, and prototypes the system via *opportunistic anchor sensing*. Testbed results confirm that our design goals could adapt to different deployment budget and service quality requirements with high scalability, e.g., from only one node to multiple nodes, and achieve significantly higher accuracy than anchor-free systems. We believe this could be a practical approach to achieve accurate localization results with very low hardware requirement and deployment costs, also scalable for applications with fine-grained resolution demand. To that end, we outline several configurations and propose solutions that hold promise in this new sensing paradigm.

The rest of the chapter is organized as follows. Section 6.2 introduces a system overview. Section 6.3 presents the displacement and direction estimation via INS. Section 6.4 proposes the location optimization approaches for less anchors. Section 6.5 investigates the fine-grained mobile phone localization via delay-constraint robust semidefinite programming. Section 6.6 evaluates the system performance via experiment. Section 6.7 summarizes related work. Section 6.8 concludes this paper.

## 6.2 System Overview

### *6.2.1 Motivation*

Achieving high accuracy and high robustness in location estimation, anchor nodes (existing or needing deployment) need to be utilized to provide location reference. Our previous approaches [57] need at least three synchronized anchors for trilateration, which is hard to meet in real cases without dense deployment. **Lowering the minimum anchor number requirement** is essential for system scalability and low-complexity.

If one anchor-based system could adapt to different deployment budgets with high scalability, e.g., from only one node to multiple nodes, and achieve significantly higher accuracy than an anchor-free system, service operators would have very high enthusiasm to deploy this system for various service quality requirements. The potential application of this proposed system could include indoor check-in, proximity, and location detection services if only one anchor is deployed; multiple anchors could be useful for shopping mall managers, factory administrators, and hospital doctors to monitor the flow of product, equipment, and inventory movement, or to trace patients' movement. It could also be used to provide turn-by-turn guides with foot-level accuracy to a destination, such as a book in a library, a toy in a store, or a butterfly sample in a museum.

## 6.2.2 System Design

The overview of our proposed solution is illustrated in Fig. 6.1.

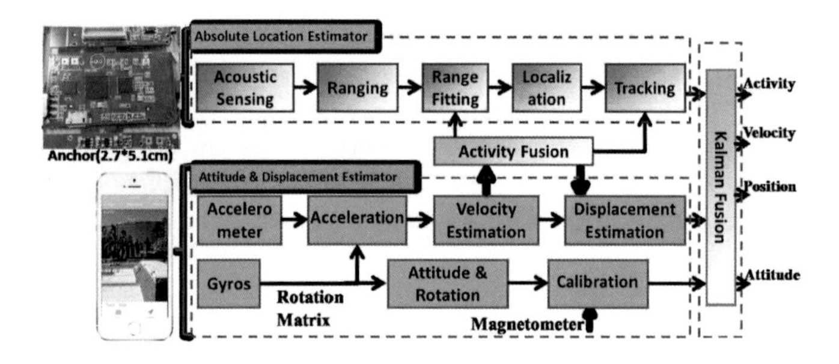


**Figure 6.1: System architecture.**

**Fewer Anchors**. For the most simple case, one anchor node provides basic check-in services, and coarse-grained location estimation. If, however, the sensing data contains: 1) the ranging information for the anchor node and smartphone pair; 2) displacement and moving direction of the smartphone, a much better accuracy could be obtained by leveraging these measurements. We utilize the acoustic RSS and TOA ranging for distance measure, INS for displacement and direction estimation, and leverage the activity measurement for error mitigation. To support multiple users simultaneously, all the ranging process is in one-way passive mode, in which the smartphone only needs to receive beacons. Thus, the highly accurate TOA result is the pseudorange between the smartphone and anchor pair. The inertial sensors, i.e., accelerometer and gyroscope, are utilized for accurate displacement and direction estimation.

If multiple anchor nodes could be sensed by smartphone, i.e., multiple ranging and relative direction results, the accuracy of the location estimation result could be further improved.

**Trilateration via Semidefinite Programming.** When the TOA measurements from accessed anchors are sufficient for trilateration ($N^A > 3$), a more accurate algorithm should be utilized. To solve the practical problem that some available anchors contain large ranging errors, and avoid over-fitting in conventional least square (LS) based approaches, we use semidefinite relaxation to convert the original localization problem into a convex problem. The delay-constraint and Huber penalty function is applied with semidefinite programming (SDP) for better accuracy and robustness. The global optimal solution of the mobile phone position can be ensured by using SDP along with post-position fusion with displacement results.

**Powerful Cloud-Based Algorithms with Low Cost.** Using the beacon data gleaned from the smartphone, the device works with an app and performs complex calculations to figure out where it is within the map. However, current powerful cloud-based processing architecture makes complex computation not a problem. All this could encourage us to utilize advanced algorithms for better accuracy, with less worrying about the implementation detail.

## 6.3 Displacement and Motion Estimation

The recent advances in the construction of micro-machined electromechanical systems (MEMS) sensor chips have now made it a reality to embed IMUs in almost every smartphone. Although MEMS sensors are not as accurate as the large mechanical or optical devices used in military applications, they are significantly cheaper, portable, and durable, thus making this technology ubiquitously available. The IMUs built in current smartphones typically contain three orthogonal gyroscopes, accelerometers, and magnetometers. Typical mobile platforms, e.g., iOS, Android, have the complete framework/package support for programming and accessing all these sensors. Industry's strong emphasis on the IMUs could stimulate future context-ware applications, and this chapter also falls into this category.

### 6.3.1 *Principle of Inertial Navigation*

#### *6.3.1.1 Coordinate System*

We model the smartphone as a rigid body in the *body* coordinate ($b$-frame), with three axes $(x, y, z)$ perpendicular with each other, and the origin point $O(0, 0, 0)$ is defined in the mass center of the smartphone. The $x$-axis is defined as horizontal and points to the right (a.k.a. horizontal-right); the $y$-axis is defined in the vertical-up direction; the $z$-axis is pointing to the user and perpendicular to the screen.

The *body* coordinate is a local coordinate of the smartphone with its orientation and displacement changes when we move the smartphone. We assume all the INS measurements in a smartphone are resolved in the $b$-frame.

To characterize the absolute movement of the smartphone with respect to the indoor/outdoor environment, we define the *navigation* coordinate ($n$-frame) as a local geographic frame that we want to navigate, e.g., stationary with respect to the indoor/outdoor map that is presented to the user. For most cases, especially for navigation over small areas, the *navigation* coordinate is fixed to the east-north-up (ENU) coordinate with shifted origin over reference point $O_R$, where $x$ is the east axis, north is $y$, and Up is $z$.

### 6.3.1.2 Rotation Estimation

The attitude, or orientation, of a smartphone relative to a local reference frame can be tracked by integrating its angular velocity measured by Gyros. The quantity results from Gyros is the rotation rate in radians/second, and can be denoted as $\omega_t^b = (\omega_t^{b_x}, \omega_t^{b_y}, \omega_t^{b_z})^T$, where $t$ is the sample timestamp; $b_x$, $b_y$ and $b_z$ means the 3-axis component in the *body* framework. The sign follows the right hand rule. To calculate the rotation angle ($\theta_t^b$) within the *body* coordinate, the integration process could be utilized as $\theta_t^b = \int_0^t \omega_t^b$.

Current Mobile Operation System all provide direct access to the attitude and rotation value in terms of Euler angles (roll, pitch, yaw), quaternions and direction cosine matrix (DCM). We could use the DCM as the rotation matrix $\mathbf{R}$, since the elements of this matrix are the cosines of the unsigned angles between the *body* axes and the *navigation* axes. The obtain $\mathbf{R}$ is actually

The orientation of the device can be specified by using several different representations, including Euler angles (roll, pitch, yaw), quaternions and direction cosine matrix (DCM). All these different representations can be converted from each other. For better illustration, we utilize the Euler angles, i.e., roll ($\phi$), pitch ($\theta$), yaw ($\psi$), in the body coordinate to illustrate the smartphone attitude. We also use the DCM as the rotation matrix $\mathbf{R}$, since the elements of this matrix are the cosines of the unsigned angles between the *body* axes and the *navigation* axes.

To derive the Euler angles, the Euler rates ($\begin{bmatrix}\phi & \dot{\theta} & \dot{\psi}\end{bmatrix}$) based on the initial Euler angles and angular velocity ($\omega_t^b$) can be derived as

$$\begin{bmatrix}\dot{\phi} \\ \dot{\theta} \\ \dot{\psi}\end{bmatrix} = \begin{bmatrix}1 & \tan\theta\sin\psi & \tan\theta\cos\psi \\ 0 & \cos\phi & -\sin\theta \\ 0 & \sin\psi\sec\theta & \cos\psi\sec\theta\end{bmatrix} \begin{bmatrix}\omega_t^{b_x} \\ \omega_t^{b_y} \\ \omega_t^{b_z}\end{bmatrix} \tag{6.1}$$

With the Euler rates available, a linear integration scheme can be used to calculate the Euler value in the current time slot as

$$\begin{aligned}\phi &= \phi + \dot{\phi} \cdot T_{ins} \\ \theta &= \theta + \dot{\theta} \cdot T_{ins} \\ \psi &= \psi + \dot{\psi} \cdot T_{ins}\end{aligned} \tag{6.2}$$

where $T_{ins}$ is the update time interval of the INS sensor. When the smartphone stands still, i.e., without undergoing any rotation, the output of the Gyros is not perfect zeros. Assume the constant bias error is $\epsilon$, the random noise is a zero-mean uncorrelated random variable with a finite variance of $\sigma^2$. The result of integrating the random noise in (6.2) is a random walk in angle, whose mean value is zero but with increasing standard deviation. The error model of the Gyros measurement can be formulated as

$$\varepsilon = \varepsilon_{\mathbf{b}} + \varepsilon_{\mathbf{r}} + \mathbf{w_g} \tag{6.3}$$

where $\varepsilon_{\mathbf{b}}$ is the constant bias error, where $\dot{\varepsilon_{\mathbf{b}}} = 0$; $\varepsilon_{\mathbf{r}}$ is the random walk error, where $\dot{\varepsilon_{\mathbf{r}}} = \frac{1}{T_g}\varepsilon_{\mathbf{r}} + \mathbf{w_r}$; $T_g$ is the correlation time of the Gyros. For the bias error, it causes an angular error that grows linearly when integrated in the rotation angle calculation. In system calibration, $\varepsilon_{\mathbf{b}}$ can be subtracted via the estimated value after taking a long term average when the Gyros is not undergoing rotation.

#### *6.3.1.3 Coordinate Transformation*

When a smartphone is placed in the initial reference frame, the $z$-axis is perpendicular to the body of the device, with its origin at the center of gravity and directed toward the bottom of the device.

Rotation and relative translation, i.e., $\mathbf{R}$ and $\mathbf{t}$, describe the relative motion of the smartphone body to the initial reference.

For one POI $\mathbf{x}_i^n$ in the *navigation* coordinate, its relative position in the *body* coordinate can be written as

$$\mathbf{x}_i^b = \mathbf{R}_n^b \mathbf{x}_i^n + \mathbf{t} \tag{6.4}$$

(6.4) relates the conversion between $b$-frame and the real physical world, i.e., $n$-frame.

### 6.3.2 Displacement Estimation

To estimate the relative translation ($\mathbf{t}$) of the smartphone, we could utilize the Accelerometer on board by measuring its acceleration force, and infer the displacement by double integrating the acceleration. The quantity result from the Accelerometer is the acceleration rate in $m/s^2$, and can be denoted as $\mathbf{f}_t^b = (f_t^{b_x}, f_t^{b_y}, f_t^{b_z})^T$, where $f_t^{b_{x,y,z}}$ is the measured force in 3-axis directions in the *body* frame.

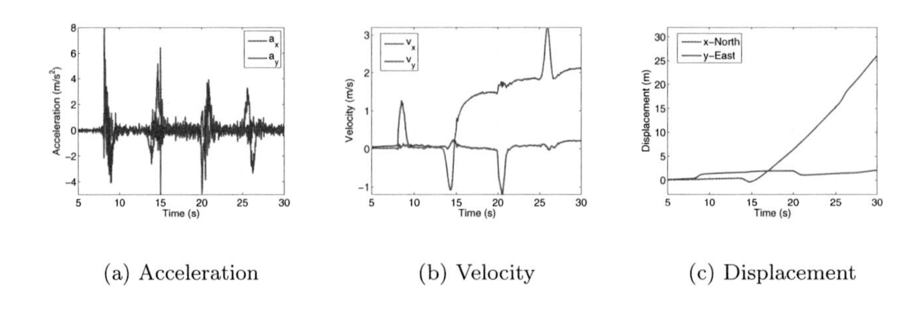


(a) Acceleration (b) Velocity (c) Displacement

**Figure 6.2: Motion estimation result via conventional method: (a) acceleration, (b) velocity, (c) displacement.**

Direct integrating $\mathbf{f}_t^b$ obtains the displacement in the *body* coordinate, which is not related to the real geodesic displacement. To convert the obtained acceleration of the smartphone to the local *navigation* coordinate, we could apply rotation and translation over $\mathbf{f}_t^b$ by

$$\mathbf{f}_t^n = \mathbf{R}_b^n \mathbf{f}_t^b + \mathbf{e}^n \tag{6.5}$$

where $\mathbf{e}^n$ is the error of the force that is applied to the smartphone. To obtain the acceleration caused by the applied forces, gravity should be subtracted by $\mathbf{a}_t^n = \mathbf{f}_t^n - \mathbf{g}$, where $\mathbf{g} = [0, 0, g]$ is the gravity vector. The measured acceleration result after gravity subtraction is shown in Fig. 6.2(a).

After the denoising process, the velocity of the smartphone can be obtained by $\mathbf{v}_t^n = \mathbf{v}_0^n + \int_0^t \mathbf{a}_t^n$ as shown in Fig. 6.2(b). From Fig. 6.2(b), we know that the velocity is drift even when the user is stationary.

The displacement can be calculated by $\mathbf{s}_t^n = \mathbf{s}_0^n + \int_0^t \mathbf{v}_t^n$, where $\mathbf{v}_0^n$, and $\mathbf{s}_0^n$ are the initial velocity and displacement. The result of estimated displacement in the $x$ and $y$ directions is shown in Fig. 6.2(b) with large drift.

The process of obtaining relative displacement $\mathbf{s}_t^n$ is contributed by double integration, in which the measurement noise, i.e., $\int \int \mathbf{e}^n$, is also integrated and amplified. The white noise in acceleration measurements is integrated *twice* and causes a second-order random walk in displacement of the smartphone. As a result, bias errors cause errors in position that grow proportional to $t^2$. The error of the accelerometer measurement can be modeled as

$$\dot{\delta} = -\frac{1}{T_a}\delta + \mathbf{w_a} \tag{6.6}$$

where $T_a$ is the correlation time of the accelerometer. The value of the $T_a$ differs from different devices, and needs to be estimated prior to calibration. $\mathbf{w_a}$ is the modeled Gaussian white noise.

## 6.3.3 Mitigating Displacement Estimation Error

### 6.3.3.1 Apply Constraint via Activity Results

One significant problem of using integration is that the estimated displacement is drifting even when the smartphone is stationary. As shown in Fig. 6.2, the estimated velocity and displacement from the noisy acceleration contains significant bias and drift.

Performing activity detection before integration, and using the estimation results as a constraint, could be feasible approaches to calibrating the drift.

To estimate the smartphone's activity before direct integration, we could perform decision-based detection for the acceleration data with multi-hypothesis tests, where the activity level $H_0$ means stationary, and $H_1$ to $H_j$ means different activity levels ($j = 1, \ldots, J$). The activity detection process can be modeled so as to detect the $H_j$ level from the noisy measurements.

Since each individual sample of acceleration $a_t^n$ can be modeled as a Gaussian random variable with the noise component as $n_t^g$, the first moment of $a_t^n$ when the smartphone is stationary ($H_0$ condition) can be written as $\mathbf{E}(a_t^n; H_0) = \varepsilon_b$. We use the mean absolute value of $|a_t^n|$ as the decision vector, which has a folded normal distribution.

The probability that the sample crosses the threshold $\hat{\eta}_{act}$ when the smartphone is stationary ($H_0$ condition) is the false alarm rate $P_{fa}^j$ ($j = 0$). Using the *Neyman-Pearson* criterion, we set a constant value of $P_{fa}^j$ and perform CFAR detection. The CFAR threshold can be determined by

$$\hat{\eta}_{act}^j = \sqrt{\pi/2}\mathbf{E}(z_i; H_j)Q^{-1}(P_{fa}^j/2) \tag{6.7}$$

For other activity levels, we need to set different $P_{fa}^j$. One key component in CFAR detection is the value of $\mathbf{E}(z_i; H_j)$, i.e., the estimated ground truth. We know that the adjacent samples from the accelerometer have strong correlation, and $\mathbf{E}(z_i; H_j)$ could not obtain an accurate value without sufficient sampling points. Here we construct two detection-aided windows to estimate $\mathbf{E}(z_i; H_j)$, i.e., the forward window ($w_f$) and backward window ($w_b$). The adaptive threshold could be obtained by estimating the future and history trends via $w_f$ and $w_b$. The decision process could be realized by comparing the adaptive threshold to the cell under test. We call it adaptive-threshold based activity detection (ATAD).

Using the detected activity level $H_j$, we could map each activity level to one weighting coefficient $w_j$. During the integration process of displacement calculation, we could apply $w_j$ for the acceleration value $a_t^n$. This process could minimize the error during integration, and obtain significantly better accuracy. For example, if the smartphone is kept stationary, we need to force the velocity component to zero instead of performing integration of the noisy acceleration measurements. For different activity levels, applying different $w_j$ could be very effective in drift reduction.

#### 6.3.3.2 Gaussian Derivative Decomposition

After using ATAD and denoising, the moving trace could become significantly better than direct integration. However, inaccurate traces still exist due to the imperfect denoising or thresholding, especially for small movements.

Performing moving pattern extraction before integration could be a feasible way. The normal movement model could not be applied for estimating the human's moving, e.g., we cannot walk at constant velocity (CV) or constant acceleration (CA) like a vehicle or a plane. Human walking or movement has its own pattern, and we need to "accelerate" and "decelerate," then "accelerate" for another footstep. Here we use the "start-moving-stop" movement model.

Performing "start-moving-stop" pattern decomposition could help us estimate the displacement in a more meaningful way. Using start, acceleration, deceleration, and stop as one basic step, the velocity changes (from zero to top to zero) can be modeled as a Gaussian shape $g_v(\mu, \sigma) = \pm \exp\{-(x-\mu)^2/(2\sigma^2)\}$, where + means moving forward; − means backward. The acceleration is the derivative of the velocity, i.e., $g_a(\mu, \sigma) = \pm(x-\mu)/\sigma^2 \exp\{-(x-\mu)^2/(2\sigma^2)\}$. Using $g_a(\mu, \sigma)$ as the *kernel* function, we can decompose the acceleration measurement $\mathbf{a}_t^n$ into a series of $\eta g_a(\mu, \sigma)$ with different parameters $\eta$, $\mu$, and $\sigma$. Then the decomposed series of acceleration is $\sum_{i=1}^{n} \eta_i g_a(\mu_i, \sigma_i)$, $\eta_i$ is the amplitude of each Gaussian derivative pulse. The fitting process can be modeled as

$$\{\eta_i, \mu_i, \sigma_i\} = \min_{\{\eta_i, \mu_i, \sigma_i\}} ||\mathbf{a}_t^n - \eta_i g_a(\mu_i, \sigma_i)|| \tag{6.8}$$

To reduce the number of parameters during the fitting process, we extract the feature points of $\mathbf{a}_t^n$, e.g., peak position and width, by thresholding the peak maximum and rising edge. Then we use the number of peaks found and the peak positions and widths to fit the specified peak model. This combination yields better and faster computation, and deals with overlapped peaks as well. During the decomposition and fitting process, the sign of $\eta_i$ is determined by comparing the remaining error of using positive and negative results.

## 6.4 Absolute Location Estimation with Fewer Anchors

For convenience, the key notations used in this section are listed in Table 6.1.

### 6.4.1 Background and Basic Procedures

In our previous approach [57], we designed an acoustic anchor-based indoor localization system from scratch with centimeter-level accuracy. However, the trilateration process needs at least three anchors for 2D location estimation.

**Table 6.1: Notation**

| | |
|---|---|
| $M$ | number of anchor nodes |
| $\mathbf{a}_m$ | location for the $m$-th anchor node ($m = 1, \ldots, M$) |
| $N$ | number of mobile phones |
| $\hat{\mathbf{p}}_n^{(k)}$ | the $k$-th initial location for the $n$-th mobile phone |
| $\mathbf{p}_n^{(k)}$ | the $k$-th optimized location for $n$-th mobile phone |
| $\hat{r}_{n,m}^{(k)}$ | the $k$-th RSS ranging result for $n$-th mobile phone |
| $\tilde{r}_{n,m}^{(k)}$ | the $k$-th TOA relative distance for the $n$-th phone |

If the anchors are deployed at different heights, at least four anchors need to be accessed for proper pseudo 3D localization.

To enable location estimation with fewer anchors, we revisit several key steps of the acoustic-based indoor localization system.

**Signal Detection.** The total number of anchor nodes is $M_A$, where $m$ is the index with $m = 1, \ldots, M_A$. Each anchor node broadcasts its own unique pseudocode sequence $\mathbf{p_m} = [p_j]$ with length $L$ for the $m$-th anchor node ($j = 1, \ldots, L$). The symbol duration of the acoustic beacon is $T_s$, i.e., the total time for each beacon is $LT_s$.

Assume the sampling rate of the smartphone is $F_s$, and the received acoustic signal sample is $g(k)$. Decode the $\hat{p}_j$ associated with the current symbol ($\hat{p}_j$ is the estimated version of $p_j$, with the vector term as $\hat{\mathbf{p}}$). Performing code matching with the pre-stored pseudocode sequence $\mathbf{p_m}$, we could obtain the station id $m$ for the $j$-th symbol.

**Ranging.** The basic process of ranging is to measure the flight delay ($t_j$) of the first one in all multipaths, i.e., $r = v_s \times t_j$, where $v_s$ is the acoustic sound speed. In the discrete sample domain, we estimate the sampling point of TOA path $\hat{k}_j$ for the $j$-th symbol as a TOA value of $\hat{t}_j = \hat{k}_j / F_s$. In this process, we associate the TOA measurement $\hat{k}_j$ to the $m$-th anchor node and $l$-th index of the pseudocode $\mathbf{p_m}$. We also convert $\hat{t}_j$ into the base symbol time ($j = 1$) and add it into the vector of TOA measurement $\mathbf{k}_m = [\hat{k}_j - jT_sF_s, \ldots]$, and obtain ranging measurements as $\hat{\mathbf{r}}_m = \mathbf{k}_m v_s / F_s$.

Due to the one-way passive ranging mode utilized for multi-user simultaneous access, the distance measured by TOA estimation is *pseudorange*, with unknown bias $\delta_r$. To solve this unknown bias, we need to synchronize all the $M$ anchor nodes, and make the $\delta_r$ fixed for every node.

**Localization:** Define $\mathbf{y}^n$, $\mathbf{x}_m^n$ as the positions of the smartphone and $m$-th anchor in the $n$-frame. Using $M$ pseudoranges $\hat{\mathbf{r}}_m$ and the preconfigured coordinates of anchor nodes $\mathbf{x}_m^n$, we could estimate the 3D position of the smartphone $\mathbf{y}^n$ by minimizing the quadric term of the remaining error

$$\varepsilon_m = ||\hat{\mathbf{r}}_m - (||\mathbf{y}^n - \mathbf{x}_m^n||_2 + \delta_r)||_2 \tag{6.9}$$

where $\delta_r$ is the unknown delay that compensates the difference between the pseudorange and real distance. The unknown bias ($\delta_r$) can be estimated during the localization process with sufficient anchor numbers, e.g., solving a 3D location $(x, y, z)$ needs four equations (anchor nodes) instead of three.

## 6.4.2 Improving the Location Accuracy via Constraints

With the acoustic ranging results and the displacement and direction estimation in Section 6.3, we could perform location optimization to the initial location even with a single anchor.

### 6.4.2.1 Initial Location

The initial location of the smartphone could be accessed by using the API provided by the mobile operating system. The obtained initial location is in geodetic coordinates (latitude $\phi$, longitude $\lambda$, height $h$), e.g., WGS 84 datum. To convert the geodetic coordinates to the *navigation* coordinate, we first convert it to the earth-centered earth-fixed (ECEF) coordinate, then convert the ECEF to the ENU frame. By subtracting the reference point $O_R$, the GPS location is mapped to the *navigation* coordinate ($n$-frame) for more intuitive and practical analysis. Define the POI's 3D position as $\mathbf{x}_i^n = [x_i^n, y_i^n, z_i^n]^T$, where the superscript $n$ denotes the position value in the *navigation* coordinate; $i$ denotes the $i$-th POI in $M$ POIs ($i = 0, M - 1$). The current location of the smartphone is defined as $\mathbf{p}^n = [x^n, y^n, z^n]^T$.

### 6.4.2.2 Measurements

Assume the position coordinate of the anchor node is $\mathbf{a}_m \in \mathbb{R}^d$, where $m$ is the index of total $M$ anchor nodes. For 2-D coordinate ($d = 2$), $\mathbf{a}_m$ is $[x_m, y_m]^T$, $m = 1, \ldots, M$. Denote the location coordinate of the $n$-th user as $\mathbf{p}_n$, $n = 1, \ldots, N$.

To refine the user's location, ranging information is utilized as a constraint. Assume the initial position coordinate of a user obtained by smartphone is $\hat{\mathbf{p}}_n$, which is direct from location API and low-accurate compared to the location of anchor nodes ($\mathbf{a}_m$). Defining the RSS ranging measurement between the user and anchor pair is $\hat{r}_{n,m}$; the estimated relative TOA distance is $\tilde{r}_{n,m}$. The real distance $r_{n,m}$ from the $n$-th mobile phone to the $m$-th anchor node is written as $r_{n,m} = ||\mathbf{p}_n - \mathbf{a}_m||_2$, where $||\cdot||_2$ calculates the 2-norm and obtains the Euclidean distance. The vector form of the RSS ranging observation from $m$-th anchor to $n$-th mobile phone can be written as

$$\hat{\mathbf{r}}_{n,m} = ||\mathbf{p}_n - \mathbf{a}_m||_2 + \hat{\mathbf{n}}_{n,m} \tag{6.10}$$

where $\hat{\mathbf{n}}_{n,m}$ is the measurement noise; $m = 1, \cdots, N^B$. Then, the TOA distance measurement from $m$-th anchor to $n$-th mobile phone is

$$\tilde{\mathbf{r}}_{n,m} = ||\mathbf{p}_n - \mathbf{a}_m||_2 + \delta_{n,m} + \mathbf{n}_{n,m} \tag{6.11}$$

where $\mathbf{n}_{n,m}$ is the TOA measurement noise, which is lower than $\hat{\mathbf{n}}_{n,m}$ in (6.10). $m = 1, \cdots, N^A$, where $N^A \leq N^B$. $\delta_{n,m}$ is the unknown bias between the $m$-th anchor and $n$-th mobile phone pair due to the unsynchronized clock. Thus, TOA result (6.11) is the relative distance measured between the smartphone and anchor.

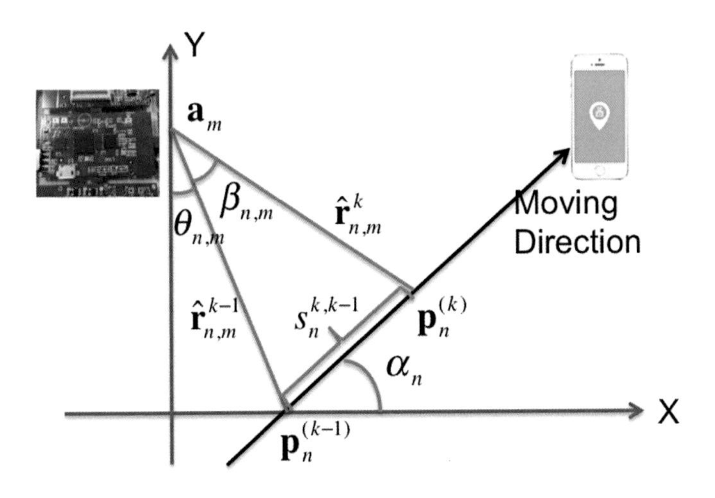


**Figure 6.3: Location estimation via single anchor.**

Fig. 6.4(a) and Fig. 6.4(b) show the TOA ranging results when the user is moving away (from 1.5 to 17 meters) and moving close to the anchor, respectively. The unit of the coordinate in Fig. 6.4(a) is the sampling point (bin) of the distance and time, respectively. When the user is standing still, the slope of $\tilde{r}_{n,m}$ can be approximated as the clock drift between the mobile phone and anchor node. When the user is moving away, the slope of $\tilde{r}_{n,m}$ in Fig. 6.4(a) shows increased distance trend, while Fig. 6.4(b) shows decreased trend. The slightly increased ranging variance when the user is far away from the anchor is caused by the signal attenuation and distortion.

The obtained displacement could be another measurement that contributes to the location optimization. For $k$ and $k+1$ measurements, the displacement can be written as

$$s_n^{k,k-1} = ||\mathbf{p}_n^{(k)} - \mathbf{p}_n^{(k-1)}||_2 + \mathbf{n}_s \tag{6.12}$$

The direction of the motion trace obtained from the attitude value is assumed as $\alpha_n$ as shown in Fig. 6.3. With the RSS ranging results and TOA relative ranging results from the anchor to the smartphone, the anchor-related

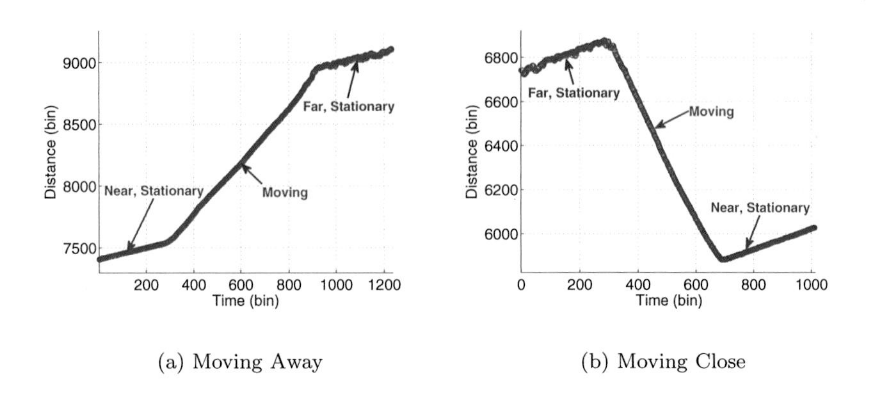


(a) Moving Away (b) Moving Close

**Figure 6.4: TOA ranging results in two different situations: (a) user is moving away from the anchor; (b) user is moving close to the anchor (*bin* is the sampling points).**

measurement could be written as

$$\hat{\mathbf{r}}_{n,m}^{k-1}\cos(\theta_{n,m}) - \hat{\mathbf{r}}_{n,m}^{k}\cos(\theta_{n,m}+\beta_{n,m}) = s_n^{k,k-1}\sin(\alpha_n) + \mathbf{n}_A \tag{6.13}$$

where the angle $\beta_{n,m}$ could be calculated by the law of cosines. The geometric relation is shown in Fig. 6.3. $\theta_{n,m}$ can be estimated by the location difference of $\mathbf{p}_n = [p_x, p_y]$ and $\mathbf{a}_m = [a_x, a_y]$, with its $x$ and $y$ coordinates related by $p_x = a_x + \hat{\mathbf{r}}_{n,m}^{k-1}\sin(\theta_{n,m})$ and $p_y = a_y - \hat{\mathbf{r}}_{n,m}^{k-1}\cos(\theta_{n,m})$.

### *6.4.2.3 Location Optimization*

Thus, the location optimization problem when the anchor node number is insufficient for trilateration can be defined by using (6.10), (6.11), (6.12), and (6.13) to obtain a refined result of $\mathbf{p}_n$.

For the $k$-th iteration, the position refinement process is achieved by minimizing the error term of adjacent measurements $\mathbf{p}_{n,m}^{(k)}$ and $\mathbf{p}_{n,m}^{(k-1)}$ for all the received anchor nodes as

$$\mathbf{p}_n^{(k)} := \arg\min_{\mathbf{p}_n^{(k)}\in\mathbb{R}} \sum_{m\in\xi_M} \mathbf{e}\left(\mathbf{p}_n^{(k)}, \mathbf{p}_n^{(k-1)}\right) \tag{6.14}$$

where $\xi_M$ is the set of all the received anchor nodes; when there is only one anchor node in the coverage area, then $m = 1$. The error term $\mathbf{e}(\mathbf{p}_n^{(k)}, \mathbf{p}_n^{(k-1)})$ illustrates the residual error between the measured distance and the calculated distance of the position coordinates (anchor and mobile phone). The introduced term $\mathbf{p}_n^{(k-1)}$ is to improve accuracy by leveraging the highly accurate relative TOA measurements and estimated moving direction. By refining the search region of $\mathbf{p}_n^k$ within the region $\mathbb{R}$, some local optima values that are outside the real region could be avoided. Such refinement could significantly minimize large errors caused by insufficient and inaccurate measurements.

Considering all the available measurements, the error constraints between the current and previous $(k-1)$-th term $\mathbf{e}(\mathbf{p}_n^{(k)}, \mathbf{p}_n^{(k-1)})$ can be written as

$$\mathbf{e}(\mathbf{p}_n^{(k)}, \mathbf{p}_n^{(k-1)}) = (\gamma_1 \mathbf{e}_S + \gamma_2 \mathbf{e}_D + \gamma_3 \mathbf{e}_M + \gamma_4 \mathbf{e}_A) \tag{6.15}$$

where $\mathbf{e}_S$ and $\mathbf{e}_D$ means the remaining error term of the RSS and relative TOA in (6.10) and (6.11). $\mathbf{e}_M$ and $\mathbf{e}_A$ are the remaining errors of (6.12) and (6.13), respectively. $\gamma_1$, $\gamma_2$, $\gamma_3$, and $\gamma_4$ are weighting coefficients that control the contribution of different measurements.

Performing the gradient operation $\nabla$ to the error residues $\mathbf{e}(\mathbf{p}_{n,m}^{(k)}, \mathbf{p}_{n,m}^{(k-1)})$ with respect to the anchor node $m$, the refined position can be updated via the steepest descent approach by

$$\mathbf{p}_n^{(k)} := \hat{\mathbf{p}}_n^{(k)} + \alpha \nabla \sum_{m \in \xi_M} (\gamma_1 \mathbf{e}_S + \gamma_2 \mathbf{e}_D + \gamma_3 \mathbf{e}_M + \gamma_4 \mathbf{e}_A) \tag{6.16}$$

where $\alpha \in (0, 1]$ is the update step size to control the convergence rate; $\xi_M$ is the total received anchor nodes, where the number is insufficient for trilateration.

Substituting measurements into (6.16), $\mathbf{p}_n^{(k)}$ can be optimized and updated by leveraging the RSS and TOA ranging measurements. (6.16) starts with the initial coarse-grained location result $\hat{\mathbf{p}}_n^{(k)}$, and optimizes the location result by substituting the initial value of $\mathbf{p}_n^{(k)}$ with $\hat{\mathbf{p}}_n^{(k)}$.

## 6.5 Trilateration via Semidefinite Programming

When there are sufficient TOA measurements for trilateration calculation, a fine-grained location result could be obtained by leveraging the TOA measurements from multiple anchors.

To prevent the location estimation algorithm from converging to the local optimality, the concept of relaxation onto convex sets has been proposed [68]. Without the requirement of performing inverse operation on the Jacobian matrix in LS-based approaches, the SDP-based approach achieves better computational efficiency by leveraging existing SDP packages, which is especially important when the Jacobian matrix in the LS problem is badly scaled or close to singular. In this section, we propose an optimized real-time SDP algorithm in mobile phone location estimation. The proposed algorithm is robust in the presence of outliers by leveraging the delay-constraint and Huber M-estimator [102].

### 6.5.1 Min-Max Criterion

The location estimation process is a *nonconvex* optimization problem, while the semidefinite programming (SDP) technique can be used to relax the initial *nonconvex* problem into convex one. Among existing relaxation criteria, *min-max* approximation and semidefinite relaxation can find the global minimum value without the "inside convex hull" requirement [68]. To utilize the SDP relaxation, we modify the problem formulation by rewriting (6.9) into $\hat{\mathbf{r}} - \delta_r = ||\mathbf{p} - \mathbf{a}_m||_2 + \mathbf{n}$. Performing square operations on both sides leads to

$$(\hat{\mathbf{r}} - \delta_r)^T \mathbf{\Sigma}^{-1} (\hat{\mathbf{r}} - \delta_r) = (||\mathbf{p} - \mathbf{a}_m||_2 + \mathbf{n})^2 \tag{6.17}$$

where the right side of (6.17) is $||\mathbf{p}-\mathbf{a}_m||_2^2 + 2\mathbf{n}^T||\mathbf{p}-\mathbf{a}_m||_2 + \mathbf{n}^T\mathbf{n}$. By adopting the *min-max* criterion [68, 98], (6.17) can be formulated as

$$\mathbf{p} = \arg\min_{\mathbf{p}} \max_{m=1,\ldots,M} \underbrace{\left| ||\mathbf{p} - \mathbf{a}_m||_2^2 - (\hat{\mathbf{r}} - \delta_r)^T \mathbf{\Sigma}^{-1} (\hat{\mathbf{r}} - \delta_r) \right|}_{\xi} \tag{6.18}$$

where the term $\xi$ can be viewed as the residual error. (6.18) calculates $\mathbf{y}$, which corresponds to the minimum value of the maximum residual error. Compared with (6.9), (6.18) remains nonconvex, but it is comfortable for the following semidefinite relaxations.

The first term in $\xi$ can be written into a matrix form of

$$\begin{aligned} ||\mathbf{p} - \mathbf{a}_m||_2^2 &= \begin{bmatrix} \mathbf{p}^T & 1 \end{bmatrix} \begin{bmatrix} \mathbf{I}_d & -\mathbf{a}_m \\ -\mathbf{a}_m^T & \mathbf{a}_m^T \mathbf{a}_m \end{bmatrix} \begin{bmatrix} \mathbf{p} \\ 1 \end{bmatrix} \\ &= \text{trace} \left\{ \begin{bmatrix} \mathbf{p} \\ 1 \end{bmatrix} \begin{bmatrix} \mathbf{y}^T & 1 \end{bmatrix} \begin{bmatrix} \mathbf{I}_d & -\mathbf{a}_m \\ -\mathbf{a}_m^T & \mathbf{a}_m^T \mathbf{a}_m \end{bmatrix} \right\} \\ &= \text{trace} \left\{ \begin{bmatrix} \mathbf{P} & \mathbf{p} \\ \mathbf{p}^T & 1 \end{bmatrix} \begin{bmatrix} \mathbf{I}_d & -\mathbf{a}_m \\ -\mathbf{x}_m^T & \mathbf{a}_m^T \mathbf{a}_m \end{bmatrix} \right\} \end{aligned} \tag{6.19}$$

where $\mathbf{P} = \mathbf{p}\mathbf{p}^T$, trace$\{\cdot\}$ calculates the trace of the matrix, and $\mathbf{I}_d$ is an identity matrix of order $d$. Following the same process as in (6.19), the second term in $\xi$ can be written into

$$\begin{aligned} &(\hat{\mathbf{r}} - \delta_r)^T \mathbf{\Sigma}^{-1} (\hat{\mathbf{r}} - \delta_r) \\ &= \text{trace} \left\{ \begin{bmatrix} \delta & \delta_r \\ \delta_r^T & 1 \end{bmatrix} \begin{bmatrix} \mathbf{\Sigma}^{-1}\mathbf{I}_d & -\mathbf{\Sigma}^{-1}\hat{\mathbf{r}} \\ -\hat{\mathbf{r}}^T\mathbf{\Sigma}^{-1} & \hat{\mathbf{r}}^T\mathbf{\Sigma}^{-1}\hat{\mathbf{r}} \end{bmatrix} \right\} \end{aligned} \tag{6.20}$$

where $\delta = \delta_r \delta_r^T$.

### 6.5.2 Delay-Constraint Robust Semidefinite Programming

From (6.9), we know that the unknown parameter $\delta_r$ incorporates the unknown clock drift. The trend of $\delta_r$ is known as a line (the stationary region in

Fig. 6.4) and the future value can be directly estimated, e.g., by linear fitting. Using such prior information, the location estimation accuracy can be further improved by substituting this pre-estimated delay $\delta_r$ as a constraint; we call this approach the delay constraint (DC).

The objective function of $\xi$ can be converted to minimize $\epsilon$ at the constraint of an inequality expression $-\epsilon < \xi < \epsilon$, while $\xi$ can be written in the form of (6.19) and (6.20). However, the outliers could not be ignored during location estimation. One possible solution is to apply a penalty function to the residual error rather than only using the quartic term ($l_2$-norm) in (6.9). Specifically, we still apply $l_2$-norm on any residual smaller than a preset threshold $\sigma_{th}$, but put a linear weight (reverts to $l_1$-like linear growth) on any residual larger than $\sigma_{th}$. Using $l_1$-norm for large errors would lower the weight for outliers and improve the robustness. We choose Huber function $\theta_{hub}(\varepsilon)$ as the penalty function [11]. This penalty function can be considered as a convex approximation of other outlier penalty functions. The constraints form of (6.19) and (6.20) are convex, but the equality constraints of $\mathbf{P} = \mathbf{p}\mathbf{p}^T$ and $\delta = \delta_r\delta_r^T$ are nonconvex. Using semidefinite relaxation, these two equalities can be relaxed to inequality constraints of $\mathbf{P} \succeq \mathbf{p}\mathbf{p}^T$ and $\delta \succeq \delta_r\delta_r^T$, respectively. The matrix form of these two equalities is

$$\begin{bmatrix} \mathbf{P} & \mathbf{p} \\ \mathbf{p}^T & 1 \end{bmatrix} \succeq 0, \quad \begin{bmatrix} \delta & \delta_r \\ \delta_r^T & 1 \end{bmatrix} \succeq 0 \tag{6.21}$$

where $\succeq$ means a positive definite (semidefinite) matrix, which is different from $\geq$.

Accordingly, the initial localization problem can be relaxed to a semidefinite programming form as

$$\begin{aligned} & \min_{\{\mathbf{p},\mathbf{P},\delta_r,\delta\}} \theta_{hub}(\epsilon) \\ \text{s.t.} \quad & -\theta_{hub}(\epsilon) < \text{trace}\left\{ \begin{bmatrix} \mathbf{P} & \mathbf{p} \\ \mathbf{p}^T & 1 \end{bmatrix} \begin{bmatrix} \mathbf{I}_d & -\mathbf{a}_m \\ -\mathbf{a}_m^T & \mathbf{a}_m^T\mathbf{a}_m \end{bmatrix} \right\} - \\ & \text{trace}\left\{ \begin{bmatrix} \delta & \delta_r \\ \delta_r^T & 1 \end{bmatrix} \begin{bmatrix} \mathbf{\Sigma}^{-1}\mathbf{I}_d & -\mathbf{\Sigma}^{-1}\hat{\mathbf{r}} \\ -\hat{\mathbf{r}}^T\mathbf{\Sigma}^{-1} & \hat{\mathbf{r}}^T\mathbf{\Sigma}^{-1}\hat{\mathbf{r}} \end{bmatrix} \right\} < \theta_{hub}(\epsilon), \\ & m = 1, \ldots, M, \\ & \begin{bmatrix} \mathbf{P} & \mathbf{p} \\ \mathbf{p}^T & 1 \end{bmatrix} \succeq 0, \quad \begin{bmatrix} \delta & \delta_r \\ \delta_r^T & 1 \end{bmatrix} \succeq 0 \\ & \hat{\delta}_r(1-\alpha) < \delta_r < \hat{\delta}_r(1+\alpha) \end{aligned}$$

where $\hat{\delta}_r$ is the estimated delay value based on historical data of $\delta_r$; $\alpha$ is predefined and used to relax the delay-constraint (DC). The $n$-th mobile phone position $\mathbf{p}$ can be extracted from the optimal solution of $\{\mathbf{p}, \mathbf{P}, \delta_r, \delta\}$. This delay-constraint robust SDP problem can be solved by some standard convex optimization packages, e.g., SeDuMi and CVX package [28]. By using the steepest descent approach in (6.16), the estimation error can be further reduced by performing a local search above the global optimized value obtained by SDP.

### 6.5.3 Improving Indoor Localization via Sensor Fusion

Compared with INS measurement, the localization result from acoustic anchors is an absolute result and does not drift over time. However, the result is only available under sufficient anchor coverage, and suffers from sound interference and blockage in normal environments.

Based on the distinct features of INS and acoustic-based localization, we propose a sensor fusion approach, and demonstrate that the error during estimation could be significantly minimized.

To fuse the INS sensor with the indoor absolute location estimator (ALE), we can write all the measurements into one state-space model, and utilize the model in Kalman filter for the fusion process. The position estimation result from the INS sensor is $\mathbf{s}_t^n$, the second-order moving model can be written as

$$\begin{aligned} \mathbf{s}_{t+1}^n &= \mathbf{s}_t^n + \dot{\mathbf{s}}_t^n T_{ins} + \ddot{\mathbf{s}}_t^n \frac{T_{ins}^2}{2} \\ \dot{\mathbf{s}}_{t+1}^n &= \dot{\mathbf{s}}_t^n + \ddot{\mathbf{s}}_t^n T_{ins} \end{aligned} \tag{6.22}$$

where $\dot{\mathbf{s}}_t^n$ and $\ddot{\mathbf{s}}_t^n$ are the moving velocity and acceleration, which are derived from the obtained displacement; $T_{ins}$ is the sampling time of the INS. Its relationship with the gyros and accelerometer measurements can be summarized as

$$\ddot{\mathbf{s}}_t^n = \mathbf{R}_{b,t}^n(\mathbf{f}_t^b - \varepsilon_{\mathbf{b}} - \varepsilon_{\mathbf{r}}\mathbf{w}_{\mathbf{g}}) - \mathbf{g} \tag{6.23}$$

where $\mathbf{R}_{b,t}^n$ denotes the rotation matrix from the *body* coordinate to the *navigation* coordinate. Assuming the output of the gyros is $\hat{\omega}_t^b$, the update function is

$$\omega_t^b = \hat{\omega}_t^b - \delta - \mathbf{w}_a \tag{6.24}$$

From (6.22), (6.23), and (6.24), we can summarize the state vector as

$$\mathbf{X_I}(t) = [\theta_{\mathbf{t}}^{\mathbf{b}}, \mathbf{v}_{\mathbf{t}}^{\mathbf{n}}, \mathbf{s}_{\mathbf{t}}^{\mathbf{n}}, \varepsilon_{\mathbf{b}}, \varepsilon_{\mathbf{r}}, \delta] \tag{6.25}$$

where $\theta_{\mathbf{t}}^{\mathbf{b}}$ is the rotation angle obtained from the gyros; $\mathbf{v}_{\mathbf{t}}^{\mathbf{n}}$ is the velocity from accelerometer; $\mathbf{s}_{\mathbf{t}}^{\mathbf{n}}$ is the location of the smartphone; $\varepsilon_{\mathbf{b}}$, $\varepsilon_{\mathbf{r}}$ are the bias and drift of the gyros; $\delta$ is the drift of the accelerometer in (6.6).

The location estimation of the smartphone is $\mathbf{p}^n$. Thus, the observation function can be written as $\mathbf{Z}(t) = \mathbf{p}^n$. Specially, the Kalman fusion process can be viewed as using the measurement from acoustic localization module (ALE) to calibrate the INS results.

The Kalman filter handles the different sample rates from INS and ALE by running at a high data rate (same as INS) and only updates when measures from ALE are available. After the fusion process, the update rate of the localization module has been improved to the same level as the INS sensor, while the bias problem of the INS is calibrated.

## 6.6 Experiment

### 6.6.1 System Evaluation

To quickly deploy the developed algorithm, we designed a back-end server for processing complex algorithms without worrying about the limited resources on a mobile platform. A Redis server [92] is used as a cloud key-value store for the smartphone; A Java server performs computation for the structural data in the Redis server. The smartphone performs sensing and estimation of the raw data, with result data (small size) uploaded to the server for complex algorithm processing. This approach balances computation and network consumption in a smartphone, and the introduced delay is less than $100ms$, which is ignorable for the sub-second level location update rate.

Besides the implementation in a smartphone, we designed the acoustic anchor node for indoor localization purposes. The maximum operation distance for one anchor node is nearly 20 meters. In the following system evaluation, we deployed 8 anchor nodes in one large exhibition room of our campus Art Museum as shown in Fig. 6.5. Enriching visiting experience in a museum is one of the promising applications of indoor localization.

**Figure 6.5: Anchor deployment and experiment environment.**

### 6.6.2 Displacement Estimation

To evaluate the drift reduction performance of our proposed INS displacement estimation algorithm, we will compare our proposed approaches, i.e., denoising (DN), adaptive threshold activity detection (ATAD), and Gaussian derivative decomposition (GDD). The performance metrics includes total drift, drift speed (drift per second), and drift rate (drift per meter).

**Drift Reduction**. After applying the denoising, ATAD, and GDD approaches for the results in Fig. 6.2, we could obtain better acceleration, velocity, and displacement results, as shown in Fig. 6.6. The ground truth of the moving path is like a rectangle, where the starting point and ending point are overlapped. The velocity is calibrated as a Gaussian pulse; the drift of the estimated displacement is significantly reduced (the starting point and end point show the same location value). With the $x$ and $y$ estimated together, the moving direction of the user in the 2-D *navigation* coordinate is also known.

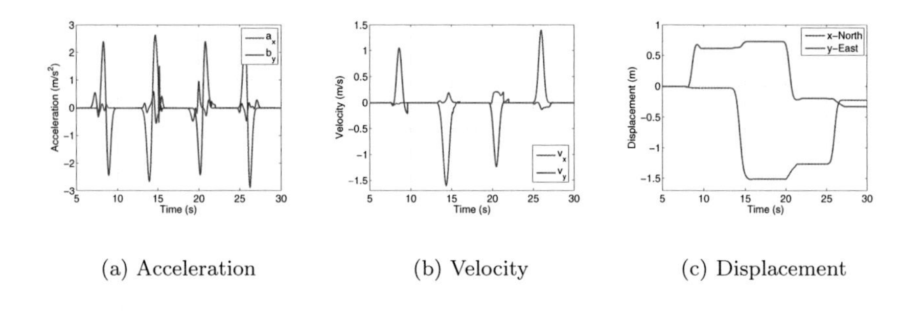


(a) Acceleration (b) Velocity (c) Displacement

**Figure 6.6: Motion estimation result via proposed method: (a) acceleration, (b) velocity, (c) displacement.**

**Drift Metrics and Experiments**. To reduce the randomness in performance evaluation, we tested more than 150 cases. The average and median value of all the metrics are shown in Table 6.2. Our proposed GDD approach achieves best performance in all metrics. The drift rate of 5.5cm per meter is significantly better than existing approaches.

## *6.6.3 Ranging Estimation*

Utilizing TOA for ranging, we define a metric *ranging rate* ($\alpha_r$) to evaluate the TOA miss-detection probability. To evaluate the false-detection, we use two metrics: *ranging error rate* and *ranging accuracy.* The rationale of using two metrics instead of a single *accuracy* is that our approaches could detect and mitigate error measurement automatically. The obtained *ranging rate* and *ranging error rate* are shown in Fig. 6.7. We observe that our proposed ranging scheme works well within 19.5m, i.e., with no apparent decline of the *ranging rate* and *ranging error rate.* The *ranging accuracy* in Fig. 6.8 also shows the *maximum operation distance* for our proposed scheme is around 15 $\sim$ 20m.

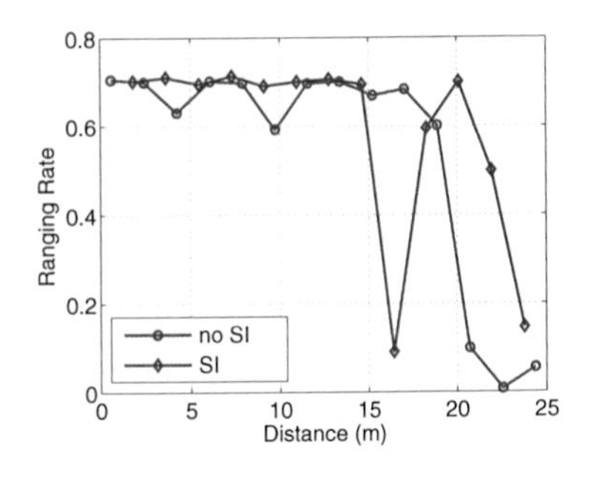


(a) Ranging Rate

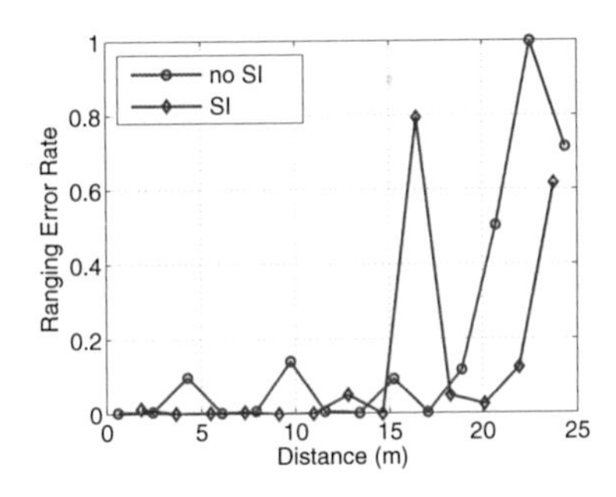


(b) Ranging Error Rate

**Figure 6.7: *Ranging rate* and *ranging error rate* in an aisle environment.**

**Table 6.2: Performance comparison with respect to different methods under specific metrics**

| Methods | Metrics | Average | Median |
|---|---|---|---|
| Basic | Total Drift (m) | 16.51 | 12.47 |
| | Drift Speed (m/s) | 0.47 | 0.40 |
| | Drift Rate (m/m) | 3.92 | 3.73 |
| ATAD | Total Drift (m) | 1.26 | 1.02 |
| | Drift Speed (m/s) | 0.04 | 0.04 |
| | Drift Rate (m/m) | 0.31 | 0.26 |
| DN+ATAD | Total Drift (m) | 0.72 | 0.55 |
| | Drift Speed (m/s) | 0.023 | 0.013 |
| | Drift Rate (m/m) | 0.18 | 0.15 |
| DN+ATAD+GDD | Total Drift (m) | 0.23 | 0.14 |
| | Drift Speed (m/s) | 0.006 | 0.0045 |
| | Drift Rate (m/m) | 0.055 | 0.043 |

### 6.6.4 Improving Location Accuracy with Fewer Anchors

To evaluate the performance improvement of our proposed algorithm when the anchor nodes are insufficient for trilateration, we conducted experiments using Apple iPhone4S in both outdoor and indoor environments. We utilized the default location management module in iOS (mainly from GPS) to obtain the initial location results without war-driving. A total of 7 different position spots are utilized to quantify the performance improvement. The ground truth of these 7 stationary position spots is aligned in a line with the step length of 4m. The coordinates of these spots are $\mathbf{p}_n = [4 \times n, 0]^T$, $n = 1, \cdots, 7$.

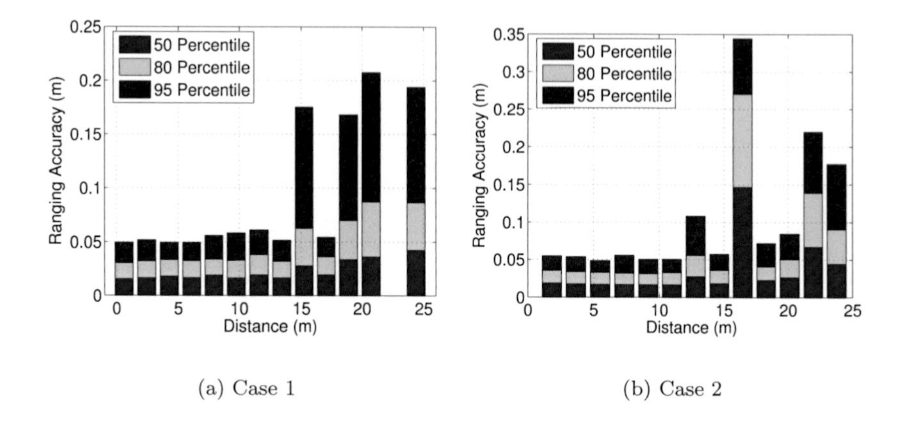


(a) Case 1

(b) Case 2

**Figure 6.8:** ***Ranging accuracy.***

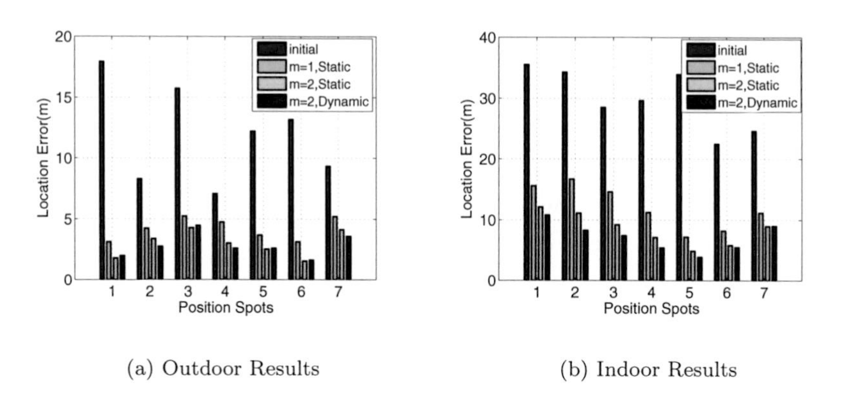


(a) Outdoor Results

(b) Indoor Results

**Figure 6.9: The location error when the anchor number is $m = 0 \sim 2$ in (a) outdoor and (b) indoor environments.**

The first anchor is placed at $\mathbf{a}_1 = [0, 0]^T$; the coordinate of the second anchor is $\mathbf{a}_2 = [19, 15]^T$. The size of the area is around $30 \times 35$ meters. The initial location update rate obtained by smartphone location API is around 1 second. By applying (6.16) for all these 9 position spots, the performance improvement is shown in Fig. 6.9(a) and Fig. 6.9(b) for outdoor and indoor environments, respectively. When the anchor number is $m = 1, 2$, our proposed approach could improve the location accuracy significantly, especially for the indoor cases with larger initial location errors. Applying the dynamic part of (6.16) by leveraging the relative TOA distance measurement and moving direction, the performance could even be improved as shown in the case of "m=2, Dynamic."

From Fig. 6.9, we know that the accuracy improvement ranges from 2 to 11 times over the initial results.

### 6.6.5 Trilateration via Semidefinite Programming

To evaluate and compare the performance of different localization algorithms when the anchor number is sufficient for trilateration, we deployed the anchor network in a typical office environment with a total of 6 anchors. This environment is polluted with voice sound and other acoustic interference.

**Performance Comparison.** The algorithms compared are: "LS-Classic" [74], "LS-PR" [57], "SDP-PR" [9], "SDP-PR-DC," and "SDP-PR-DCR." Fig. 6.10(a) and Fig. 6.10(b) show the CDF of the position error when the mobile phone is placed near $[5.13, 1.08]m$ and $[5.5, 1.4]m$, respectively. The SDP-based approaches perform better than the LS-based approaches in these two cases. By performing delay-constraint (DC) and the robust (R) approach (using Huber Estimator) during the SDP optimization, "SDP-PR-DCR" outperforms other approaches in most situations with different performance gains.

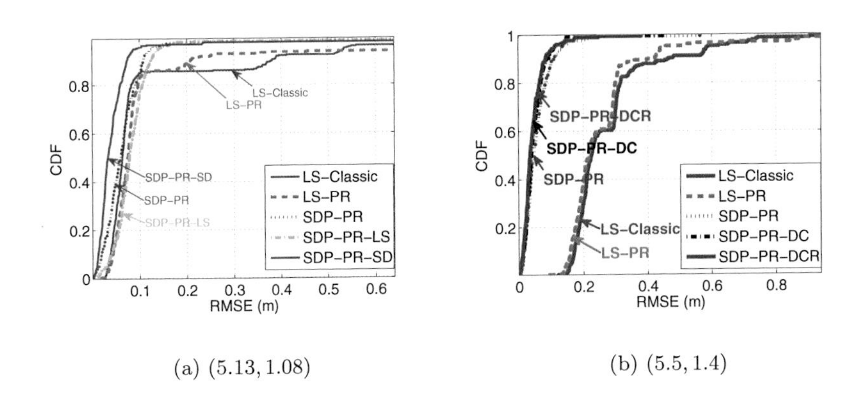


(a) $(5.13, 1.08)$ (b) $(5.5, 1.4)$

**Figure 6.10: Cumulative distribution of different algorithms when the mobile phone is in: (a) $[5.13, 1.08]$ and (b) $[5.5, 1.4]$ by using experimental data with 6 anchors.**

### 6.6.6 Location Fusion

**Ranging and Localization Results** Fig. 6.11(a) shows the ranging accuracy (cm) of smartphones from different anchor nodes (a total of 8 nodes) in 50%, 80%, and 95% percentile probability. For the 8 group bars, the left bar is the result of initial acoustic localization, while the right bar shows our improved result using Attitude and Displacement (AD) for ranging and location tracking. From the ranging error, we show significant reduction in all cases under different probability. Fig. 6.11(b) shows CDF results of the position error without AD. The achieved improvement is more than three times that of the initial result. The smartphone localization accuracy is near 2.7*cm* with 80% probability.

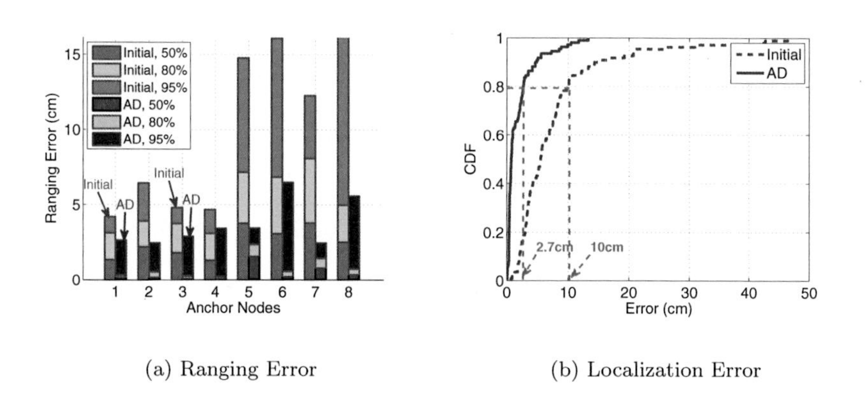


(a) Ranging Error

(b) Localization Error

**Figure 6.11: (a) Ranging results; (b) localization results.**

**Kalman Fusion Experiment**. Fig. 6.12(a) shows the estimated displacement from INS results, the obtained smartphone location via acoustic anchors, and the Kalman fusion results. As expected, the INS result has a very high update rate, but the displacement result in unknown bias to the true location. The location result via acoustic anchors (ALE) is unbiased to the ground truth, but suffers from measurement noise and outliers. After Kalman fusion, the achieved results contain no bias and show better variance in accuracy. Moreover, the fusion results show the same update rate as INS results, which is significantly higher than for ALE. Thus, the fusion process could improve the accuracy and update rate simultaneously. To test the robustness of the fusion process, we utilize the INS result with strong bias and drift as shown in Fig. 6.12(b). The fusion result is still satisfactory, with higher update rate and no bias to the true location.

## 6.7 Related Work

**Localization without anchor nodes**: Low complexity localization techniques are more convenient and popular since they do not rely on the deploy-

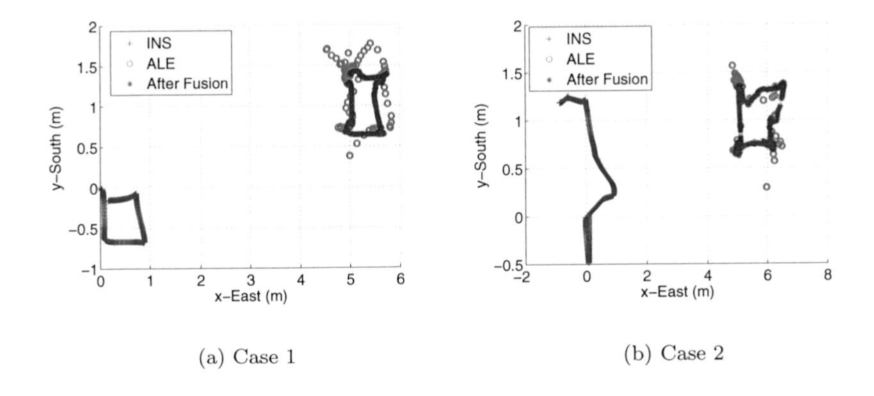


(a) Case 1

(b) Case 2

**Figure 6.12: The Kalman fusion results for location estimation in two cases.**

ment of additional anchor nodes, e.g., inertial sensor (INS)-based approaches [18, 129], fingerprinting-based approaches [8, 132, 15, 6], and radio signal strength (RSS)-based ranging approaches [20, 44, 50].

Solutions based on pervasive WiFi systems have remained the most popular theme of indoor localization without anchor node deployment, but come at the cost of accuracy and need meticulous war-driving. Recent approaches trying to eliminate the war-driving via crowdsourcing or other techniques are proposed in [19, 89, 129], but still suffer low accuracy and lack of clear incentive for crowdsourcing.

Another important issue of a WiFi-based localization system lies in its high energy consumption and long delay. Experiments show that WiFi scanning is very high cost both in time and energy, e.g., Liu et al. [52] show more than 4.5s delay for only 5 WiFi samples. Apple has even banned WiFi RSS scanner apps in its App Store since 2010.

**Localization via anchor nodes**: Conventional highly accurate infrastructure-based localization systems rely on dense anchor nodes for trilateration computations, and require special devices on the user side for ranging purposes, e.g., ultrasound [85, 10], UWB devices [54]. The inconvenience introduced by requiring additional hardware makes these approaches impractical, at least in the near future.

Recent approaches relying on the high-band of the microphone sensor introduces a convenient approach for trilateration without additional hardware attachment on a user's smartphone [81, 74, 48, 57]. Liu et al. [57] utilized low-complexity anchor nodes for broadcasting unnoticeable acoustic beacon with high accuracy. However, at least three anchor nodes are needed for one location calculation with 2-D coordinates, and more nodes are needed for covering large areas, which inhibits wide deployment.

**Localization via hybrid approaches**: Leveraging multiple sensors in a mobile phone and optimizing the location accuracy via moving traces without anchor nodes are proposed in [18, 129]. However, the moving traces obtained by accelerometer and compass are inaccurate and highly dependent on the prior information of the foot-step length. Authors [52, 55] proposed a localization optimization approach via peer-to-peer ranging. However, the two-way ranging process among all user-pairs is too time-consuming and inconvenient. Rajalakshmi et al. [74] proposed systems named EchoBeep and DeafBeep that fuse RF and acoustic-based techniques into a single framework. However, all these scenarios are based on fixed desktops without any mobility considerations and are not direct applicable to smartphones. Moreover, the requirement on the two-way ranging for EchoBeep and the triangular ranging for DeafBeep would limit the user numbers (they only support one user) and introduce complex ranging protocol and long delay, which is impractical in a smartphone-based mobile system. SAIL [65] combines physical layer (PHY) information and human motion for indoor localization using a single Wi-Fi AP. However, the required Wi-Fi AP with PHY information is not ubiquitous, and the achieved accuracy is only at the meter level.

**Proximity detection without localization**: Relying on the proximity detection, one anchor node can provide location references to the user whose accuracy depends on the density of the anchor deployment. The RFID network and the recently introduced BLE network, e.g., Apple's iBeacon [5] and Qualcomm's Gimbal proximity beacons [86], and Estimote [24], are examples of using proximity detection approaches. However, proximity-based approaches are simple but inaccurate; purely relying on the anchor density to improve accuracy is not an efficient and economic method.

## 6.8 Conclusion

We proposed location optimization approaches in a mobile phone via opportunistic anchor sensing. Using the obtained coarse-grained absolute and fine-grained relative ranging information from accessible anchors, the location accuracy achieved significant improvement even with only one or two anchors. When sufficient anchors are available for trilateration, we proposed delay-constraint robust semidefinite programming to ensure robustness in the presence of ranging outliers. The achieved results show 2 to 11 times greater performance with limited anchors and sub-second delay for supporting unlimited users, plus they achieve 80 percentile accuracy of $8cm$ with sufficient anchors. The flexibility and accuracy of the proposed approaches provide strong incentives for service operators to deploy this low-complexity system with various location resolution demands.

# *Chapter 7*

# Pushing Location Awareness to Context-Aware Augmented Reality

**Kaikai Liu**

*Assistant Professor, San Jose State University*

**Xiaolin Li**

*Associate Professor, University of Florida*

## CONTENTS

## 7.1 Introduction

Augmented Reality (AR), as a new form of connections between the virtual and physical worlds, is emerging as an innovative way of expressing and presenting information to users [7, 110]. Via estimating and displaying relevant information of the user's immediate surroundings in the physical world, AR applications present highly contextualized, spatially relevant information that enhances users' experience of mobile lives.

Although AR has been used recently to overlay digital information, e.g., websites, notes, videos, or photos, directly on top of items or point-of-interests (POIs) around us via camera view, the promise of AR is far more ambitious and transformative. Although AR technology is very desireable for wearable devices like Google Glass, smartphones are more practical, affordable, and pervasive. The popularity of these smart devices equipped with cameras, GPS, WiFi, and inertial and other sensors, has lowered the hardware requirement, making it unnecessary to build expensive special head-mounted devices from scratch.

Existing solutions of AR applications fall into two distinct directions: geographic or marker scale. Geographic-scale AR applications focus on location-based services (LBS) by adding location-related information on top of the camera view [71, 29]. They utilize GPS for location tracking, and inertial sensors (INS) for attitude estimation. The marker-scale AR applications focus on computer vision techniques, e.g., marker or non-marker based. For example, *ARToolKit* displays and attaches a virtual character on the top of a marker image [41]. However, marker-based solutions are only suitable for small-scale problems that do not consider mobility. Non-marker based approaches allow using a normal image template as a marker, but entail higher computational costs and even exceed the power of current smartphones. Leveraging the cloud infrastructure could lower the computation cost on the mobile side, but suffers from high-latency in AR view tracking.

More attractive line-of-sight, median scale, or indoor AR applications are largely missing due to several challenges: 1) **High-Accuracy Location Requirement:** GPS with 10-meter accuracy outdoors is unable to differentiate POIs in fine-grained environments and its performance is degraded dramatically in indoor environments. 2) **High-Accuracy Attitude Requirement:** Compared with almost invisible remote POIs, the line-of-sight AR view is more sensitive to the attitude estimation result. The attitude estimation error of the camera would introduce visible drift and bias between the "rendered" objects and the actual objects due to the short distance. 3) **Displacement Matters:** The displacement of a user's camera is ignored in outdoor AR applications, since its moving distance is far shorter than the POI distance. For indoor AR applications with short-range POIs, the movement would have significant impact for the screen location of the AR view.

Using only a smartphone on the user's side without any additional hardware to realize indoor AR poses significant challenges. In this chapter, we propose solutions to enable AR applications in a smartphone for various scales and mobilities. We propose activity identification, pattern decomposition, interacting multiple models, and vision fusion approaches to reduce the displacement and attitude estimation error, drift, and distortion. Leveraging our previous work on high-precision smartphone indoor localization [58], we obtain the absolute location of the smartphone via the deployed anchor network, and estimate the nearby candidate POIs to minimize the POI image matching complexity. We utilize the epipolar geometry of the image feature pairs to estimate the initial relative rotation ($\mathbf{R}$) and translation ($\mathbf{t}$) between the POI and the smartphone. We derive the AR view projecting process by utilizing the intrinsic and extrinsic matrix of the smartphone camera. Computer vision approaches are combined with INS results to improve the accuracy and sensitivity of the AR view tracking under various mobility levels and POI-camera relations. Through a suite of optimization and fusion approaches, we demonstrate that innovative indoor augmented reality applications are emerging on the horizon.

# 7.2 System Design

## 7.2.1 Application Overview

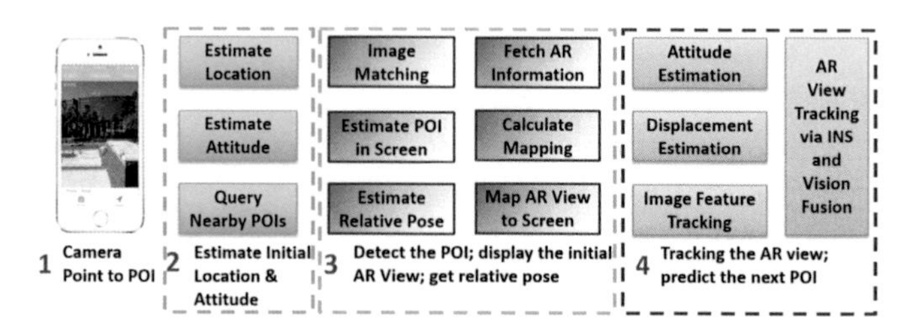


**Figure 7.1: Overview of the proposed AR application.**

A complete AR application involves localization, estimation, tracking to AR view rendering, and even 3D visualization. The localization problem alone is hard to solve. Instead of targeting all the existing harsh challenges directly, we focus on solving practical AR view tracking problems via attitude, displacement, and vision estimation. Leveraging the features of AR application itself, we redesign the AR application procedure by making it feasible and still meet the application demand.

The key procedures of our envisioned AR application are shown in Fig. 7.1. When the user pulls out the smartphone and points the camera to the POI, he intends to know the background information related to current context. The software in the smartphone would initiate the location, attitude and candidate POIs detection process as shown in step 2 of Fig. 7.1. In step 3, a) the smartphone performs image matching for the sensed camera view to detect the actual POI; b) the relative position of the POI to the smartphone will be calculated; c) the AR view will be projected near the real physical POI according to the estimated mapping relations. To minimize the high-cost image matching process, the displacement and attitude of the smartphone will be tracked under various mobilities, e.g., small shaking, or large movement. In step 4, the smartphone estimates moving displacement and performs AR view tracking via INS and vision fusion. At the same time, we can view the world in an additional dimension and makes the "projected" objects and the actual objects indistinguishable to the user. More interactive applications could be built on top of this prototype.

## 7.2.2 System Model and Terminology Definition

**Body coordinate.** The *body* coordinate is a local coordinate of the smartphone with its orientation and displacement changes when we move the smartphone.

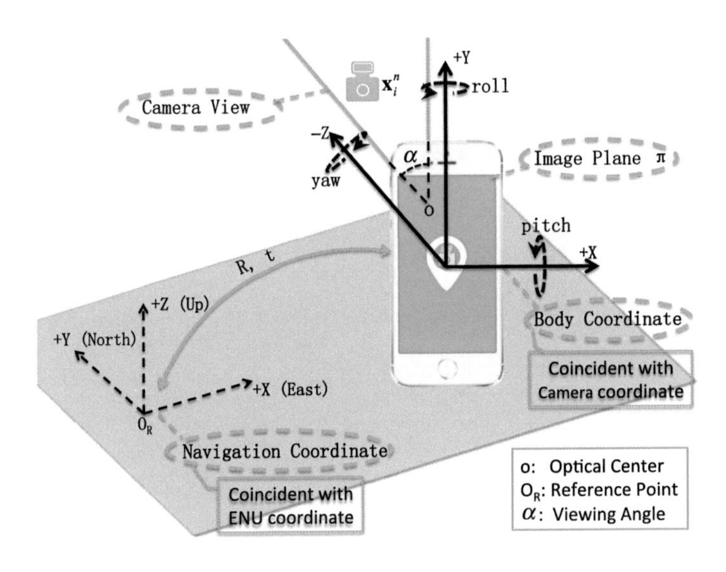


**Figure 7.2: System model and coordinate system.**

**Navigation Coordinate.** To characterize the absolute movement of the smartphone with respect to the indoor/outdoor environment, we define the *navigation* coordinate ($n$-frame) as a local geographic frame that we want to navigate, e.g., stationary with respect to the indoor/outdoor map that is presented to the user.

**Points-of-Interest (POI).** POIs ($\mathbf{x}_i^n$) are specific location points that someone may find interesting and helpful, which we should present to users in terms of a virtual layer over the camera view. We use the term POI to refer an outdoor building, or a small painting in a museum, even a virtual note or comment. The generating of POI could be manually input or via crowdsourcing, and each POI will be associated with a unique ID, a location, and other related information. The unique ID could be a random 64bit UUID; the location could be stored in terms of geodetic coordinates (latitude and longitude), or relative coordinate.

**Smartphone Attitude.** The attitude of a smartphone is the scientific term "pose," which means the orientation of an object with respect to an inertial frame of reference or another entity (the celestial sphere, nearby objects, etc.). The attitude or orientation of the device, i.e., smartphone or camera, can be specified by using several different representations, including Euler angles (roll ($\phi$), pitch ($\theta$), yaw ($\psi$)), quaternions, and direction cosine matrix (DCM), a.k.a. the rotation matrix $\mathbf{R}$. All these different representations can be converted from each other [125].

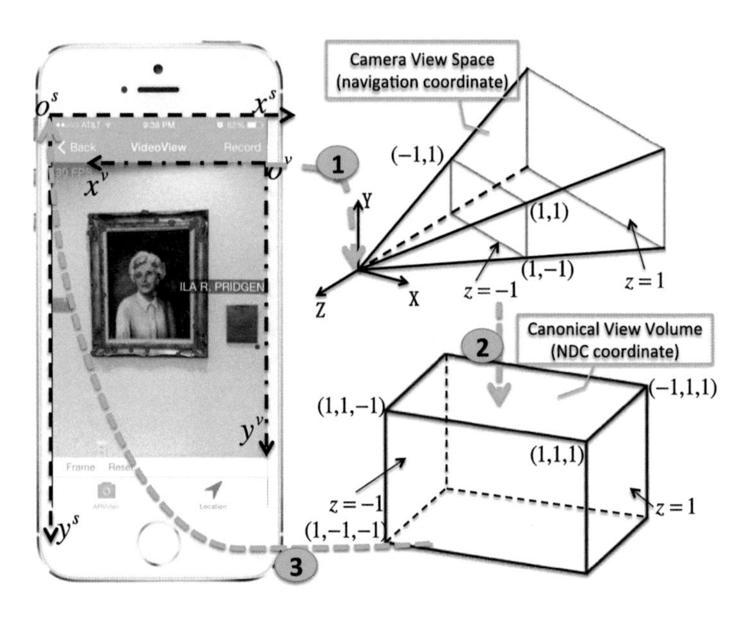


**Figure 7.3: Camera mapping system.**

**Euler Angles**. Euler angles, i.e., roll ($\phi$), pitch ($\theta$), yaw ($\psi$), could be used to illustrate the attitude in a more direct way. Roll, pitch, and yaw are defined in the body coordinate to quantify the smartphone attitude as shown in Fig. 7.2. A roll ($\phi$) is a rotation around a longitudinal axis ($Y$) that passes through the device from its top to bottom. A pitch ($\theta$) is a rotation around a lateral axis ($X$) that passes through the device from side to side. A yaw ($\psi$) is a rotation around an axis ($Z$) that runs vertically through the device.

**Rotation Matrix**. Euler angles are simple and intuitive. On the other hand, Euler angles are limited by a phenomenon called "Gimbal lock." Due to this reason, we use the DCM, a.k.a. the rotation matrix $\mathbf{R}$, to compute the orientation and coordinate transformation. The elements of matrix $\mathbf{R}$ are the cosines of the unsigned angles between the *body* axes and the *navigation* axes. When the attitude or other INS measurement information needs to be converted from the *body* coordinate to the local *navigation* frame, a $3 \times 3$ rotation matrix $\mathbf{R}_b^n$ representation is used for such transformation. Converting a vector quantity $\mathbf{v^b}$ in the body frame to the local *navigation* frame $\mathbf{v^n}$, we can apply $\mathbf{v^n} = \mathbf{R}_b^n \mathbf{v^b}$. Assuming rigid motion, two important properties of the rotation matrix are $\det \mathbf{R} = 1$ and $\mathbf{R}^{-1} = \mathbf{R}^T$. When the rotation matrix multiplies with a vector, the vector rotates while preserving its length. Thus, the inverse transformation from *navigation* coordinate to *body* is given by $\mathbf{v^b} = \mathbf{R}_n^b \mathbf{v^n} = (\mathbf{R}_b^n)^{-1} \mathbf{v^n} = (\mathbf{R}_b^n)^T \mathbf{v^n}$.

The relation between the direction cosine matrix (rotation matrix) and Euler angles can be derived as

$$\mathbf{R} = \begin{bmatrix} \cos\theta\cos\psi & \alpha\sin\phi - \cos\phi\sin\psi & \alpha\cos\phi + \sin\phi\sin\psi \\ \cos\theta\sin\psi & \beta\sin\phi + \cos\phi\cos\psi & \beta\cos\phi - \sin\phi\cos\psi \\ -\sin\theta & \sin\phi\cos\theta & \cos\phi\cos\theta \end{bmatrix} \tag{7.1}$$

where $\alpha = \sin\theta\cos\psi$, $\beta = \sin\theta\sin\psi$. $\theta$, $\psi$, and $\phi$ are the rotation angles around the axis of $y$, $z$, and $x$, respectively.

**Translation Matrix.** A translation $\mathbf{t}$ is a function that moves every point a constant distance in a specified direction in Euclidean geometry, a.k.a. displacement. Translation can be utilized to shift the origin of one coordinate system, i.e., $\mathbf{x}' = \mathbf{x} + \mathbf{t}$, where $\mathbf{t} = [x, y, z]^T$.

## 7.3 Attitude and Location Trace Estimation

One essential problem in this paper is to have fine-grained and efficient camera location and moving estimation, preferably without additional hardware on the users' side. In this section, we utilize the inertial sensor (INS) to estimate the relative attitude and motion of a smartphone, and propose approaches to reduce the drift and error. To get the absolute location results, we leverage our previous work for smartphone indoor localization and propose INS-assisted approaches to improve its dynamic performance. The overview of the algorithm structure is illustrated in Fig. 7.4.

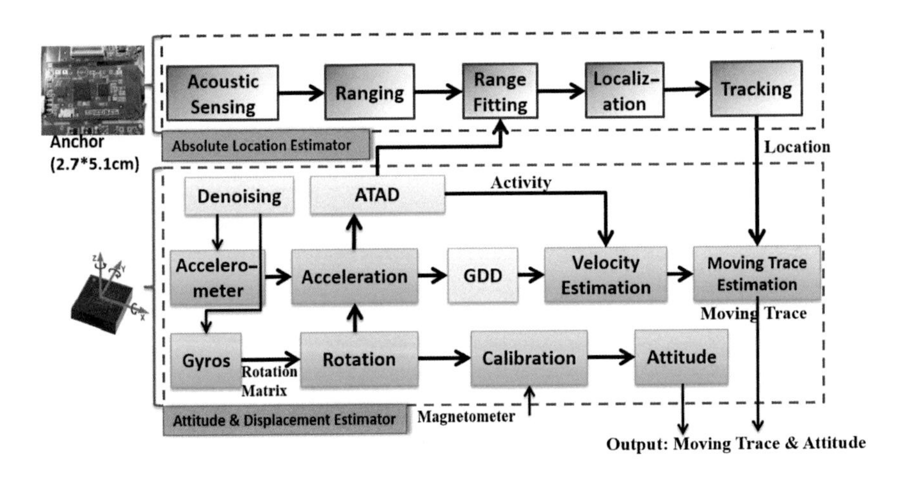


**Figure 7.4: Moving trace and attitude estimation.**

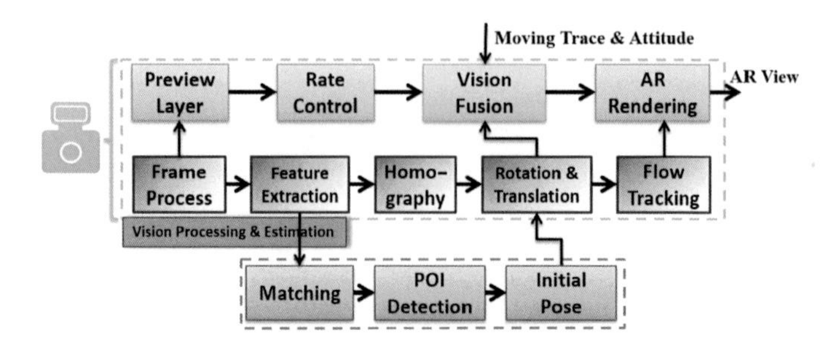


**Figure 7.5: AR view projecting and tracking.**

## 7.3.1 Attitude Estimation

### 7.3.1.1 State-of-the-Art Approaches

The INS, e.g., gyros, accelerometer, and magnetometer, are used for rotation and attitude estimations. Current mobile platforms, e.g., iOS, Android, have the complete framework/package support for programming and accessing all these sensors for attitude estimation [4]. However, the major disadvantages of current MEMS INS is that they are far less accurate than mechanical and optical devices.

State-of-the-art approaches utilize the magnetometer (compass) to opportunistically calibrate the drift. The magnetometer measures the magnetic field observed by the smartphone and derives the *angle* with respect to the global *inertial* coordinate frame (fixed in the space, i.e., it does not accelerate or rotate with respect to the rest of the universe). Both the $A^3$ in [134] and Apple's solution in [4] utilized this principle and achieved good result in heading estimation and compass applications.

The rationale is that gyroscopes and accelerometers belong to the category of strapdown inertial systems that measure the angular velocity and the specific force acted upon the smartphone, respectively. Unlike the gyroscopes and accelerometers where all measurements are made in the smartphone's own frame of reference, magnetometers measure the magnetic field observed by the smartphone and derive the *angle* with respect to the global *inertial* coordinate frame (fixed in the space, i.e., it does not accelerate or rotate with respect to the rest of the universe). Both the $A^3$ in [134] and Apple's solution in [4] utilized this principle and achieved better accuracy in heading estimation.

However, these solutions are not satisfactory for moving conditions in AR applications. The basic requirement of calibration is the identification of a "good" opportunity [134], i.e., when the measurement is stable. If the smartphone kept moving (the common case), there would be no opportunity for

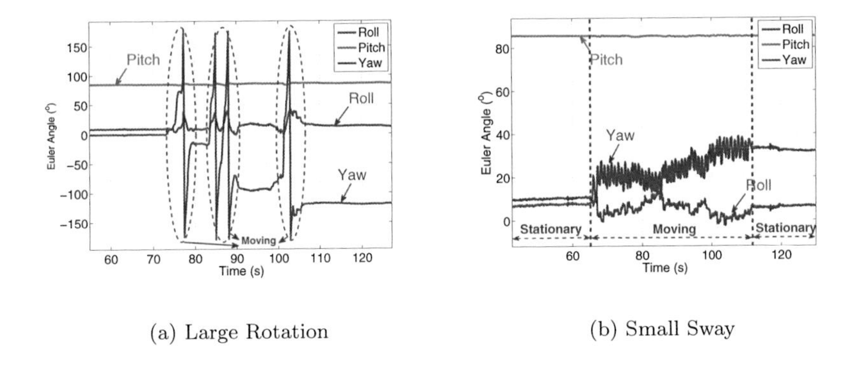


(a) Large Rotation

(b) Small Sway

**Figure 7.6: Attitude under (a) large rotation; (b) small sway.**

calibration. Fig. 7.6 shows the obtained attitude via Apple's solution under two different moving cases. Fig. 7.6(a) is the case where the smartphone is stationary in the initial stage, then performs four large rotations ($360^o$), back to the initial location with the same ground-truth attitude. The attitude drift is well controlled in the stationary stage, but not good after rotations. As shown in Fig. 7.6(a), every rotation introduces several degrees of drift (yaw angle). Fig. 7.6(b) contains stationary and small back-and-forth sway (with the same degree) in the moving part. The attitude suffers strong drift during the small movement. Overall, the existing state-of-the-art approach works well in stationary cases, but is still not good for movement due to the insufficient "good" opportunity for calibration.

### *7.3.1.2 Attitude Error Reduction*

The traditional attitude calibration approach uses the 6-state EKF to estimate the current attitude and gyro biases simultaneously [66]. When the sensor misalignment is considered, there will be a total of 12 calibration parameters.

A paramount issue of the attitude accuracy is the precision of the EKF model. If assuming the bias is constant and the resulting drift is linear, the performance of EKF will be downgraded if the assumption is not true. The worst case is the diverging of the EKF. From the experimental results shown in Fig. 7.6, the bias sometimes really changes, especially in different moving modes. For military or spacecraft applications, a higher order model, e.g., a 15-state model, is often derived to solve this nonlinear problem. However, a higher-order model does not work for low-accuracy MEMS INS in the smartphone. Estimating more unknown parameters will make the result worse in real cases.

**Gyros Error Modeling.** The output of the gyros contains small bias and random noises. Assuming the constant bias error is $\epsilon$, the random noise is a zero-mean uncorrelated random variable with a finite variance of $\sigma^2$. The error model of the gyros measurement can be formulated as $\varepsilon = \varepsilon_{\mathbf{b}} + \varepsilon_{\mathbf{r}} + \mathbf{w_g}$, where $\varepsilon_{\mathbf{b}}$ is the constant bias error, where $\dot{\varepsilon_{\mathbf{b}}} = 0$; $\varepsilon_{\mathbf{r}}$ is the random walk error (result of integrating the random noise), where $\dot{\varepsilon_{\mathbf{r}}} = \frac{1}{T_g}\varepsilon_{\mathbf{r}} + \mathbf{w_r}$; $T_g$ is the correlation time of the gyros [66]. When calculating attitude, the error term causes an angular error that grows linearly when performing integration. Even when the smartphone stands still, i.e., without undergoing any rotation, the output of the gyros is not perfect zero, i.e., it is drifting.

**Accelerometer Error Modeling.** Different from the gyros, the white noise in acceleration measurements is integrated *twice* and causes a second-order random walk in displacement. As a result, position errors grow proportional to $t^2$. The error of the accelerometer measurement can be modeled as $\dot{\delta} = -\frac{1}{T_a}\delta + \mathbf{w_a}$, where $T_a$ is the correlation time of the accelerometer. The value of $T_a$ differs from different devices, and needs to be estimated prior to calibration. $\mathbf{w_a}$ is the modeled Gaussian white noise.

The integration process causes the most important problem of estimating the displacement. The measurement noise will be accumulated and the displacement is drifting even when the smartphone is stationary. Thus, simply performing integration will cause significant errors in real situations.

**Interacting Multiple Model (IMM).** Instead of estimating all the bias and scale factors ($\varepsilon_{\mathbf{b}}$, $\varepsilon_{\mathbf{r}}$, $\delta$) continuously and simultaneously, we change the EKF model in moving conditions, i.e., Interacting Multiple Model (IMM). The basic principle is to smooth the result, i.e., reduce the noise, instead of estimating more unknown parameters in worse cases, e.g., moving cases. We estimate the bias and scale factors when the smartphone is stationary; when the smartphone is moving, we reduce the unknown parameters and directly utilize the scale factors and misalignment that were estimated in the stationary stage. The state vector of the EKF model is written as

$$\mathbf{X_I}(t) = [\theta_{\mathbf{t}}^{\mathbf{b}}, \mathbf{v}_{\mathbf{t}}^{\mathbf{n}}, \mathbf{s}_{\mathbf{t}}^{\mathbf{n}}, \varepsilon_{\mathbf{b}}, \varepsilon_{\mathbf{r}}, \delta] \tag{7.2}$$

where $\theta_{\mathbf{t}}^{\mathbf{b}}$ is the rotation angle obtained from the gyros; $\mathbf{v}_{\mathbf{t}}^{\mathbf{n}}$ is the velocity from the accelerometer; $\mathbf{s}_{\mathbf{t}}^{\mathbf{n}}$ is the location of the smartphone; $\varepsilon_{\mathbf{b}}$, $\varepsilon_{\mathbf{r}}$ are the bias and drift of the gyros; $\delta$ is the drift of the accelerometer. We reduce the $\mathbf{X_I}(t)$ by removing the drift parameters $\varepsilon_{\mathbf{r}}$ and $\delta$ when the smartphone is moving, i.e., change the model via IMM.

Another improvement is to reduce the noise. The relationship of the gyros and accelerometer measurements can be written as

$$\ddot{\mathbf{s}}_t^n = \mathbf{R}_{b,t}^n(\mathbf{f}_t^b - \varepsilon_{\mathbf{b}} - \varepsilon_{\mathbf{r}}\mathbf{w_g}) - \mathbf{g} \tag{7.3}$$

where $\mathbf{R}_{b,t}^n$ denotes the rotation matrix from the *body* coordinate to the *navigation* coordinate. Thus, leveraging the adaptive-threshold-based activity detection (ATAD) results via accelerometer as stated in Section 7.3.2.3 smooths the gyros result.

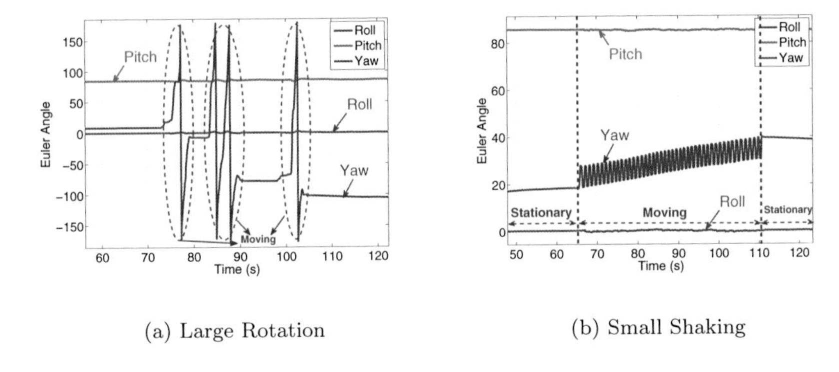


**Figure 7.7: Attitude via IMM under (a) large rotation; (b) small sway.**

Utilizing (7.3), we perform real-time wavelet denoising [90] for the update function of the gyros ($\omega_t^b = \hat{\omega}_t^b - \Delta - \mathbf{w}_a$), where $\hat{\omega}_t^b$ is the gyros output. After the smoothing process, the optimized attitudes of Fig. 7.6 are shown in Fig. 7.7. Both Fig. 7.7(a) and Fig. 7.7(b) show reduced noise with clear movement trace. Note that our approach does not reduce the drift, since we used a low-order model during movement. Our purpose is to reduce the noise. The drift reduction is achieved via INS and vision fusion as stated in Section 7.4.4.2.

## 7.3.2 Displacement Estimation

The movement and location of a smartphone or camera is critical in presenting the AR view to the right place. However, a highly accurate indoor localization approach is still missing, and remains a hot research topic. Instead of directly utilizing the absolute location of a smartphone, we leverage coarse-grained location, POI image matching, and displacement estimation to map the AR view to the right place in the physical world.

### *7.3.2.1 Challenges*

To estimate the displacement ($\mathbf{t}$) of the smartphone, double integrating the acceleration via movement model, or counting the step number from acceleration features (pedometer) are two typical approaches [97, 114]. However, step-counting requires the prior knowledge of step size, and is only suitable for applications when the number of steps are large. For indoor and small-scale AR applications, people usually linger around and do not walk with their normal step size. For example, museum visitors often take photos, record videos or stare at exhibitions, which is very subtle in movement with no apparent footsteps. Thus, fine-grained double integration is more preferred than coarse-grained step counting.

The quantity result from the accelerometer is the acceleration rate in $m/s^2$, and can be denoted as $\mathbf{f}_t^b = (f_t^{b_x}, f_t^{b_y}, f_t^{b_z})^T$, where $f_t^{b_{x,y,z}}$ is the measured force in 3-axis directions in the *body* frame. Direct integrating $\mathbf{f}_t^b$ obtains the displacement in the *body* coordinate, which is not related to the real geodesic displacement. Moreover, the user's moving path is not a line. Turns and rotations should also be estimated.

### *7.3.2.2 Existing Approaches*

To obtain the complete moving paths and turns, we convert the obtained acceleration of the smartphone to the local *navigation* coordinate (apply rotation and translation over $\mathbf{f}_t^b$) by

$$\mathbf{f}_t^n = \mathbf{R}_b^n \mathbf{f}_t^b + \mathbf{e}^n \tag{7.4}$$

where $\mathbf{e}^n$ is the error of the force applied to the smartphone. $\mathbf{f}_t^n$ is the estimated force of the smartphone in the *navigation* coordinate. To obtain the acceleration caused by the applied forces, gravity should be subtracted by $\mathbf{a}_t^n = \mathbf{f}_t^n - \mathbf{g}$, where $\mathbf{g} = [0, 0, g]$ is the gravity vector.

Using Fig. 7.8 as an example, the ground truth of moving is a rectangle with stationary stays during movements. The measured acceleration result after gravity subtraction is shown in Fig. 7.8(a). The four strong non-zero regions demonstrate the four edge movement for a rectangular shape.

After the denoising process, the velocity of the smartphone can be obtained by $\mathbf{v}_t^n = \mathbf{v}_0^n + \int_0^t \mathbf{a}_t^n$ as shown in Fig. 7.8(b), where $\mathbf{v}_0^n$ is the initial velocity. From Fig. 7.8(b), we know that the velocity drifts even when the user is in stationary.

The displacement can be calculated by $\mathbf{s}_t^n = \mathbf{s}_0^n + \int_0^t \mathbf{v}_t^n$, where $\mathbf{s}_0^n$ is the initial displacement. The result of the estimated displacement is shown in Fig. 7.8(c) with large drift. We apply different colors in Fig. 7.8(c) for different

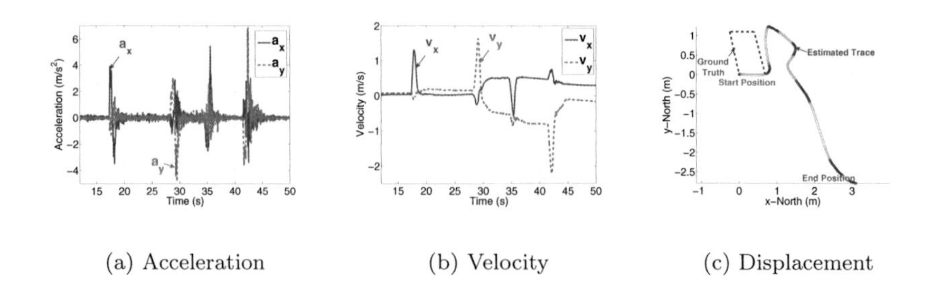


(a) Acceleration (b) Velocity (c) Displacement

**Figure 7.8: Motion estimation result via conventional method: (a) acceleration, (b) velocity, (c) displacement.**

time periods during the movement, which provides an additional views of the temporal information.

The process of obtaining relative displacement $\mathbf{s}_t^n$ is contributed by double integration, in which the measurement noise, i.e., $\int\int \mathbf{e}^n$, is also integrated and amplified. The white noise in acceleration measurements is integrated *twice* and causes a second-order random walk in displacement of the smartphone.

#### *7.3.2.3 Mitigating Displacement Error*

The portable and inexpensive features, in addition to the fast response time and low power consumption, motivates us to design better approaches to compensate for the error introduced in INS. The major disadvantages of current MEMS INS in smartphones is that they are far less accurate than mechanical and optical devices. The integration process causes the most important problem of using INS measurement to estimate the displacement. Thus, simply performing integration will cause significant errors in real situations.

**Adaptive-Threshold-based Activity Detection (ATAD).** Performing signal denoising [90] and activity detection before integration, and using the estimation results as a constraint, could be a feasible approach to calibrating the drift. One significant problem of using integration is that the estimated displacement is drifting even when the smartphone is stationary.

One important problem in activity detection is the dynamic setting of the threshold for various environments and conditions. We propose to utilize forward and backward window-aided CFAR detectors for high detection probability with constant false-alarm rate [58]. Each individual sample of acceleration $a_t^n$ can be modeled as a Gaussian random variable with the noise component as $n_t^g$. Using the *Neyman-Pearson* criterion, we set a constant value of false-alarm rate $P_{fa}$ and perform CFAR detection. Relying on the fact that the adjacent samples from the accelerometer have strong correlation, here we construct two detection-aided windows ($w_f$ and $w_b$) to estimate the environment noise and the detection threshold. The adaptive threshold is obtained by

$$\hat{\eta}_{act} = \sqrt{\pi/2}\mathbf{E}(z_i)Q^{-1}(P_{fa}/2) \tag{7.5}$$

where $\mathbf{E}(z_i)$ is the similar term for the signal-to-noise ratio, i.e., the normalized decision vector with regard to the values estimated by $w_f$ and $w_b$. The activity decision process is realized by comparing the adaptive threshold to the cell under test. We call it adaptive-threshold-based activity detection (ATAD).

Using the detected activity level as shown in Fig. 7.9(a), we map the activity level to one weighting coefficient $w_t$ via low-pass filter for boundary smoothness, as shown in Fig. 7.9(b). During the integration process of displacement calculation, we multiply $w_t$ to the velocity value $v_t^n$ as shown in Fig. 7.9(b). This process minimizes the error during integration and improves the boundary continuity. The estimated displacement is shown in Fig. 7.9(c) with significantly reduced drift compared with the initial version in Fig. 7.8(c).

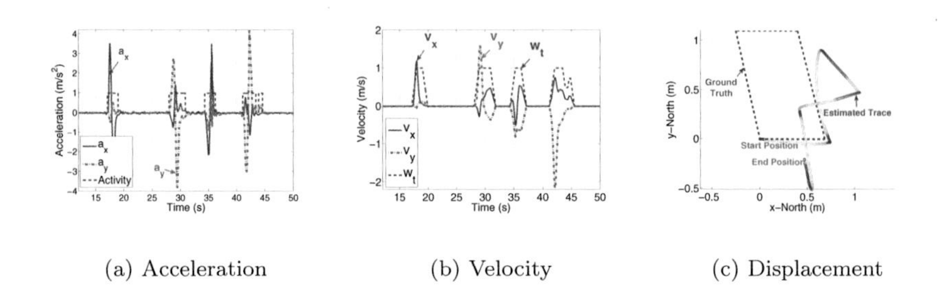


(a) Acceleration (b) Velocity (c) Displacement

**Figure 7.9: Motion estimation result via ATAD: (a) acceleration, (b) velocity, (c) displacement.**

**Gaussian Derivative Decomposition (GDD).** After performing denoising and using ATAD, the moving trace becomes significantly better than direct integration. However, inaccurate traces still exist due to the imperfect denoising or thresholding. Purely detecting the signal from noisy background without prior information of the signal pattern would not guarantee better results.

Performing moving pattern extraction before integration could be a feasible method. The normal movement model could not be applied for estimating the human's moving, e.g., we cannot walk at constant velocity (CV) or constant acceleration (CA) like a vehicle or a plane. Human walking or movement has its own pattern, and we need to "accelerate" and "decelerate," and then "accelerate" for another footstep. Thus, the acceleration pattern for one movement should look like a pulse. Here we use the "start-moving-stop" movement model.

Performing "start-moving-stop" pattern decomposition helps us estimate the displacement in a more meaningful way. Using start, acceleration, and deceleration as one basic step, the velocity changes (from zero to top to zero) can be modeled as a Gaussian shape $g_v(\mu, \sigma) = \pm\exp\{-(x-\mu)^2/(2\sigma^2)\}$, where $+$ means moving forward; $-$ means backward. The acceleration is the derivative of the velocity, i.e., Gaussian derivative pulse, as

$$g_a(\mu, \sigma) = \pm(x-\mu)/\sigma^2 \exp\{-(x-\mu)^2/(2\sigma^2)\} \tag{7.6}$$

Using $g_a(\mu, \sigma)$ as the *kernel* function, we decompose the acceleration measurement $\mathbf{a}_t^n$ into a series of $\eta g_a(\mu, \sigma)$ with different parameters $\eta$, $\mu$, and $\sigma$. The decomposed series of acceleration is $\sum_{i=1}^{n} \eta_i g_a(\mu_i, \sigma_i)$, where $\eta_i$ is the amplitude of each Gaussian derivative pulse. The fitting process can be modeled as

$$\{\eta_i, \mu_i, \sigma_i\} = \min_{\{\eta_i, \mu_i, \sigma_i\}} ||\mathbf{a}_t^n - \eta_i g_a(\mu_i, \sigma_i)|| \tag{7.7}$$

To reduce the number of parameters during the fitting process, we extract the feature points of $\mathbf{a}_t^n$, e.g., peak position and width, by thresholding the peak maximum and rising edge. Then we use the number of peaks found and the peak positions and widths to fit the specified peak model. This combination yields better and faster computation, and deals with *overlapped* peaks as well. During the decomposition and fitting process, the sign of $\eta_i$ is determined by comparing the remaining error of using positive and negative results.

After the GDD-based fitting, the fitted acceleration result is shown in Fig. 7.10(a). We calibrate the measured acceleration and mitigate the outliers by relying on the "start-moving-stop" model. The estimated velocity and displacement results after using GDD are shown in Fig. 7.10(b) and Fig. 7.10(c), respectively. Compared with the result in Fig. 7.9, Fig. 7.10 demonstrates a very clear and accurate moving rectangle (almost the same as the ground truth), which is significantly better in terms of estimation accuracy.

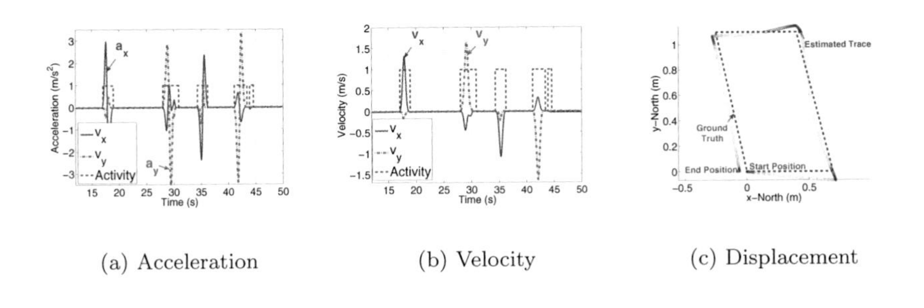


(a) Acceleration (b) Velocity (c) Displacement

**Figure 7.10: Motion estimation result via GDD: (a) acceleration, (b) velocity, (c) displacement.**

The GDD approach also works for overlapped steps or situations when the steps are continuous. We do not assume users fully stop during the movement. Continuous movement will result in several overlapped pulses, and the process of (7.7) also works. As shown in the second and fourth movement in Fig. 7.10(a), the GDD fitting process is effective when two pulses are overlapped together.

### 7.3.3 Smartphone Localization

One important problem of the AR application is to predict which POI is in the camera view, and map the physical POI into camera coordinates, rendering and tracking the AR view in an appropriate screen location. This process requires an accurate location estimation result of the camera with real-time response. However, smartphone indoor location estimation is still the subject of extensive works [74, 52, 48].

To relax this demanding requirement, we utilize a coarse-grained location result and combine it with the POI image matching to obtain the relative location. The coarse-grained location result could narrow down the searching space of the POI candidates, which significantly reduces the matching complexity. The image matching process detects which POI is in the camera view. Based on our previous work [58], we leverage a plurality of acoustic sensor nodes as a preconfigured anchor constellation to broadcast an unnoticeable acoustic beacon. The smartphone captures the location beacon, demodulates the beacon symbol, and performs time-of-arrival (TOA) estimation. To differentiate different anchor nodes, i.e., station i.d. the decoded sequence of symbols is utilized to match with the pre-stored pseudocode sequence $\mathbf{p_m}$.

Define $\mathbf{p}^n$, $\mathbf{x}_m^n$ as the position of the smartphone and the $m$-th anchor in the $n$-frame. Using $M$ pseudoranges $\hat{\mathbf{r}}_m$ and the preconfigured coordinates of anchor nodes $\mathbf{x}_m^n$, we estimate the 3D position of the smartphone $\mathbf{p}^n$ by minimizing the quadric term of the remaining error

$$\varepsilon_m = ||\hat{\mathbf{r}}_m - (||\mathbf{p}^n - \mathbf{x}_m^n||_2 + \delta_r)||_2 \tag{7.8}$$

where $\delta_r$ is the unknown delay that compensates for the difference between the pseudorange and real distance. The value of $\delta_r$ is estimated by adding an additional measurement, e.g., solving 3D location $(x, y, z)$ needs four equations (anchor nodes) instead of one. The superscript $n$ in $\mathbf{p}^n$ denotes the position value in the *navigation* coordinate (defined in Fig. 7.2).

**Range and Location Fusion via Activity Detection.** Measurements obtained from the indoor absolute location estimator (ALE) contain various kinds of jitters and outliers due to the ambient noise and multipath. Things will get even worse when blockage occurs with no ranging information at all.

To solve this problem, we create two steps of Kalman filtering to eliminate the outliers and smooth the data via the activity detection result in Section 7.3.2.3. The first step is to process the ranging data from multiple anchors before location computation, in other words, tracking before localization; the second step is to smooth and track the coordinates after location computation, i.e., location tracking.

To fuse the INS sensor with the ALE, we can write all the measurements into one state-space model, and utilize the model in the Kalman filter for the fusion process. The position estimation results from the INS sensor is $\mathbf{s}_t^n$, the second-order moving model can be written as

$$\begin{aligned} \mathbf{s}_{t+1}^n &= \mathbf{s}_t^n + \dot{\mathbf{s}}_t^n T_{ins} + \ddot{\mathbf{s}}_t^n \frac{T_{ins}^2}{2} \\ \dot{\mathbf{s}}_{t+1}^n &= \dot{\mathbf{s}}_t^n + \ddot{\mathbf{s}}_t^n T_{ins} \end{aligned} \tag{7.9}$$

where $\dot{\mathbf{s}}_t^n$ and $\ddot{\mathbf{s}}_t^n$ are the moving velocity and acceleration, which is derived from the obtained displacement; $T_{ins}$ is the sampling time of the INS.

The relationship with the INS result (7.9) and the estimated location result (7.8) can be summarized as $\mathbf{s}_{t+1}^n = \mathbf{p}_{\tau+1}^n - \mathbf{p}_0^n + \varepsilon_{\mathbf{p}}$, where $\varepsilon_{\mathbf{p}}$ is the error term, and $\tau$ is the update time period for the localization result, which is longer than $T_{ins}$.

The Kalman filter handles the different sample rates from INS ($t$) and ALE ($\tau$) by running at a high data rate (same as INS) and only calibrating an INS result when the measurements from ALE are available. After the fusion process, the update rate of the localization module has been improved to the same level as the INS sensor, while the bias problem of the INS is calibrated.

# 7.4 Projecting and Tracking the AR View

The process of projecting the AR view is summarized in Fig. 7.3. Basically, it involves four steps: 1) detect the POI; 2) map the POI into the camera view space; 3) convert to the canonical view volume, i.e., NDC (normalized device coordinate); 4) present the AR point in screen coordinate. The overall algorithm processing structure is illustrated in Fig. 7.5.

## 7.4.1 POI Detection and Relative Pose Estimation

### 7.4.1.1 POI Detection via Image Feature Detection and Matching

Using the smartphone's location to estimate a rough region for the POI image matching, the matching database is significantly reduced. When the activity detection module triggers the image feature detection process, we perform ORB feature detection [94] for current image frame $I_k^v$ as $\rho_k^v = f_{\mathcal{A}_k^v}^{ORB}(I_k^v)$. To find the best transform between two sets of feature points, the RANdom SAmple Consensus (RANSAC) approach is applied to reject outliers [32]. To determine whether two images match, the inlier rate of the matched points after RANSAC is utilized. Assuming the feature point for the POI is $\rho^m$, the inlier rate for $\rho_k^v$ is $p_k^m = Match(\rho_k^v, \rho^m)$. If the maximum $p_k^m$ among all the images in the database exceeds the threshold, the POI is matched to the smartphone's camera view. We get the POI's augmented information and its pre-stored coordinate $\mathbf{x}_i^n = (x_i^n, y_i^n, z_i^n)^T$ expressed in terms of the *navigation* coordinate.

### 7.4.1.2 Initial Relative Pose Estimation

With the detected POI and its pre-stored information, we estimate the relative rotation $\mathbf{R}$ and translation matrix $\mathbf{t} = \begin{bmatrix} t_x & t_y & t_z \end{bmatrix}$ between the POI and the smartphone. According to the epipolar geometry, the fundamental matrix can be written as $\mathbf{F} = \mathbf{K}^{-T}\hat{\mathbf{T}}\mathbf{R}\mathbf{K}^{-1}$ [32], where $\mathbf{t} = \begin{bmatrix} t_x & t_y & t_z \end{bmatrix}$, $\mathbf{K}$ is the camera intrinsic matrix. Using the result of image matching, we obtain the fundamental matrix $\mathbf{F}$. With $\mathbf{F}$ available, relative rotation $\mathbf{R}$ and translation matrix $\mathbf{t}$ are inversely calculated. The ambiguity values of $\mathbf{R}$ and $\mathbf{t}$ can be ruled out via the frustum constraints, i.e., the POI is faced to the camera and cannot be in the back of the camera view (depth value $z_i^{ndc} < 0$).

The estimated matrix from fundamental matrix $\mathbf{F}$ are actually camera's initial rotation $\mathbf{R}_o$ and translation $\mathbf{t}_0$ that relative to the POI in the *navigation* coordinate. When the camera moves, further readings of attitude and displacement from INS, i.e., the rotation matrix $\mathbf{R}_n^b$ and the translation matrix $\mathbf{t}_n^b$, reflect the physical dynamics. Combing the initial $\mathbf{R}_o$ and $\mathbf{t}_0$ with $\mathbf{R}_n^b$ and $\mathbf{t}$, the POI's location in the smartphone's *body* coordinate is written as

$$\mathbf{x}_i^b = \mathbf{R}_n^b(\mathbf{R}_0\mathbf{x}_i^n + \mathbf{t}_0) + \mathbf{t}_n^b \tag{7.10}$$

(7.10) relates the conversion between the $b$-frame and the real physical world, i.e., $n$-frame. Converting the POI location into the smartphone's *body* coordinate is the prerequisite for the POI projection process in Fig. 7.3.

### *7.4.1.3 Homography Relation*

For two consecutive image frames, using homogeneous representation could represent the projective transformation by means of simple matrix multiplications [32]. Epipolar geometry holds when translation exists, i.e., the baseline of two cameras is larger than 0. For our continuous video frame capture, the translation between each frame is small. For this zero epipolar geometry condition, homography holds [32]. Assuming a pinhole camera model, the feature points in consecutive planar images $\rho_k^v$ and $\rho_{k+1}^v$ are related by a 3 *homography* matrix [32]. For a single feature point $\rho_k^v = [x_k^v, y_k^v, 1]^T$ and $\rho_{k+1}^v = [x_{k+1}^v, y_{k+1}^v, 1]^T$, the homography relation can be modeled by

$$\rho_k^v = H_k^{k+1}(w'\rho_{k+1}^v) \tag{7.11}$$

where $w'$ is the added dimension in homogeneous coordinates [32]. The homography relation $\mathbf{H}_k^{k+1}$ is calculated by finding the best matched transform between two sets of points $\rho_k^v$ and $\rho_{k+1}^v$ after RANSAC [32].

The camera rotation ($\mathbf{R}_k^{k+1}$) and translation ($\mathbf{t}_k^{k+1}$) between two images can be extracted from an estimated homography matrix $\mathbf{H}_k^{k+1}$. This information is useful for rendering the correct AR perspective in the next frame $(k+1)$ and appear to have been part of the POI point (scene).

The predicted POI point in image $k$ is $\mathbf{p}_{i,k}^{ndc}$, and its related pixel point in the $v$-frame will be $\mathbf{p}_{i,k}^v$ via $\mathbf{f}_{ndc}^v$. Then, the predicted next position of the POI point via homography is

$$\mathbf{p}_{i,k+1}^v = \mathbf{R}_k^{k+1}(\mathbf{p}_{i,k}^v - \mathbf{C}) = \mathbf{R}_k^{k+1}\mathbf{p}_{i,k}^v + \mathbf{t}_k^{k+1} \tag{7.12}$$

where $\mathbf{C}$ is the original location of image $k$. The related screen coordinate can be calculated by $\mathbf{p}_{i,k+1}^s = \mathbf{f}_{ndc}^s\mathbf{f}_v^{ndc}\mathbf{p}_{i,k+1}^v$. If iterate via (7.12) from $k = 1$, we could obtain the whole moving trace of the POI point in the image plane.

### 7.4.2 Mapping the POI to the Screen

To address the problem of how and where to present the overlay view to the right position on the screen, the mapping process between the rendered AR view and real physical POI should be derived.

To model the process of calculating the screen point position $\mathbf{x}_i^{\pi}$ for the POI point ($\mathbf{x}_i^b$), the projection process can be modeled as $\mathbf{x}_i^{\pi} = \mathbf{T}_b^{\pi}\mathbf{x}_i^b$, where the image point is $\mathbf{x}_i^{\pi} = (x_i^{\pi}, y_i^{\pi}, z_i^{\pi})^T$, and $\mathbf{T}_b^{\pi}$ determines the relationship between a point in the image plane and a POI point. Converting $\mathbf{x}_i^{\pi}$ to the point in pixel format $\mathbf{x}^p$ involves two steps. The first step is to change the scale from metric units to pixels. The second step is to translate the origin of the coordinate frame from the principal point to the video capturing ($v$-frame) original point as shown in Fig. 7.3. Thus the transformation can be described as $\mathbf{x}_i^p = \mathbf{T}_{\pi}^p\mathbf{T}_b^{\pi}\mathbf{x}_i^b = \mathbf{K}\mathbf{x}_i^b$.

For AR applications, we need to render a synthetic scene in the screen via OpenGL. While the world of OpenGL is slightly different, our $3 \times 3$ intrinsic camera matrix $\mathbf{K}$ needs three modifications to make it compatible with OpenGL. Thus, the intrinsic and extrinsic matrix combined could illustrate the full-perspective model that describes the relationship between a 3D point $\mathbf{x}_i^n$ expressed in the *navigation* frame and its projection on the screen. The overall relationship is written as

$$\begin{aligned}\begin{bmatrix} x_i^p \\ y_i^p \\ 1 \end{bmatrix} = \mathbf{K}^G\mathbf{D}\begin{bmatrix} x_i^n \\ y_i^n \\ z_i^n \\ 1 \end{bmatrix} &= \begin{bmatrix} f/\alpha & 0 & 0 & 0 \\ 0 & f & 0 & 0 \\ 0 & 0 & -\frac{far+near}{far-near} & -\frac{2far \times near}{far-near} \\ 0 & 0 & -1 & 0 \end{bmatrix}\begin{bmatrix} \mathbf{R}_n^b & \mathbf{t} \\ 0 & 1 \end{bmatrix}\begin{bmatrix} x_i^n \\ y_i^n \\ z_i^n \\ 1 \end{bmatrix} \\ &= \mathbf{P}\begin{bmatrix} x_i^n \\ y_i^n \\ z_i^n \\ 1 \end{bmatrix}\end{aligned} \tag{7.13}$$

where $\mathbf{D}$ is the camera's *extrinsic matrix* that describes the camera's location coordinate transformation. The added one dimension is to utilize the homogeneous representation [32]. $\mathbf{K}^G$ is the new projection matrix, where $f = \cot(fovy/2)$; $fovy = 2\arctan(height/2(n_y f))$; pixel aspect ratio $\alpha = (n_y width)/(n_x height)$; $near$ and $far$ is the near and far plane of the OpenGL camera system. Using Apple iPhone 5S as an example, the $fovy = 60$ degrees; $near = 0.25$ meters; $far = 1000$ meters; $width = 320$ pixels; $height = 180$ pixels. The projection matrix $\mathbf{K}^G$ allows you to represent all of the intrinsic camera parameters, e.g., focal length, principal point, pixel aspect ratio, and axis skew. The matrix $\mathbf{P}$ denotes the perspective projection matrix, which transforms a POI point in the *navigation* coordinate to the screen of a smartphone as shown in the first step of Fig. 7.3.

### 7.4.3 Projecting the AR View

To summarize the procedures of projecting the AR view to the smartphone screen, there are several basic steps:

- Get the initial absolute location of the smartphone via Section 7.3.3; query all the nearby POI positions.
- Detect the actual POI on screen via the image matching; estimate the initial relative pose via fundamental matrix in Section 7.4.1.2.
- Obtain the smartphone body attitude in rotation matrix $\mathbf{R}_n^b$ and relative translation matrix $\mathbf{t}$ via INS.
- Update the camera projection matrix $\mathbf{P}$ via (7.13); iterate over all the local POI positions, and calculate their projected homogeneous position $[\mathbf{x}_i^p \quad w_i^p]$ in view frustum as shown in step 1 of Fig. 7.3.
- Collapse the 4 component vector $\mathbf{x}_i^p$ in view frustum to 3-dimensional space. The achieved normalized device coordinate (NDC) is $\mathbf{x}_i^{ndc}$, where $x_i^{ndc} = x_i^p/w_i^p$, $y_i^{ndc} = y_i^p/w_i^p$, $z_i^{ndc} = z_i^p/w_i^p$. Through this process, the POI position is mapped into the *NDC* coordinate as shown in step 2 of Fig. 7.3, where $x, y, z$ is in the range of $[-1, 1]$ for the points on the screen. Since $x$ and $y$ coordinates are being biased from the range $[-1, 1]$, to match up with the UIKit coordinate system in iOS, we convert it to range $[0, 1]$ by $x_i^{ndc} = (x_i^{ndc}+1)/2$, $y_i^{ndc} = (y_i^{ndc}+1)/2$.
- The z component, $z_i^{ndc}$, should be checked to see if it is greater than 0. If $z_i^{ndc} < 0$, then the POI is in the first half of the projection frustum, i.e., in front of the camera.
- After the third step in Fig. 7.3, the overlay view location in *screen* coordinate for the $i$-th POI is obtained via $\mathbf{f}_{ndc}^s$ as

$$\begin{bmatrix} x_i^s \\ y_i^s \end{bmatrix} = \mathbf{f}_{ndc}^s \begin{bmatrix} x_i^{ndc} \\ y_i^{ndc} \\ 1 \end{bmatrix} = \begin{bmatrix} W^s & 0 & 0 \\ 0 & -H^s & H^s \end{bmatrix} \begin{bmatrix} x_i^{ndc} \\ y_i^{ndc} \\ 1 \end{bmatrix} \tag{7.14}$$

  where $H^s$ and $W^s$ are the height and width of the screen frame.

### 7.4.4 AR View Tracking

#### 7.4.4.1 Problems

Users' movement is highly dynamic, from hand vibration to fast walking. They may hold their smartphones to view the AR with slight hand movement or vibration. The relative displacement estimation obtained in Section 7.3 is accurate for major movement, but cannot detect vibration or other slow changes. Only using INS to estimate the displacement and rotation is not sufficient.

#### *7.4.4.2 Vision Tracking and Adaptive Rate Control*

It is possible to perform feature detection in the same order as the refresh rate of the AR view, and then apply the matching process between two consecutive image frames. However, the computation demand of performing high rate feature detection will kill the battery of smartphones very quickly and slow down the update frame rate. To balance the tracking accuracy and efficiency, the feature detection process should only be performed when necessary. The low-complexity optical flow tracking approach should be applied for consecutive frames.

**Optical Flow Tracking.** KLT (Kanade-Lucas-Tomasi) feature tracking has been extensively studied and applied [112, 47]. For the $k$-th frame ($I_k^v$) in the video source, KLT uses image alignment, and solves image displacement via nonlinear optimization by minimizing the remaining error of

$$e_k^v = \sum \left[ f_{\mathcal{A}_k^v}^{ORB}(I_k^v) - (f_{\mathcal{A}_k^v}^{ORB}(I_k^v) + \mathbf{d}_k^v) \right]^2 \tag{7.15}$$

Using window $\mathcal{A}_k^v$ to represent the major region rather than detecting and searching the whole image frame can lower the computation complexity. $\mathbf{d}_k^v$ is the pixel displacement between two video frames. After the KLT tracking process, the feature points ($\rho_{k+1}^v$) in the next image frame are estimated.

**Adaptive Rate Control for AR View Tracking.** Not every image frame is needed for feature detection and matching, since users may shake their smartphone with no purpose. Utilizing the activity detection results, we only need to perform feature detection for images when the smartphone detected the POI, which means the user is intended to know the POI in its camera.

During the KLT feature tracking process, the frame rate could also be controlled. If the frame rate is too high, and the smartphone is not undergoing movement, performing KLT feature tracking for all the frames is not economical. However, the frame rate cannot be too low since KLT tracking is vulnerable to complex movement.

## 7.5 Evaluation

We implemented all the proposed algorithms on the iOS platform. Several libraries are utilized for fast implementation, e.g., OpenCV [12], CMMotion [4], and OpenGL. The first photo in Fig. 7.11 illustrates the indoor AR application scenario: when a user wants to know the details about one painting, he could just point the camera view to the painting. The second figure demonstrates feature detection and matching, and the retrieval of the information and POI coordinate of the painting. If the POI is detected, the augmented information is presented and aligned to the real painting, e.g., the name of the painting. The third figure shows the tracking of the POI when the camera moves without repeating the image matching process. For simplification, the rendered

AR information for the painting is just one text label; the POI database is small, i.e., matching 10 POI images. More complex views and realistic POI databases will be included in our future work.

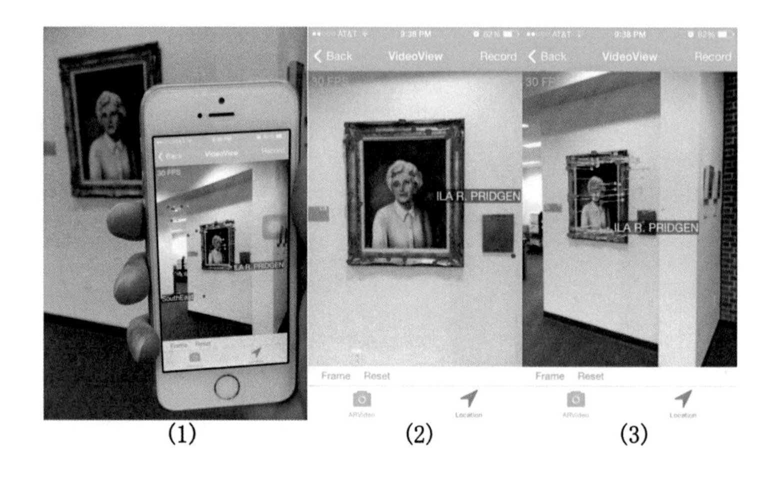


**Figure 7.11: Indoor AR App and screenshots.**

Besides the implementation in the smartphone, we deployed 8 anchor nodes in one large exhibition room of our campus Art Museum, as shown in Fig. 7.12. Enriching visiting experience in a museum is one of the promising applications of our proposed AR application. Based on our previous work [58], we leveraged the deployed anchor network for indoor location estimation.

### *7.5.1 Displacement Estimation*

To evaluate the drift reduction performance of our proposed INS displacement estimation algorithm, we compared our proposed approaches, i.e., adaptive threshold activity detection (ATAD), and Gaussian derivative decomposition (GDD).

Fig. 7.13 shows the INS displacement estimation results in different cases when moving in line (forward and backward style) patterns. The ground truth is like a rectangle, the start point and end point are overlapped. These estimated moving traces show very small drift after series of movements.

To quantify the performance of drift reduction, we utilize the metrics of: total drift (meters), drift speed (drift per second), and drift rate (drift per meter). To reduce the randomness in performance evaluation, we tested more than 150 cases of moving trace from one point to another, and then moving back. We calculate the drift by measuring the difference between the estimated

**Figure 7.12: Anchor deployment and experiment environment.**

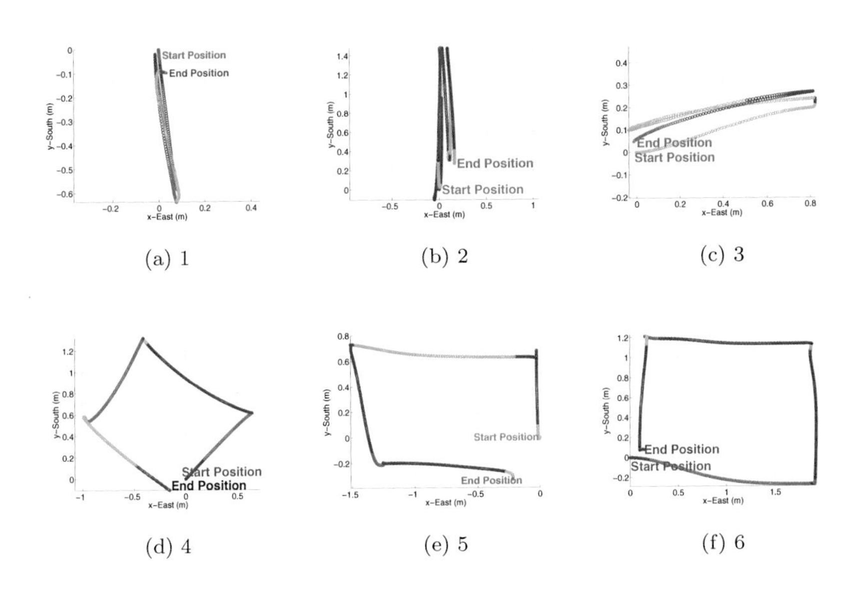


**Figure 7.13: Estimated moving traces for line and rectangular moving patterns.**

end point and the ground truth. The average and median value of all the metrics are shown in Table 7.1. Our proposed GDD approach achieves best performance in all metrics. The drift rate of 5.5cm per meter is significantly better than rates for existing approaches.

**Table 7.1: Performance comparison with respect to different methods under specific metrics**

| Methods | Metrics | Average | Median |
|---|---|---|---|
| Basic | Total Drift (m) | 16.51 | 12.47 |
| | Drift Speed (m/s) | 0.47 | 0.40 |
| | Drift Rate (m/m) | 3.92 | 3.73 |
| ATAD | Total Drift (m) | 1.26 | 1.02 |
| | Drift Speed (m/s) | 0.04 | 0.04 |
| | Drift Rate (m/m) | 0.31 | 0.26 |
| GDD | Total Drift (m) | 0.23 | 0.14 |
| | Drift Speed (m/s) | 0.006 | 0.0045 |
| | Drift Rate (m/m) | 0.055 | 0.043 |

**Displacement Estimation via Vision.** Displacement estimation via vision is highly sensitive, and suitable for slow movement when the INS is impossible to detect. Fig. 7.14 shows the estimated moving trace for POI using vision tracking. The smartphone moves (translates) to the left, and then moves back without rotation. By decomposing the homography matrix into rotation and translation matrix, we could get the full movement trace of the POI point in screen coordinate as shown in Fig. 7.14(b). The clear moving path demonstrates the AR view displacement estimation capability.

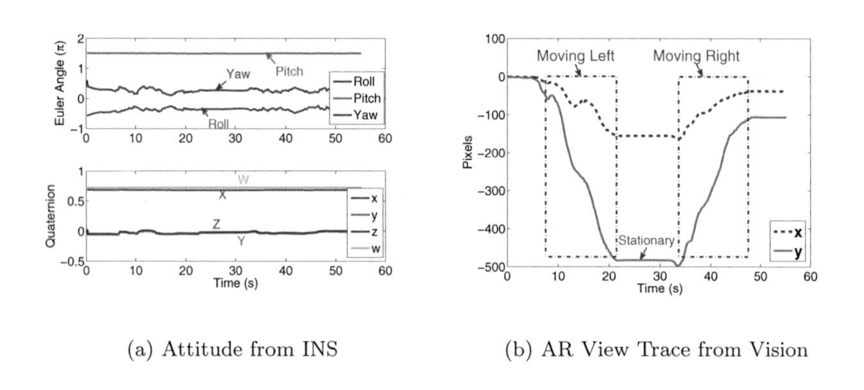


(a) Attitude from INS (b) AR View Trace from Vision

**Figure 7.14: The attitude and AR view trace results.**

## 7.5.2 Absolute Location Estimation

Fig. 7.15(a) shows the estimated displacement from INS results; the obtained smartphone location via acoustic anchors (ALE), and the fusion results. As

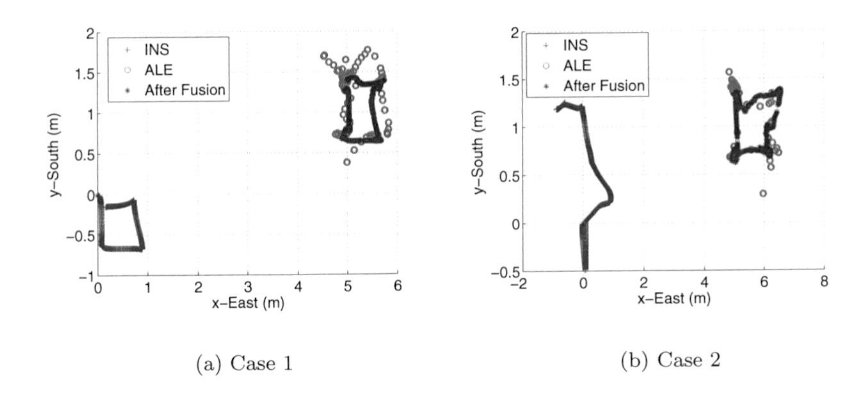


(a) Case 1

(b) Case 2

**Figure 7.15: The fusion results for moving location estimation in two cases (the ground truth is a rectangle).**

expected, the INS result has a very high update rate, but the displacement result has an unknown bias to the true location. The location result via ALE is unbiased to the ground truth, but suffers from measurement noise and outliers. After fusion, the achieved results show better variance in accuracy with higher update rate. To test the robustness of the fusion process, we utilize the INS result with strong bias and drift as shown in Fig. 7.15(b). The fusion result is still satisfactory with higher update rate and no bias to the true location.

### 7.5.3 Attitude Estimation

The greatest problem of state-of-the-art approaches in attitude estimation lies in the moving conditions. To demonstrate the error and noise reduction performance of our proposed approach, we perform small sway with equal amplitudes. As shown in Fig. 7.16(a), the Euler Angle is noisy and the overall trend is drifting. After using noise-reduction and vision fusion, Fig. 7.16(b) shows clear moving angle change with no apparent drifting.

### 7.5.4 POI Image Matching and Relative Pose Estimation

Table 7.2 illustrates the measured average execution time of image matching and relative pose estimation. The training process in Table 7.2 can be done offline and does not impact the real-time performance. For the image matching process, the iPhone5S takes nearly 0.055 seconds for one image frame, which can only reach 18 frames per second (FPS) for real-time processing. To enable real-time image matching, we utilize the rate control module in Fig. 7.5 to

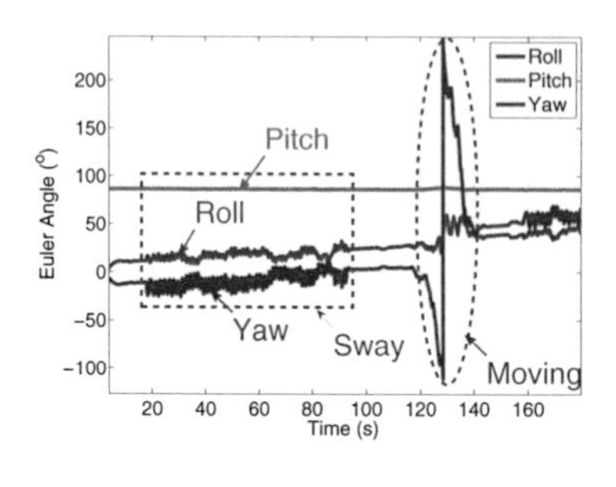


(a) Initial Euler Angle

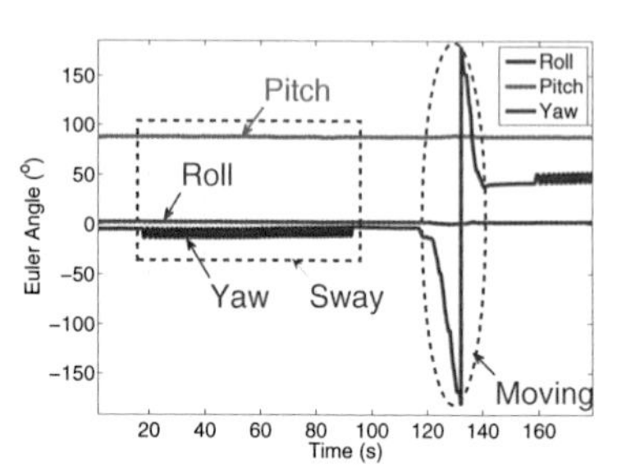


(b) Optimized Euler Angle

**Figure 7.16: The comparision of a) initial Euler angle, and b) optimized Euler angle.**

**Table 7.2: Comparison of execution time**

| Device | Training (s) | Matching (s) | Refined Match (s) | Estimate Pose (s) |
|---|---|---|---|---|
| iPhone5S | 0.232 | 0.055 | 0.128 | 0.003 |
| Server | 0.055 | 0.027 | 0.050 | 0.001 |

lower the frame rate to < 18FPS and only perform image matching when the detected activity level is stationary. The refined image matching takes longer than the normal matching process; however, the refined matching does not impact the real-time performance since this process is only performed once for the initial relative pose estimation. Compared with the execution time of the iPhone5S and the server, the difference is not significant, especially considering the network delay. That's why we choose local processing instead of using the delay-unbounded server for computation offloading.

Fig. 7.17 shows the relative pose estimation results via epipolar geometry proposed in Section 7.4.1.2. The template image is illustrated as "0"; the estimated rotation and translation matrix from the sensed image ($i = 1 \sim 6$) are $\mathbf{R}_i$ and $\mathbf{t}_i$, respectively. The relative 3D viewpoint locations for the 6 images in Fig. 7.17 are estimated via $\mathbf{R}_i\mathbf{x} + \mathbf{t}_i$. The estimated viewpoint relations in Fig. 7.17 are consistent with the real image view, which demonstrates the accuracy of our proposed approach in initial pose estimation.

### *7.5.5 Mapping the POI to the Screen*

When the smartphone moves, the AR view of the POI should be dynamically rendered at screen locations that make the AR view attached to the real physical object. This process requires mapping the POI to the appropriate

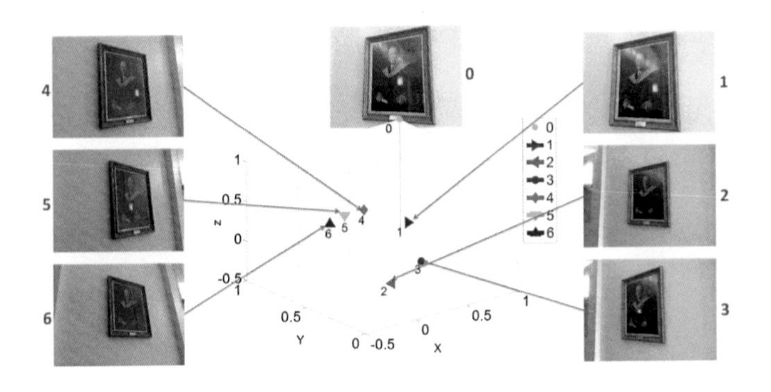


**Figure 7.17: Illustrating relative poses for different images in 3D space.**

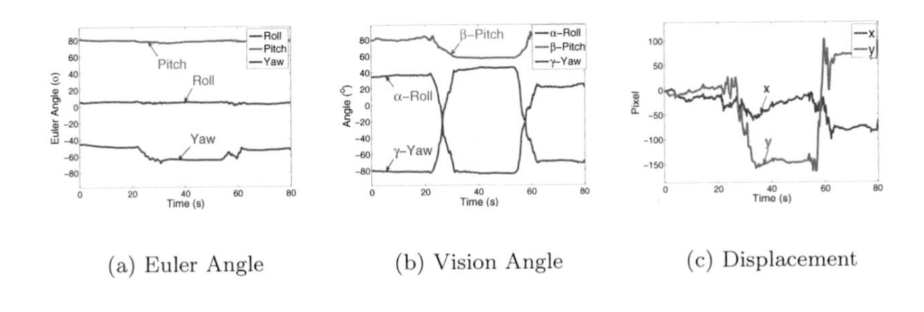


(a) Euler Angle (b) Vision Angle (c) Displacement

**Figure 7.18: Displacement in vertical.**

screen location by utilizing the estimated attitude and displacement. Fig. 7.18 and Fig. 7.19 show two cases of moving the smartphone forward and back in vertical and horizontal direction, respectively. We intentionally move the smartphone at a very low speed to test the sensitivity of the proposed POI tracking approach. As shown in Fig. 7.18(a) and Fig. 7.19(a), the euler angle of the smartphone shows very small movement, i.e., vertical and horizontal. The relative vision angles between the POI and the smartphone axis are shown in Fig. 7.18(b) and Fig. 7.19(b) with very apparent and clear changes during movement. The screen location changes (x and y coordinates) of the POI are shown in Fig. 7.18(c) and Fig. 7.19(c), which is consistent with the real physical movement.

### 7.5.6 On-Screen AR View Detection and Tracking

When there is no POI covered by the camera view, we only utilize the INS for the attitude and displacement estimation to save energy. Specifically, the high-cost vision processes (feature detection, matching, and tracking) are only

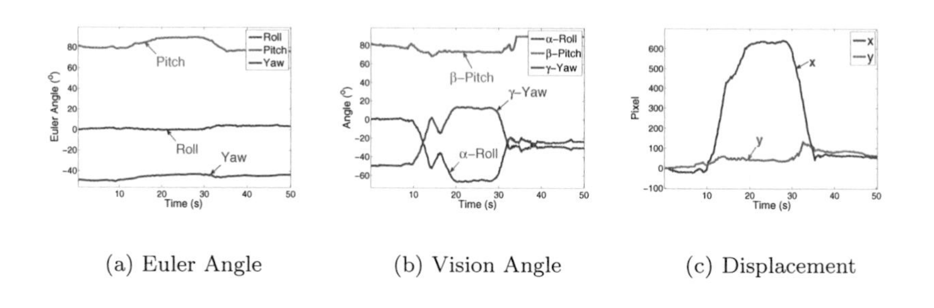


(a) Euler Angle (b) Vision Angle (c) Displacement

**Figure 7.19: Displacement in horizontal.**

launched when detected within the on-screen POI, i.e., the NDC coordinates of the POI are located within [0, 1]. However, only using the INS results of the estimated NDC coordinates for launching the vision processes is not reliable. In a real system, we add two engineering treatments: 1) Calibrate the INS results periodically via magnetometer [93], which reduces large drifts; 2) leave headroom $\alpha$, i.e., change the on-screen NDC coordinates range to $[0-\alpha, 1+\alpha]$, where the value of $\alpha$ balances the accuracy and energy cost (we choose $\alpha = 0.1$ in our implementation).

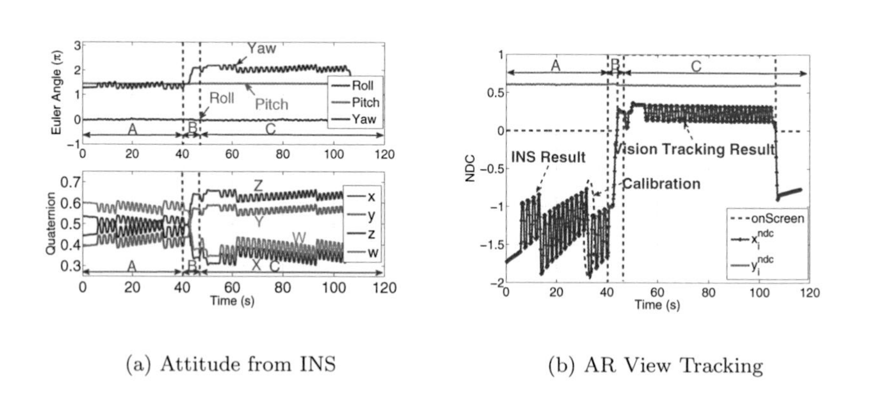


(a) Attitude from INS (b) AR View Tracking

**Figure 7.20: The attitude and AR view tracking result.**

Fig. 7.20 shows the complete evaluation process by moving the camera view. During the $0 \sim 40s$ time period, i.e., the period labelled A in Fig. 7.20, the POI is outside the camera view. We sway the smartphone via tripod with the same angle to test its attitude and displacement drift. The estimated attitude results (Fig. 7.20(a)) and POI's locations (Fig. 7.20(b)) show near periodical movement but with apparent drift over time. The two sudden jumps of the *NDC* location near $15s$ and $33s$ are caused by the magnetometer calibration process (to prevent the POI drift into the screen).

At the time point near $42 \sim 47$s (period B), we moved the smartphone, and the POI is covered by the camera view. When the *NDC* location value of POI is within $[0, 1]$, i.e., on-screen, the vision tracking process is started immediately for better accuracy. We still perform the same periodical movement (50s and later) as with the $0 \sim 40s$ period (period C). Fig. 7.20(b) shows very stable and accurate estimation of the POI's ndc location; while the INS results of the attitude in Fig. 7.20(a) suffer from apparent drift. The whole process illustrates the accurate detection of the AR view within the camera's viewpoint, and performs vision tracking to improve the accuracy.

### 7.5.7 AR View Tracking via Adaptive Frame Rate

We propose solutions for the AR view tracking and make it suitable for various movements from small hand shakes to large rotations, with high energy efficiency. After detected on-screen POI, i.e., there exists one NDC coordinates that located within $[0 - \alpha, 1 + \alpha]$, vision processes (feature detection, matching, and tracking) are launched for better accuracy. Due to the inaccuracy of the INS estimation, errors exists in detecting the on-screen POI. The cost of the error detection is mainly the computational power. Thus, we add a self-stop scheme to terminate the vision processes after the feature matching stage when no POI feature matches.

To further reduce the computation cost and energy consumption in the vision processes, we propose an adaptive frame scheme to select critical video frames that are necessary for the overall performance. To evaluate the effectiveness, we conduct two real-time experiments with similar movement. The whole movement is divided into different stages: 1) A and E: equal-angle sway via tripod when POI is not under the camera view; 2) B: large movement when the POI is not on-screen; 3) C: equal-angle sway when POI is under the camera view; 4) D: large movement when the POI is on-screen.

When using all the image frames, the AR view trace and execution time for each frame are shown in Fig. 7.21(a) and Fig. 7.21(b), respectively. Fig. 7.22 shows the logged energy trace of the smartphone via Apple's Xcode Instrument. When the POI is not on-screen, the execution time is very low since only INS is utilized. In stage "C," i.e., when POI is on-screen with small movement, the execution time is small since only vision tracking is needed. In stage "D," there is a large movement with multiple POIs entering and leaving the camera view, and the execution time is longer due to the frequent image feature detection and matching process.

After performing adaptive frame selection, the performance of Fig. 7.23 is similar to Fig. 7.21 with no apparent difference. When comparing the logged energy trace as shown in Fig. 7.24, the reduction in terms of energy usage and CPU activity is significant.

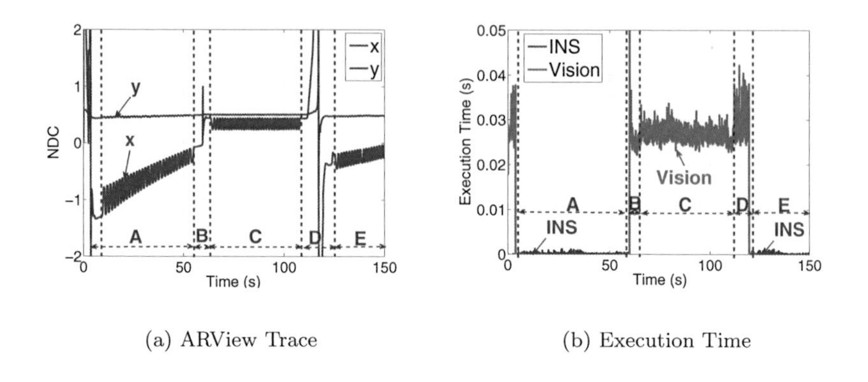


(a) ARView Trace

(b) Execution Time

**Figure 7.21: a) Estimated AR view trace; b) execution time for each frame.**

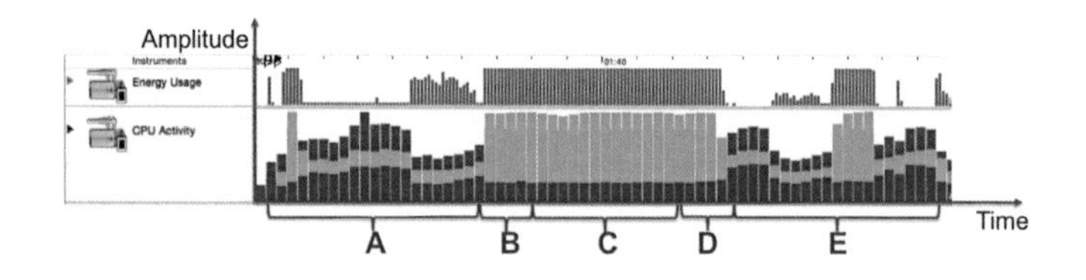


**Figure 7.22: Energy and CPU activity profile with high frame rate.**

Several narrow peaks in Fig. 7.21(b) and Fig. 7.23(b) are caused by the image matching process that takes longer. The execution time of the normal image matching and refined image matching for initial pose estimation is illustrated in Table 7.2.

## 7.5.8 Comparison with Existing Apps

**Energy Consumption.** Fig. 7.25 shows the profiled results of our indoor AR App (InAR) vs. other competitors in terms of energy usage, and CPU activity via Apple's Xcode Instruments running on an iPhone5S [75]. We launch different apps at different time slots: 1) "InAR" (30FPS); 2) An app performs ORB feature detection at full frame rate (30FPS); 3) An app performs SURF feature detection (near 20FPS); 4) iPhone-AR-Toolkit [120]; 5) Mixare [71]; 6) Layar [29]; 7) Qualcomms Vuforia [87]. Our app is less resource intensive than others. For cases 2 and 3, the smartphone's CPU activity is nearly 100%; the powerful iPhone5S cannot run in full frame rate ($<$ 30FPS).

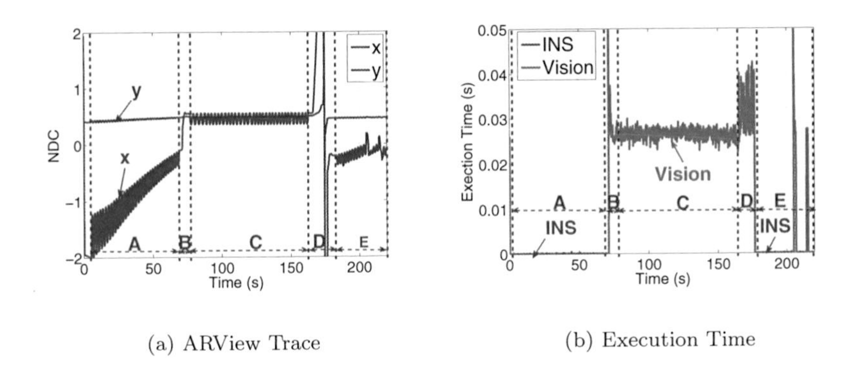


(a) ARView Trace

(b) Execution Time

**Figure 7.23: a) Estimated AR view trace; b) execution time for each frame.**

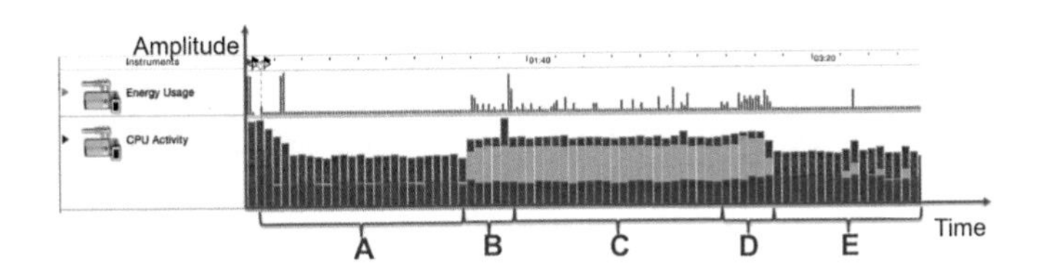


**Figure 7.24: Energy and CPU activity profile with adaptive frame rate.**

**Accuracy.** Among existing solutions, e.g., iPhone-AR-Toolkit, Mixare, and Layar, all these are focused on AR applications and directly utilized the system-provided API for attitude and location estimation, i.e., the Apple's solution that we compared in Section 7.3 and Section 7.5. We select the Layar (INS-based) [29] and Vuforia (vision-based) [87] as two representative examples. The screen-shots of the tracking results are presented in Fig. 7.26. From Fig. 7.26, the problem of Layar is drift after movement (the same AR view is displayed in different screen positions); vision-based solutions like Vuforia suffer from short distance when the letter-sized image template is slightly far away (the rendered teapot has disappeared). Our proposed solution, i.e., InAR, does not have these problems and achieves the best accuracy in AR view tracking under different movements.

Fig. 7.27 shows the AR view estimation results from INS-based, vision-based, and our proposed approach. The movement pattern is the same in all three cases, i.e., small sway and large rotation. Our proposed solution, i.e., InAR, does not drift during small sways, and also supports large movements.

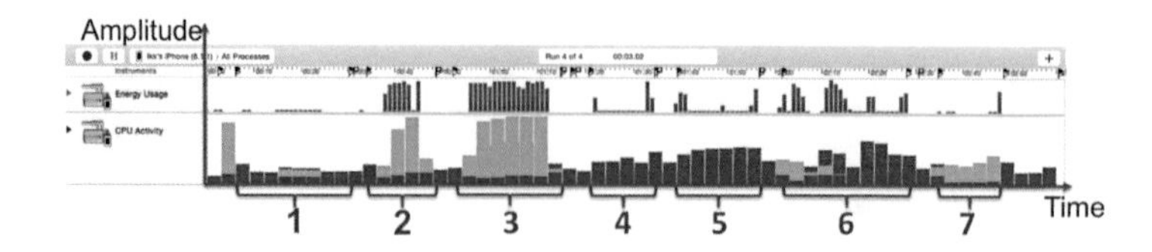


**Figure 7.25: The comparison of energy usage, and cpu activity for different apps.**

**Figure 7.26: The drift problem of Layar, and the short distance problem of Vuforia.**

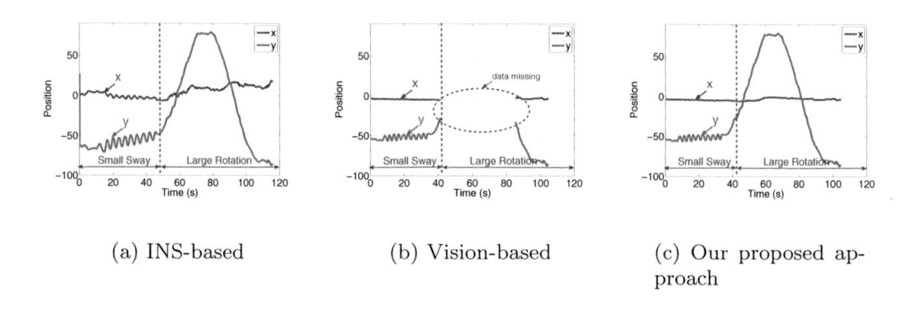


(a) INS-based

(b) Vision-based

(c) Our proposed approach

**Figure 7.27: AR view tracking results via different methods.**

## 7.6 Related Work

**Related AR Applications**. AR applications put a childlike spin on the world and help users interact with the AR-enabled physical world. Dating back to 2001, a special head-mounted device was built by Newman to enable AR applications [76]. Although the device looks bulky and is not ready for massive adoption, the pursuit of AR devices and applications never stops. With the wide popularity of sensor-enabled smartphones, the device required for enabling the AR application is ready. *Layar* and *Wikitude* are good examples of geo-AR applications using GPS [29]. *Golfscape GPS Rangefinder* is an AR range finder for Golf lovers; *Cyclopedia* adds Wikipedia information to your camera view.

Another category of AR applications is using computer vision techniques to render a character or a model on top of a reference image (marker), e.g., *ARToolKit* [41]. *SnapShop Showroom* provides a preview when you select furniture or redecorate a room. Authors in [73] even use markers for localization purposes. However, marker-based approaches focus on short distance applications (you can clearly see the marker in the camera view), and are not scalable to handle large-scale navigation. Thus, using marker a is not suitable for our intended application for displaying POI information in indoor/outdoor environments with high freedom.

**Attitude Estimation.** The estimation of the camera attitude/pose could be INS-based [125] or vision-based [14]. Chandraker et al. [14] proposed a fully mobile, purely vision-based attitude tracking system using artificial landmarks. The most strongly related works of attitude estimation via INS in the literature are based on an extended Kalman filter (EKF) for military or spaceship applications [125, 66], where the device and movement model is accurate and stable. For less-accurate MEMS INS sensors embedded in a smartphone, a high-order EKF model may make things worse since users' mobility is highly dynamic and unpredictable. Zhou et al. [134] presented $A^3$ for smartphone-based attitude estimation via opportunistic calibration. They achieved significant performance improvement in applications like heading estimation. As we discussed in Section 7.3.1.1, the calibration performance is downgraded when the smartphone constantly moves. For AR applications, we are more interested in the dynamic performance of attitude estimation.

**Displacement Estimation.** Existing displacement estimation approaches are mostly in the area of indoor localization and navigation. UnLoc [114] leverages distinct pattern on a smartphone's accelerometer for indoor localization; Walkie-markie [97] realizes indoor pathway mapping via INS and Wi-Fi. Most of these solutions for localization rely on step counting of the accelerometer signatures. Step counting works in cases when the step size could be averaged over time. For AR applications, small-scale movement is common; fine-grained model-based double integration is more appropriate. However, the double integration process causes severe drifts and errors.

Visual odometry (VO) could be another sensing solution for displacement [25]. Authors in [63] proposed object localization via structure-from-motion (SfM).

**Indoor Localization** Smartphone/camera localization techniques, especially indoors, remain an open problem. Radio-fingerprinting-based approaches [129, 52] are not sufficient for fine-grained indoor AR applications. The authors in [35] leveraged ultra-wideband (UWB) and INS for sensor fusion. However, it is not suitable for the smartphone application, since UWB devices are not available and too expensive. Solutions using acoustic anchors [58] could achieve high resolution in indoor localization, but are missing essential attitude estimation and tracking techniques needed for AR application.

**Sensor Fusion Approach** Hybrid approaches using sensors and vision would allow us to alleviate the shortcomings of individual approaches. The authors in [67, 43] utilized IMU-Vision synchronisation for egomotion estimation in autonomous unmanned aerial vehicle (UAV) navigation. In [37], an improved KLT feature tracking via INS assistance is proposed. A depth-camera could also be combined with INS for tracking [23]. Authors in [83] proposed a hybrid tracking system consisting of an active marker made by an infrared (IR) light-emitting diode (LED), and two Nintendo Wii Remotes as vision tracking devices. Ribo et al. [93] presented a hybrid system combing gyro, compass, and vision sensors for self-contained tracking.

**Summary of Differences** With the popularity of smartphones, our by using low-complexity sensor nodes and off-the-shelf smartphones for localization and viewpoints tracking is more convenient. To further improve its location accuracy in different mobilities, i.e., small vibration or large moving, we propose INS and vision fusion approaches with adaptive frame rate for high-energy efficiency. Based on existing state-of-the-art solutions for image matching and attitude estimation, we complete the whole loop of the indoor AR application with improved smoothness and sensitivity in AR view tracking.

## 7.7 Conclusion

One of the most exciting aspects of indoor AR is that it stirs the curiosity of customers. Service operators could create a situation in which customers will keep looking and exploring wherever they get excited about the hidden information. Leveraging the distinct features of the sensor-based and vision-based approaches, we present a hybrid approach for AR applications with better accuracy and efficiency for various mobilities. Experimental results demonstrate significantly reduced AR view tracking errors with modest CPU utilization. Our future work is to address other essential problems in AR applications that are not covered in this paper, e.g., POI database generation, impact of the POI number, various moving speed, information visualization, camera calibration (intrinsic matrix) for different devices, and enabling apps for Google Glasses. With all these key components readily available, AR applications can be realized, which has significant implications for potential innovative AR-related applications.

# *Chapter 8*

# Towards Location-Aware Mobile Social Networks with Missions

**Kaikai Liu**

*Assistant Professor, San Jose State University*

**Xiaolin Li**

*Associate Professor, University of Florida*

## CONTENTS

## 8.1 Application Description

The ToGathor App and mobile social network platform is designed for users coming, keeping, and working together. The hybrid cyber-physical social group formed could enable interactive, reactive, and proactive interaction/activities beyond pure physical or online social networks. We focus on two important features of being together, i.e., location and communication, which highly relate to the "mobile" and "social."

The location tag, either absolute or relative, will be dynamically created and updated according to users' mobility and intention. Leveraging device-to-device communication and opportunistic sensing, location-based check-in services could be universally available in both an active and a passive way. Check-in will not just be limited to static places like stores or restaurants, or need intentional manual operation; any social group or user could finish the check-in process hands-free. The only thing you need to do is to enable the location beacon in the app, and you can be a check-in point for your social groups or families. The app will push immediate feedback when your member joined or got lost. You could also perform automatic attendance check for your social group without special attendance check devices; for example, checking students attending classes or monitoring employees' daily arrival and departure times to the office. You will never be too busy staring to prevent the loss of a friend, child, or pet.

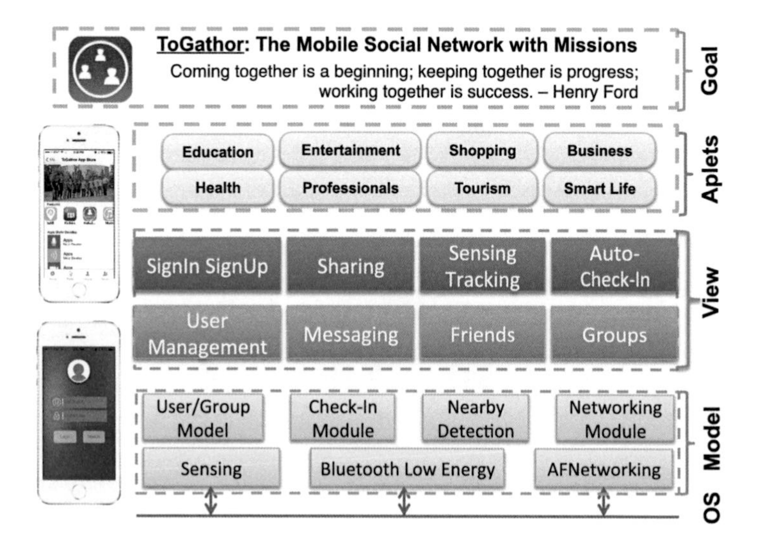


**Figure 8.1: System architecture.**

The communication between group members could be realized in an opportunistic way. You can even share and communicate with your nearby friend without Internet connection. We combine device-to-device communication and Internet communication in an opportunistic way, such that you can send and receive message/media or any other information with your nearby friends without worrying about the intermediate Internet connection loss. Your friend status will be extended from simple "online" and "offline" to multiple levels: nearby active, nearby inactive, online and nearby, online but not nearby, offline with meet notification, nearby with lost notification, and other combinations. Sharing your business card with a new friend, reminding you of the nearby old friend, or exchanging photos/videos taken during events will become super easy with the nearby "sense" of the app. The messaging module will keep you and your friends connected before/during/after the event; the cloud storage module will memorize all of your exciting moments.

## 8.2 Innovation and/or Uniqueness in Marketing

- **Check-In-Related Apps**: Foursquare, ShopKick, Gowalla, Google Latitude, et al. These are mobile apps that allow users to check in to any fixed locations — cafes, stadiums, shopping malls, restaurants. Sometimes they offer game-like "rewards" for users who perform check-in. Our app extends the check-in concept for any places and groups with/without real location information. You can create a check-in point, and your friends or families could check-in to join social events together; schoolteachers could open the app and set up a check-in point in class for the students who attend; company managers could monitor their employees' arrival and departure times.
- **Mobile Message Apps**: Whats-App, iMessage, Google Talk, Skype, WeChat, et al. These messaging apps connect you with your friends, for sharing your exciting moments. We extend Internet-based messaging to nearby location, and enable fine-grained status checks for your friends. We deliver your exciting news in a best-effort way; we store your happy moments in the cloud for you to access at any time.

## 8.3 Key Design Features

ToGathor is designed from the ground up to help you socialize and interact with friends better. It is a unique hybrid cyber-physical mobile social network with the following salient features: connecting friends socially and physically, mission-oriented groups, tracking nearby friends or team members, or children/elders.

### 8.3.1 User Management

The first part of the ToGathor App is user management. We provide interface to easily sign people up to our cloud service. In the Sign Up view, as shown in Fig. 8.2, we only need a user's email address and user name information for registration.

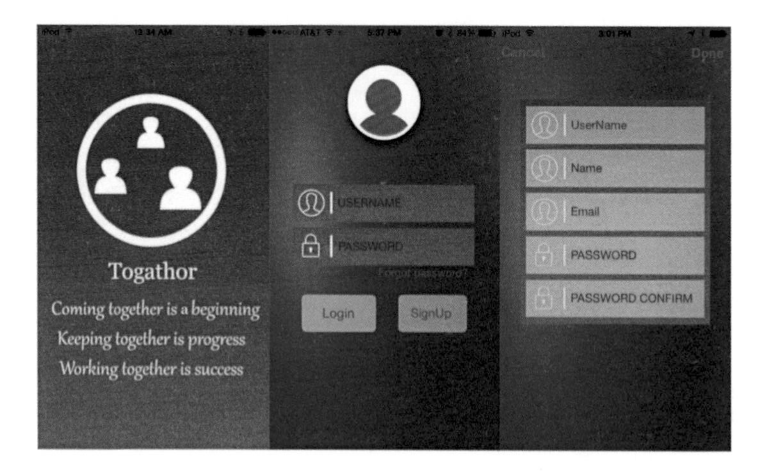


**Figure 8.2: User management.**

### 8.3.2 Connecting Friends and Messaging

We provide a dedicated Contact view for users to manage their contacts, as shown in the first page of Fig. 8.3. To add or search for friends, we only need to click the upper-right button to explore online friends.

By clicking the friend cell in the Contact view, more details of the friend will be presented. We could click friends with similar interests, and click "add" to add a friend into our contact list. We also could click "Start Conversation" to start messaging with friends via opportunistic ad hoc links that can survive without WiFi/4G connections. Currently, we could send text message and location data to our friends. Other media data, e.g., audio, video, photo, file, will be implemented later. A timeline view of events will also be implemented in the near future.

### 8.3.3 Managing Social Groups

Coming together is not only for two friends, but for a group. The concept of group is not limited to online social groups, but also for physically co-located

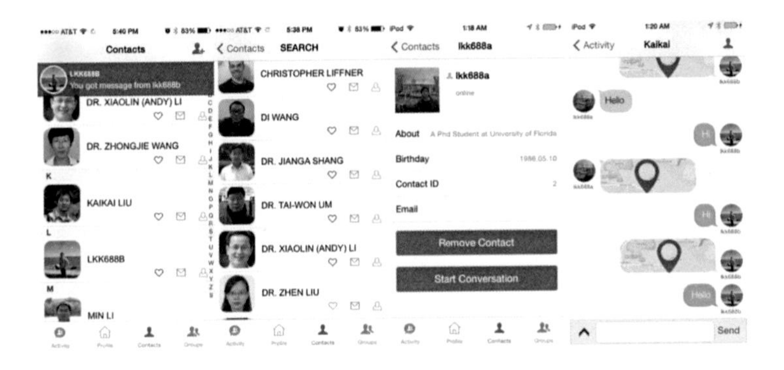


**Figure 8.3: Contacts and mobile messaging between friends.**

groups, e.g., company employees, students attending the same class. The group detail view as shown in Fig. 8.4 contains group members, and other group properties. Adding and removing a group is easy and just requires clicking the first button on the bottom. Users can search groups online and create a new group.

When creating a new group, the group can have a password, and request other users to enter the password when joining the group. One important feature is to enable auto check-in and create location-enabled groups. A normal group without auto check-in is similar to current online groups.

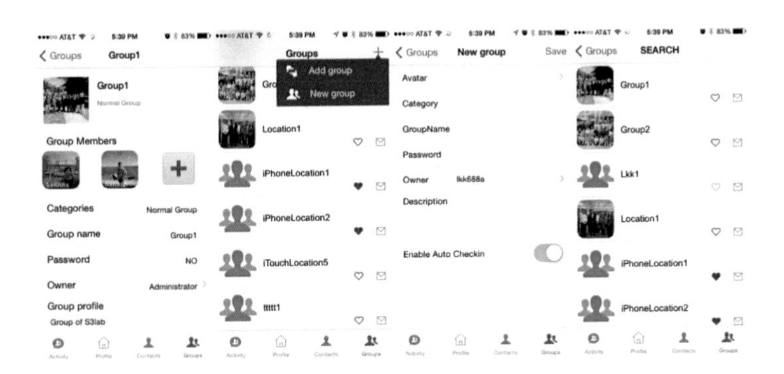


**Figure 8.4: (a) Group detail; (b) add and search group; (c) new group; (d) search groups.**

### 8.3.4 Group Auto Check-In

For location-enabled groups, the group owner could enable beacon broadcast to advertise the existence of this group by turning on the switch as shown in Fig. 8.5. Group members can enable group auto check-in by enabling the "Auto Check-in" switch in the group detail view page.

When the "Auto Check-in" switch is turned on, the user's smartphone would perform low-power and low-cost Bluetooth-Low-Energy (BLE) scanning by following the Apple iBeacon specification. This enables beacon detection even when the smartphone is in the background. When the user's smartphone has detected the group owner's beacon signal, it will push notification to the user and automatic send a remote check-in message to the group's owner so that users and the group owner could know who is already checked in to the group, i.e., nearby members in proximity.

The applications of the group auto check-in can be automatic member check-in for students in a class, tourists in a group, friends at a party, or any other scenarios that involve a group of friends, making our life smarter.

**Figure 8.5: (a) Enable group beacon; (b) enable auto check-in; (c) auto post check-in to group; (d) sharing location.**

### 8.3.5 Sensing and Tracking Nearby Friends

ToGathor not only connects you with friends online, but also provides physical proximity and location sensing capability. Imagine when we go outside with our friends or family members, we have to pay attention to our group members' locations to prevent members getting lost, especially for children who are more likely to wander around. With this design goal in mind, we enable BLE ranging and tracking in a smartphone, and trigger alerts when a member is just about

out of sight. One important requirement is that the ranging and tracking process should also work when ToGathor is not active, and without Internet connection.

As shown in Fig. 8.6, we designed and implemented the BLE background tracking functionality into ToGathor. We can scan our nearby friends via BLE beacon, and request friend tracking by sending a request via the BLE communication channel. After receiving acceptance from the friend, our smartphone tracks our friend and pushes notification when our friend is lost. Ranging information, e.g., the BLE RSS, is utilized to estimate the distance and perform lost alerts.

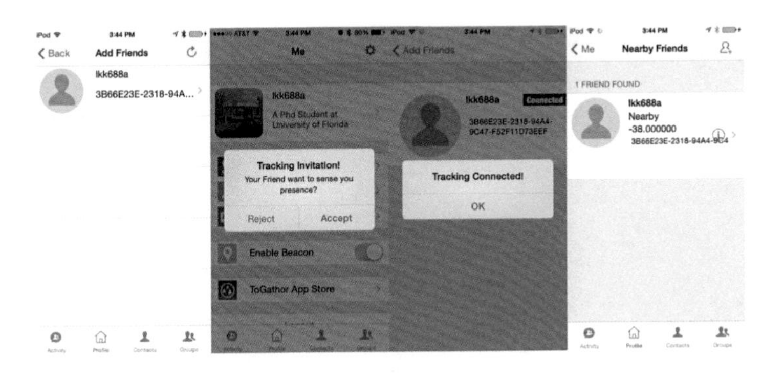


**Figure 8.6: (a) Sense friend nearby; (b) friend received request; (c) rReceived acknowledgement from friend; (d) tracking friend.**

## 8.4 Conclusion

The ToGathor app is designed from the ground up for users coming, keeping, and working together with common interests or missions. The *five* basic functionalities of the current version are user management, connecting friends and messaging, managing social groups, group auto check-in, and sensing and tracking friends. ToGathor-enabled hybrid cyber-physical social groups could dramatically enhance interactive, reactive, and proactive interaction/activities, e.g., sensing and auto-check-in, that are beyond current online social networks. Focusing on two important features of being together, i.e., location and communication, we provide basic modules that are dedicated for "mobile" and "social." ToGathor acts as a platform for future mobile social network applets (through in-app "purchase" or third-party plugins) with support in terms of friend and group management, in-app messaging and notification, and location-based auto-check-in and nearby sensing.

# *Chapter 9*

# Wearable Localization via Mobile Crowd Sensing

**Kaikai Liu**

*Assistant Professor, San Jose State University*

**Xiaolin Li**

*Associate Professor, University of Florida*

## CONTENTS

## 9.1 Introduction

Losing their beloved child is the worst nightmare for every parent. Sometimes after you turn around for just a few seconds, your child is gone when you turn back. If you are at home or some less-crowded small regions, you probably can find your child in some corner or in the immediate vicinities. But for public areas, like shopping malls, streets, or even your child's favorite Disney World, it is hard to find your child in crowds when lost.

There are so many reasons for your child to get distracted and wander, then get lost. In places like Disney theme parks, there is simply so much to see, and so many people attending, especially for events like fast-paced parades. Even if your child doesn't typically wander, kidnapping could happen, making it even harder to find your child. Guarding a child in crowded places full of attractions is nontrivial; locating your lost child is mission impossible.

Normal approaches include writing your phone number on a shoe tag or sticker of your child; or go to your designated meeting place, if you have one or your child could do that. You also can look at the closest locations that are of interest, or go to the baby center to locate a lost child found by others. However, this manually blind searching is not efficient. Giving your "big kids" a cell phone could be a high-tech approach, but most cases they are not old enough to carryone.

To find your child quickly if you are separated, lots of systems and approaches have been developed. One kind of approach is using GPS locator that is installed on your child's shoes or clothes, e.g., Amber Alert GPS, PocketFinder, AT&T Family Locator [21]. This kind of device includes GPS and cellular communication modules. One problem for this kind of locator is that it is high cost and bulky. GPS and cellular communication are all expensive and power hungry, especially when they work in continuous mode. Providing sufficient battery life for one day's use could result in a bulky and heavy device, not suitable for little kids to carry. For indoor places like castles and shopping malls, GPS may suffer significant performance degradation or even not work due to the signal blockage.

Another category of approaches relies on devices with peer-to-peer communication capabilities. The transmitter and receiver pair, or smartphone and peripheral pair, are carried by parents and child, respectively. If the child goes out of the communication range or predetermined threshold, parents will get an alert. This kind of approach leverages the existing low-power communication standard and could be made with high efficiency in power, and portable in size, e.g., Toddler Tag Child Locator, Keeper 4.0, Chipolo [17]. A drawback of this approach lies in its lacking absolute location information, e.g., GPS location. It is impossible for parents to locate their child when the child goes out of the communication range.

In this chapter, we propose crowd sensing solutions for locating the lost child, using low-power and miniature wearable devices without high cost cellular and GPS modules. To solve the coverage and localization problems of wearable devices, we focus on the investigation of nearby "crowds" of smartphones for transparent ranging and locating via peer collaboration. State-of-the-art approaches leverage connection information to detect the presence of wearable devices in a specific region near to one participator. However, the resulting searching area of the obtained location resolution still makes it hard for parents to pinpoint their children in crowds, e.g., 20-meter peer-to-peer distance could exaggerate the initial location error surface to thousands of square meters. Instead of just relying on the single connection information, we propose opportunistic localization approaches to derive the absolute location of the wearable tag via multi-hop assistance for covering more participators. Even with no sufficient measurements from the immediate surroundings, joint estimation solutions via semidefinite programming (SDP) could achieve a global relaxed optimal solution for the lost child (wearable tag). To solve the bias error and unsolvable problems caused by sparse crowd measurements, conventional SDP problems are reformulated by jointly optimizing the multi-hop ranging and coarse-grained location measurements. We introduce a *virtual anchor* from participators with better measurements, and utilize this "anchor" location to assist the localization of the true target, i.e., wearable tag on a child. Detailed simulation and experimental results are presented to evaluate the performance. In summary, our main contributions are as follows:

- We propose a system architecture for smartphone-based transparent crowd localization with high-energy efficiency and location resolution; we develop a portable and low-cost wearable tag based on Bluetooth Low Energy (BLE) that could be embedded in kids' clothing or shoes with long operating time.

- To solve the problem of insufficient measurements available among immediate nearby collaborators, we perform joint location optimization via multi-hop opportunistic communication and ranging.

- We propose semidefinite relaxation for location optimization under various unreliable sources in crowd sensing, which could overcome the over-fitting problem and large errors, and ensure best-effort solutions.

- By evaluating the connection topology of the wearable tag (target) and nearby peers, we identify the localizable peers as virtual anchor, and perform back propagation to assist the target localization.

## 9.2 System Overview

### *9.2.1 Design Considerations*

**Despite substantial research on the localization, why is locating a child still an unsolved problem?** The popularity of GPS-enabled devices and smartphones makes us take locating any object for granted, although we have been somewhat surprised and exhausted when we cannot locate our beloved ones. When the GPS and cellular communication-enabled devices are not cost-effective and prevalent, tracking any objects without bulky devices and a cellular subscription fee is impossible. Tiny devices based on peer-to-peer communication, e.g., BLE, Zigbee, WiFi, cannot obtain location information without infrastructure or fingerprinting data from site survey.

From the literature, most papers on localization have attempted to solve this problem in terms of accuracy improvement under specific settings, i.e., without infrastructure. Both settings have their distinct drawbacks, especially when the measurement comes from various unreliable sources. Solutions using acoustic anchors [58] could achieve centimeter-level resolution at low cost, but only work in small areas with infrastructure. For real cases without infrastructure, the central server cannot even collect sufficient measurements, not to mention perform localization via trilateration. Fingerprinting-based approaches [129, 52] cannot work without site survey, and suffer the problem of collecting RSS fingerprinting data when the child is already lost. Liu et al. [56] propose location optimization approaches via peer-to-peer ranging without infrastructure. However, this approach requires initial location from the GPS module, which does not exist in our application.

**Why not give kids GPS-enabled smartphones?** For most cases, GPS-enabled smartphones with cellular subscription to prevent them from getting lost are expensive. Even if you can purchase one for your child, it is not practical for every family member including your pets. Current smartphones are still too bulky for embedding in your kid's clothing or shoes. Your child may not be old enough to carry one, or cannot perform correct operations when they get lost.

**Why use BLE tags; what about other devices?** BLE seems; like it is becoming the most promising solution for connecting peripherals to your smartphone with low-cost and high energy efficiency. WiFi or Cellular solutions are too heavy for low-data-rate applications. Zigbee and RFID are also low-cost solutions, but not popular in current smartphones.

**Why do you need crowd sensing; what's the incentive for people participating in finding lost children?** For locating a lost child in crowds without infrastructure and site survey, collecting measurement data via nearby ubiquitous sensor-rich smartphones becomes a flexible and cost-effective solution. The powerful computing/communication capacities of nearby smartphones, huge population in crowds, and the inherent mobility makes mobile crowd sensing (MCS) a fast-growing consumer-centric sensing paradigm.

Incentive mechanism design is one of the key components in MCS. Participatory sensing should be performed in a transparent, energy- and privacy-preserving way; otherwise users are reluctant to release sensor data. We propose solutions for the background processing of sensory data only for a limited time period. The uploaded data from a smartphone utilizes pseudo-ID that is irrelevant to users' personal identification. Moreover, helping others locate a lost child earns "credit," which you can spend when you need crowd sensing services in the future. This could be a beneficial environment for locating the child, and in turn provides incentives for participation.

**Why do you need multi-hops and opportunistic connection?** We cannot assume that there are enough participators nearby with sufficient measurements, e.g., GPS location and ranging results. To enable target localization in real cases, we need to leverage multiple sources and perform opportunistic connection. One-hop or multi-hop assistance could make the target localizable and provide sufficient performance improvement.

## 9.2.2 System Design

Fig. 9.1 shows the key building blocks and connections underlying the "FindingNemo" system for kids or other family members, including pets. The overall "FindingNemo" system has three main modules, namely: (1) Wearable Tag installed on the child's belongings, (2) Smartphone App for social networking and crowd sensing, and (3) Cloud Server for aggregating all the crowd-sensed measurements and performing global location optimization.

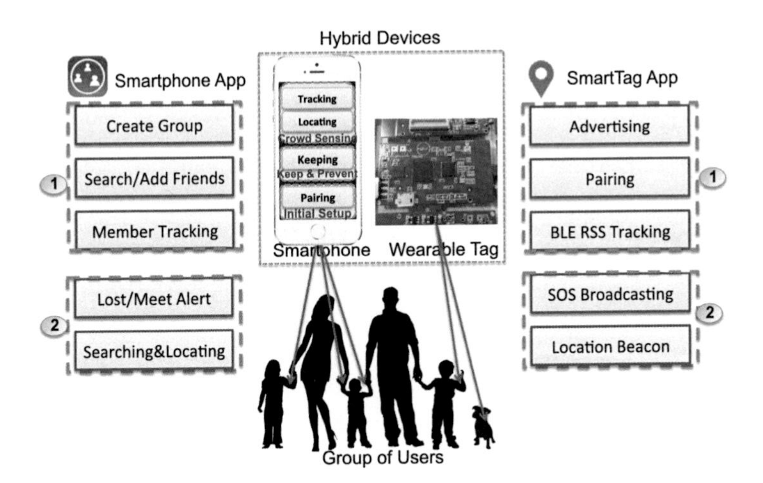


**Figure 9.1: System architecture.**

Leveraging the low-power and portable features of the wearable tag, parents could place one tag on their beloved ones, paired with their smartphone. If the child goes outside of the warning threshold, the smartphone could wake up automatically and post an alert immediately to prevent the child from getting lost. If the child is already lost, our developed App "FindingNemo" installed in nearby users' smartphones could receive notifications when the "lost" child is passing or nearby. To locate the child, opportunistic communication and ranging is performed with nearby peers without GPS location in a transparent way, without disturbing the users. Depending on the available measurements, one-hop or multi-hop configuration is automatically selected for the best-effort localization.

**Wearable Tag.** We designed a wearable tag from scratch based on the BLE technique, as shown in Fig. 9.1, which is dedicated for child or family member tracking and locating. BLE devices could directly communicate with users' smartphones without additional hardware attachments. It is convenient for users to use their personal smartphones to locate the BLE peripherals attached to their child's clothing or shoes. The energy efficiency and miniaturization of BLE peripherals make it perfect for tracking the child continuously without significant degradation of the battery life of parents' smartphones. Compared with existing solutions, e.g., Keeper 4.0, Chipolo [17], we provide additional features that include loss detection, dual mode operation (connection and broadcast), heterogeneous power source, portability, low-power via auto-sleep, UUID hiding for preserving privacy, and secure pairing and authentication.

**FindingNemo App.** We designed a mobile social network from the ground up to enable social crowd sensing. The user base and incentive are the two major problems of crowd sensing/sourcing. With a better-designed mobile social network app or framework that could be embedded in other apps, e.g., Disney's official app, or Facebook, sufficient users could be accumulated and leveraged to enable crowd sensing. The dedicated mobile social network could make the social collaboration process more convenient.

The FindingNemo App continuously tracks the wearable tag even in sleep mode, with a low-duty cycle for better energy efficiency. If the feedback signal from a wearable tag is lost for a period of time, the app will be launched to the front and the "lost" alert will be sent to the user. Performing background BLE scan and tracking in an app is lightweight. Evenif the app is switched off from the background, the OS could still launch the app automatically when the BLE notification is received. The native support from modern mobile OS provides an essential reason for using BLE for child tracking. This feature could make the participatory sensing process transparent to users, and perform all the tasks in the background when necessary and as notified by the beacon. We apply constraints in terms of total active interval and power cost when the participator's app is being woken up, and perform localization.

**Cloud Server.** Due to the inconsistent, error-prone, and opportunistic features of crowd sensing, one central cloud server for optimizing all the available measurements is necessary. Most algorithms proposed in this chapter run on this cloud server. The cloud server consists of the pub/sub messaging module (Kafka), the stream processing module (Storm), and the persistent datastore with efficient writing/reading and flexible query mechanisms (Cassandra) [101, 39, 13].

# 9.3 Mobile Crowd Sensing via Multi-Hop Assistance

## *9.3.1 Models and Protocols*

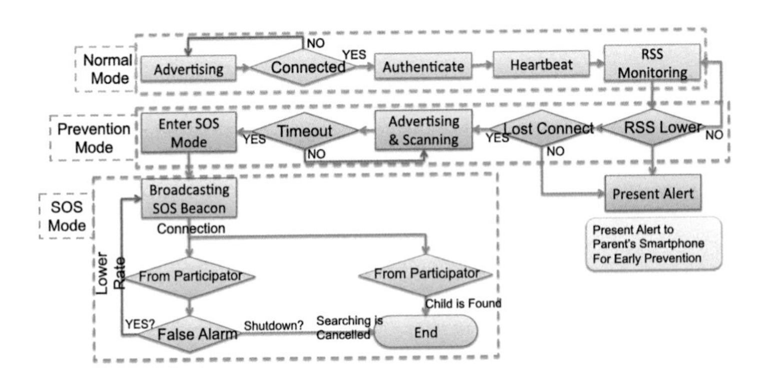


**Figure 9.2: FindingNemo protocol.**

**Family Group** As shown in Fig. 9.1, we define the family group as a virtually connected unit via peer-to-peer communication channels. The smartphone and the wearable tag are necessary components for connecting this group. The family's leader could initialize this group by adding all of its nearby devices via pairing. The family leader needs to use the smartphone with BLE functionality, and install our mobile social network app. The family member could use either a smartphone (with BLE functionality and "FindingNemo" app) or a wearable tag. Users could give big kids a smartphone, and give wearable tags to little kids and pets.

**Normal Mode** In the initial stage, each group member advertises its unique UUID and waits to be connected, as shown in Fig. 9.3. The group leader could add all of its members via BLE connection and authentication. After all the members are joined in the group created by the group leader, all the members will transmit its unique UUID along with meta data as the "heart beat" signal to the leader via BLE connection mode. The smartphone of the group leader could receive this "heart beat" signal and reply with a response. The received signal strength (RSS) is monitored as the closeness indication. If the RSS value is higher than the predefined threshold, no members are lost. The RSS monitoring will continue and the system state is kept in normal mode.

**Prevention Mode** If the RSS value is lower than the threshold, the "FindingNemo" app will enter into prevention mode. The wireless environment is highly unstable, and the RSS value's being low or lost does not necessary mean one of the group members is lost. The app needs further information to determine the current status of the family member. Outlier detection with RSS historical values is performed to minimize false alarms. If the RSS value continues to decrease or even is lost, the app running in the group leader's smartphone will push alert notification to prevent the member from being lost as shown in Fig. 9.3. This immediate alert could remind the leader to check their members around, and prevent some members going further away.

**SOS Mode** For the wearable tag or the member's smartphone, it will re-scan the nearby BLE beacon and try to re-connect to the leader's smartphone. If the lost BLE connection to the group leader could not be re-established, and none of the other members' BLE heartbeats scanned and captured within a pre-defined time period, it will enter into SOS mode to broadcast its existence using special SOS UUID.

### 9.3.2 Balance the False-Alarm and Detection Rates

**Early Losts Prevention.** In Fig. 9.3, the leader's smartphone would push alert notification if one of its members' BLE beacons is lost or the RSS is too weak. However, such alerts should not be too frequent since the highly unstable BLE communication channel may cause false alarms. It will be very annoying for users when a false-alarm happens. To minimize false alarms while at the same time keeping the detection rate, low-complexity detection approaches should be applied.

Predict current RSS value ($\hat{r}_i$) and confidence interval $\alpha_i$, if the measured RSS value ($r_i$) is outside the range of $[\hat{r}_i - \alpha_i, \hat{r}_i + \alpha_i]$, then the current measured RSS is an outlier, further measurement is needed to determine the closeness of group members instead of direct push alert.

**Enter, Delay, and Cancel the SOS Mode** For the wearable tag or group member's smartphone, the software program needs to determine when to enter the SOS mode and search for help. The SOS mode could trigger nearby participators' smartphones to perform assistance. This triggering process involves a high cost and could waste the battery power of nearby participators. If the tag or smartphone is not highly confident that it is lost, the SOS mode should be re-examined by using further information.

If the reply message from the leader's smartphone is missing or the connection is lost, the tag or smartphone should perform scanning again and wait for re-connection. Only after a pre-set timeout without any connection from the group leader will it enter into SOS mode, and broadcast an SOS beacon.

One issue of the wearable tag is its lack of Internet connection. If the group leader did not report the child/pet/tag lost, the transmission frequency of the SOS beacon should be lowered to minimize the energy cost both to itself and nearby participators. If the group leader canceled the searching/finding process, we need to design one mechanism to make the lost tag aware of the situation even without Internet connection and direction communication from group leader and other members.

As shown in Fig. 9.3, we designed a crowd sensing approach that let nearby participators relay the message from the server. When the wearable tag is in the SOS mode, it also accepts connection from nearby participators. If one participator received the SOS beacon, the mobile OS will wake up the FindingNemo App and report the SOS beacon to the server. If the identification of this SOS beacon matches one item in the "Finding" table, i.e., the group owner reported the member as lost, the participator will get a response from the server to begin the localization process to locate the lost tag. Other cases are: 1) If the group owner does not report the member loss, the participator will get a response and send this information to the detected wearable tag and let it decrease the frequency of the SOS beacon. 2) If the participator received a "cancel" command from the server, it will connect to the tag and send "shutdown" information to this tag. For the wearable tag, if the information from the nearby participator is verified, it will follow the instruction to delay or cancel the SOS beacon.

### 9.3.3 Crowd Localization

When the wearable tag enters into SOS mode, the nearby users with the "FindingNemo" app will receive the SOS beacon and wake up to assist the tag localization process. The crowd localization process is illustrated in Fig. 9.3. The lost tag tries to reach as many participators as possible, especially the participators with GPS information, so that the location of the tag (child) could be estimated and reported to the server and the group leader (parent).

**User Participatory Model** Participators can choose to allow GPS localization assistance, or only allow communication assistance. The communication consumes far less energy than the GPS involvement. It is up to the user whether to allow GPS to participate in the crowd sensing process. As shown in Fig. 9.3, the users with "location" icon means the GPS location is available for this participator, a.k.a., Anchor. Other users without the "location" icon could participate via communication and ranging, and the sensed information could upload to the central server opportunistically.

We model the smartphone as a navigator in the *navigation* coordinate ($n$-frame). The $n$-frame is used to characterize the absolute movement of the smartphone with respect to the indoor/outdoor environment, which is a local geographic frame. The location from the GPS module of a smartphone is at geodetic coordinates (latitude $\phi$, longitude $\lambda$, height $h$), e.g., WGS 84 datum. To convert the geodetic coordinates to the *navigation* coordinate, we first convert it to the earth-centered earth-fixed (ECEF) coordinate, then convert the ECEF to the ENU frame via the formula in [56]. By subtracting the reference point $O_R$, the GPS location is mapped to the *navigation* coordinate ($n$-frame) for more intuitive and practical analysis.

Assume the position of the wearable tag in the $n$-frame is $\mathbf{y}^n \in \mathbb{R}^d$, i.e., 2-D coordinate ($d = 2$) of $\mathbf{y}^n$. For notation simplicity, we refer to $\mathbf{y} = [x, y]^T$ as $\mathbf{y}^n$ in the navigation coordinate without the superscript. When the child is lost, the wearable tag will broadcast the BLE beacon in "SOS" mode with the communication range as $R_B$. Assume the nearby $m$-th participator within the range $R_B$ is $a_m$, where $m$ is the index number in total $M$ participators. The location of the $m$-th participator (a.k.a. the smartphone's location) are denoted as $\mathbf{x}_m \in \mathbb{R}^d$. For most of the cases, the dimension could be simplified as $d = 2$, thus each element of $\mathbf{x}_m$ is a 2-D coordinate as $[x_m, y_m]^T$, $m = 1, \ldots, M$. The location of other participators $\mathbf{x}_m$ is unknown for most cases. If the participator allows the app to access the GPS location, then $\mathbf{x}_m$ could be estimated by the GPS result as $\hat{\mathbf{x}}_m$, with the total number as $M_a$. Since the GPS module is power hungry, and computationally heavier than the BLE communication, we could not require every participator to open their GPS location. Thus, the location of the $m$-th participator is denoted as $\hat{\mathbf{x}}_m$ only if available, and the set of GPS-enabled participators is defined as a subset of $\mathbb{R}^d$ as $\mathbb{R}^d_a$.

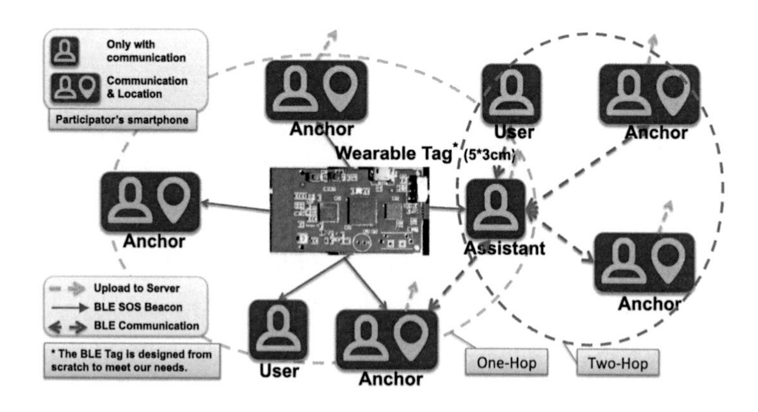


**Figure 9.3: System configuration.**

**Accuracy Requirement for Localization.** The objective of finding the lost child via crowd sensing is to estimate $\mathbf{y}$ if the BLE tag goes out of the range of the parents' smartphone. The high accuracy of $\mathbf{y}$ is more desirable, which could result in a smaller searching region for parents. The measurements available are the BLE RSS distance measure and the GPS location available in the subset of participators $\mathbb{R}^d$. Since the wearable tag is only equipped with radio communication, the estimation of $\mathbf{y}$ without GPS is challenging.

If the wearable tag could reach one participator with location information ($\mathbf{x}_m$), the value of $\mathbf{y}$ could be roughly estimated via the $\mathbf{x}_m$ and the ranging value $\hat{\mathrm{r}}_m$, where $\mathrm{r}_m$ denotes the ground truth distance; $\hat{\cdot}$ denotes the estimated value. The accuracy of $\mathbf{y}$ depends on the distance and the accuracy of ranging and location. For example, if the GPS location accuracy of $\mathbf{x}_m$ is 8 meters, the resulting search area is around $201m^2$. For the lost child, if $\mathrm{r}_m = 20$m and the ranging accuracy of $\hat{\mathrm{r}}_m$ is around 5m, the resulting search area is around $3421m^2$, which is a huge area for parents to search.

To narrow down the search area and make it acceptable in terms of efforts, significantly higher accuracy of $\mathbf{y}$ should be obtained. Multiple participators should be leveraged to improve the location estimation. The detailed accuracy requirement depends on the specific environment. Typically, the accuracy of $\mathbf{y}$ within several meters of tens of meters should be sufficient for parents to perform manual search.

To improve the localization accuracy, multiple participators should be utilized to locate the lost child, e.g., via trilateration. The distance measure from multiple participators can be written as

$$\hat{\mathrm{r}}_m = ||\mathbf{y} - \mathbf{x}_m||_2 + \delta_m + n_m \tag{9.1}$$

where $\delta_m$ is the drift or bias for the BLE tag and smartphone ranging pair, $n_m$ is the measurement noise. Each element of $\mathrm{r}_m$ is $r_m = ||\mathbf{y} - \mathbf{x}_m||_2$, i.e., from the BLE tag to the $m$-th participator, where $|| \cdot ||_2$ calculates the 2-norm and obtains the Euclidean distance. Since $r_m$ should be non-negative, we use $\hat{\mathrm{r}}_m = |\hat{\mathrm{r}}_m|$ to prevent the negative value due to the noise. We define the unknown parameter vector as $\theta = [\mathbf{y}]^T$. The localization process of relying on multiple participators is to estimate $\theta$ by using approaches like Bayesian or maximum-likelihood (ML) estimation techniques.

For 2-d coordinates, the number of participants with GPS location should be $M_p \geq 3$ for localization, a.k.a. trilateration. Since the BLE communication only covers limited areas, the available participants with GPS location may not meet this requirement. For the extreme cases, if none of the participants allows releasing GPS location, the lost child could never be located. How to locate the child with sufficient accuracy without enough measurements poses a stringent challenge.

**Reaching More Participators via Multi-Hop Assistance.** We need to reach more participators to get a better location accuracy, e.g., larger than three for trilateration; however, it is hard to obtain enough measurements in crowd sensing, especially when the nearby participators are sparse.

One possible solution is using multiple hops of communication and connection to cover more participants. Specifically, if the immediate surroundings of the BLE tag does not have enough participants with location information, relying on the participants' nearby as one-hop could provide a higher probability of location access. More hops could be involved for better localization probability. Using one-hop via $m'$-th participant as an example, the accessible participants with locations for the $m'$-th participant can be denoted as $a_{m',n}$, where $n = 1, \cdots, N_a$. The available location measurements for these $N_a$ participants is included in vector $\mathbf{x}_{m',n}$. Among all these second-hop participants, the set of participants with GPS location can be written as $\mathbb{R}^d_{a,m'}$. Thus the measurement model can be rewritten as

$$\begin{aligned} \hat{\mathrm{r}}_m - ||\mathbf{y} - \mathbf{x}_m||_2 &= \delta_m + n_m, \quad m \in \mathbb{R}^d_a \\ \hat{\mathrm{r}}_{m'} - ||\mathbf{y} - \mathbf{x}_{m'}||_2 &= \delta_{m'} + n_{m'}, \quad m' \notin \mathbb{R}^d_a, m' \in \mathbb{R}^d \\ \hat{\mathrm{r}}_{m',n} - ||\mathbf{x}_{m'} - \mathbf{x}_{m',n}||_2 &= \delta_{m',n} + n_{m',n}, \quad n \in \mathbb{R}^d_{a,m'} \end{aligned} \tag{9.2}$$

where the second added equation in (9.2) means the ranging between the BLE tag and the $m'$-th participant that does not contain GPS location; the third added equation means the ranging measurements between the $m'$-th participant and its own nearby accessed participants with locations. Through this one-hop via $a_{m'}$, more location-enabled participants are available; the probability of locating the target $\mathbf{y}$ is increased.

## 9.4 Improving Crowd Sensing Accuracy via Semidefinite Programming

To improve the localizability and accuracy of the lost child $\mathbf{y}$, we obtain more sensing results of (9.2) by leveraging multi-hop assistance to reach more participators. However, using all these new measurements in the localization process is more complex than the conventional trilateration process. The involved multi-hop assistance and sparse ranging measurement make the location estimation problem highly nonlinear, nonconvex, and hard to solve.

### 9.4.1 Challenges

**Solving Non-Convex Problem.** Solving the localization problem via multi-hop measurements in (9.2) is different from the conventional trilateration approaches that use least squares (LS)-based methods. As shown in Fig. 9.4, LS-based trilateration utilized the known location of anchor nodes and anchor in relation to mobile node ranging measurements to obtain approximate solutions of maximum likelihood (ML) by linearizing the initial *non-linear* problem. However, crowd sensing-based localization does not have enough one-hop anchor nodes, and the available anchor locations are not accurate.

When the problem becomes complex, i.e., a self-localization application, as shown in Fig. 9.4, LS-based approaches may converge to the local optimal solutions when the measurement noise is not Gaussian or the initial value in iteration is not good. Converting the initial non-convex problem to a convex one is a typical solution to avoid the local optimal. Among existing solutions, semidefinite programming (SDP) has been demonstrated to have a tight bound with the initial *non-convex* problem and tends to achieve optimal global results [117]. The SDP approach has been widely used in the sensor network localization areas, i.e., self-localization as shown in Fig. 9.4, where the dense inter-node ranging information is utilized to infer the location relative to the anchor node. Usually two anchor nodes are need to resolve the 2D coordinates of the whole network without ambiguity.

For the crowd sensing-based localization as shown in Fig. 9.4, the available ranging measurements are significantly sparser than in conventional self-localization problems. The available distance constraint is not sufficient when applying the conventional SDP for localization.

**Beyond Conventional Semidefinite Programming.** Although conventional SDP is not suitable for crowd sensing scenarios, it does provide a technical tool for solving this *non-convex* localization problem. Different from the LS-based method that linearizes the initial *non-linear* problem and requires a good initial value to ensure convergence, the SDP approach is insensitive to large errors and could eliminate the problem of converging to the local optimal. However, two stringent challenges should be solved: 1) Insufficient ranging information may render the SDP problem unsolvable; 2) amplified location bias error when mapping back to the physical space due to the requirement of two accurate anchor locations as base-line.

To solve the problem of insufficient ranging information when applying SDP to the crowd sensing setting, we reformulate the SDP problem by jointly leveraging the available sparse ranging and coarse-grained location information. Instead of relying on two anchor locations as baseline, we derive the mapping during the location estimation by optimizing the global bias between all the available participator locations.

The unknown location of the assistant $\mathbf{x}_{m'}$ in multi-hop relay is jointly estimated with the target $\mathbf{y}$. Denote the residue vector in (9.2) as $\varepsilon_m = [\delta_m + n_m, \cdots]$, $\varepsilon_{m'} = [\delta_{m'} + n_{m'}, \cdots]$, and $\varepsilon_{m',n} = [\delta_{m',n} + n_{m',n}, \cdots]$, respectively. We define $\varepsilon$ as a summation of $\varepsilon = [\varepsilon_m, \varepsilon_{m'}, \varepsilon_{m',n}]$. The vector of unknown parameter is $\theta = [\mathbf{y}, \mathbf{x}_{m'}]$.

### 9.4.2 *Cases of Insufficient Measurements*

For one extreme case, if the number of participators with GPS location near the lost child is zero, then it is unable to obtain the location $\mathbf{y}$ via conventional approaches. This case can be denoted as $\mathbb{R}_a^d = \emptyset$, and the first constraint in (9.10) cannot be used. If there are no participators $a_{m'}$ in $\mathbb{R}^d$ that accessed GPS-enabled participators, i.e., $\mathbb{R}_{a,m'}^d = \emptyset$, $\mathbf{y}$ via (9.10) is unsolvable. Only when $M_a + N_a \geq 1$, could $\mathbf{y}$ be located. Here we focus on the case of $M_a + N_a \geq 1$, and improve the location accuracy.

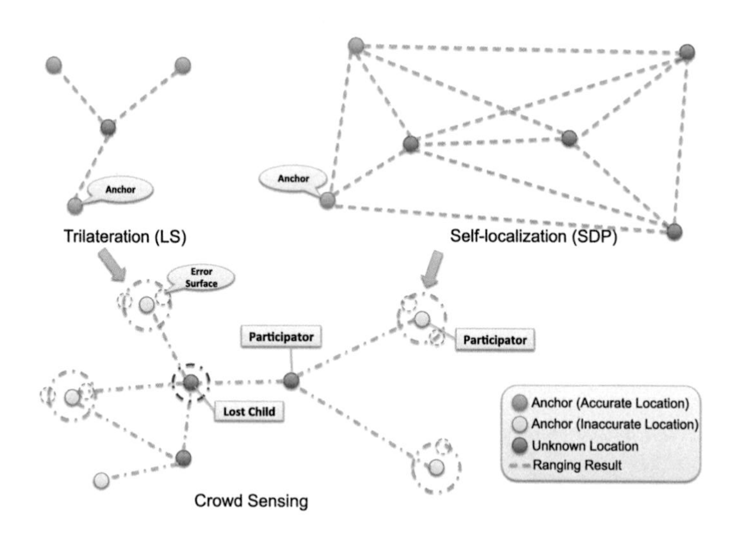


**Figure 9.4: The comparision of crowd sensing to trilateration and self-localization.**

**Different value of $M_a$.** If $M_a = 0$, the BLE Tag is only localizable via participator $a_{m'}$, which has $N_a \geq 1$. In (9.10), only the measurements $\hat{r}_{m',n}$ and $\hat{r}_{m'}$ are actually utilized.

The lowest resolution case is $N_a = 1$, assuming $x_{m',n_1}$ has location available with accuracy as $\sigma^{GPS}_{m',n_1}$. Then the achievable accuracy of $\mathbf{y}$ is on the order of

$$\sigma^{\mathbf{y}} = \sigma^{GPS}_{m',n_1} + \hat{r}_{m',n_1} + \sigma^{r}_{m',n_1} + \hat{r}_{m'} + \sigma^{r}_{m'} \tag{9.3}$$

where $\sigma^{r}_{m',n_1}$,$\sigma^{r}_{m'}$ is the ranging accuracy between $a_{m'}$ and $a_{m',n_1}$, $y$ and $a_{m'}$, respectively. Through this one-hop cooperation via $a_{m'}$, we could locate $\mathbf{y}$. However, the accuracy of (9.3) is very high, but not sufficient for finding a lost child in crowds.

Increasing the number $M_a$, the accuracy of $\mathbf{y}$ could be significantly improved. When $M_a = 1$, the $\sigma^{\mathbf{y}}$ could be reduced to $\sigma^{\mathbf{y}} = \sigma^{GPS}_{m} + \hat{r}_m + \sigma^{r}_{m}$. When $M_a = 2$, the possible positions of $\mathbf{y}$ are reduced to two spots, with the accuracy in each spot depending on $\sigma^{GPS}_{m} + \sigma^{r}_{m}$. When $M_a \geq 3$, $\mathbf{y}$ could be localized even without the one-hop assistance, where the accuracy depends on $\sigma^{GPS}_{m}$ and $\sigma^{r}_{m}$.

**Different value of $N_a$.** If $N_a = 0$, there will be no benefit from using this one-hop assistance. If $0 < N_a < 3$, the performance improvement from this one-hop assistance is significant when $M_a$ is insufficient for trilateration. If $N_a \geq 3$, $a_{m'}$ could be localized by trilateration that is more reliable, which could in turn provide performance improvement, especially for cases when $M_a < 3$.

### 9.4.3 Problem Formulation for Crowd Sensing SDP

The location objective function of $\theta$ for jointly estimating the vector of unknown parameter $\theta = [\mathbf{y}, \mathbf{x}_{m'}]$ could be written as

$$\hat{\theta}_{ML} = \arg\min_{\theta}( \sum_{\forall m, m', (m', n)} \varepsilon^2) \tag{9.4}$$

$$= \arg\min_{\theta} \left( \sum_{\forall m \in \mathbb{R}_a^d} \varepsilon_m^2 + \sum_{\forall m' \in \mathbb{R}^d \cap \bar{\mathbb{R}}_a^d} \varepsilon_{m'}^2 + \sum_{\forall n \in \mathbb{R}^d_{a,m'}} \varepsilon_{m',n}^2 \right)$$

where the three different constraints leverage all the sparse ranging and coarse-grained location information. The solution $\hat{\theta}_{ML}$ of (9.4) is optimal in the ML sense and highly nonlinear with constraints in (9.2). Instead of using the square errors in (9.4) that result in a non-linear problem, we modify the error formulation as a matrix linear operation before SDP relaxation. Using the first equation in (9.2) as an example, we can rewrite it as $\hat{\mathrm{r}}_m = ||\mathbf{y} - \mathbf{x}_m||_2 + \varepsilon_m$. Performing square operation on both sides will lead to $\hat{\mathrm{r}}_m^2 = (||\mathbf{y} - \mathbf{x}_m||_2 + \varepsilon_m)^2$, where the right side is $||\mathbf{y} - \mathbf{x}_m||_2^2 + 2{\varepsilon_m}^T ||\mathbf{y} - \mathbf{x}_m||_2 + {\varepsilon_m}^T \varepsilon_m$. $\varepsilon_m$ is the variance vector of the ranging error. Assuming ranging measurements are independent, we will have ${\varepsilon_m}^T \varepsilon_m = 0$; $2\varepsilon_m ||\mathbf{y} - \mathbf{x}_m||_2$ will be the new noise term as $\varepsilon'$. Thus, for all three equations in (9.2), the location objective function can be rewritten as

$$\mathbf{y} = \arg\min_{\mathbf{y}} \max_{m, m'} \underbrace{\left\{ \left| ||\mathbf{y} - \mathbf{x}_m||_2^2 - \hat{\mathrm{r}}_m^2 \right| + \left| ||\mathbf{y} - \mathbf{x}_{m'}||_2^2 - \hat{\mathrm{r}}_{m'}^2 \right| \right\}}_{\xi_1} \tag{9.5}$$

$$+ \arg\min_{\mathbf{y}} \max_{m', n} \underbrace{\left| ||\mathbf{x}_{m'} - \mathbf{x}_{m',n}||_2^2 - \hat{\mathrm{r}}_{m',n}^2 \right|}_{\xi_2}$$

(9.5) calculates $\mathbf{y}$ by minimizing the maximum residual error, where the $\sum$ in (9.4) becomes max operation. *Minimax* approximation plays a key role in linearizing the initial problem into linear matrix operations for semidefinite relaxation.

### 9.4.4 Location Optimization via Semidefinite Programming

The objective function in (9.5) can be converted to minimize $\epsilon$ at the constraint of an inequality expression $-\epsilon < \xi_1 + \xi_2 < \epsilon$, while $\xi_1$ and $\xi_2$ are the residual error in (9.5) for the first and second term.

The term $||\mathbf{y} - \mathbf{x}_m||_2^2$ in $\xi_1$ of (9.5) can be written into a matrix form of

$$\begin{aligned}||\mathbf{y} - \mathbf{x}_m||_2^2 &= \begin{bmatrix} \mathbf{y}^T & 1 \end{bmatrix} \begin{bmatrix} \mathbf{I}_d & -\mathbf{x}_m \\ -\mathbf{x}_m^T & \mathbf{x}_m^T\mathbf{x}_m \end{bmatrix} \begin{bmatrix} \mathbf{y} \\ 1 \end{bmatrix} \\ &= \text{tr}\left\{ \begin{bmatrix} \mathbf{y} \\ 1 \end{bmatrix} \begin{bmatrix} \mathbf{y}^T & 1 \end{bmatrix} \begin{bmatrix} \mathbf{I}_d & -\mathbf{x}_m \\ -\mathbf{x}_m^T & \mathbf{x}_m^T\mathbf{x}_m \end{bmatrix} \right\} \\ &= \text{tr}\left\{ \begin{bmatrix} \mathbf{Y} & \mathbf{y} \\ \mathbf{y}^T & 1 \end{bmatrix} \begin{bmatrix} \mathbf{I}_d & -\mathbf{x}_m \\ -\mathbf{x}_m^T & \mathbf{x}_m^T\mathbf{x}_m \end{bmatrix} \right\}\end{aligned} \tag{9.6}$$

where $\mathbf{Y} = \mathbf{y}^T\mathbf{y}$, $\text{tr}\{\cdot\}$ calculates the trace of the matrix, and $\mathbf{I}_d$ is an identity matrix of order $d$. From step 1 to step 2 in (9.6), we utilized the property of matrix trace $\text{tr}\{\mathbf{x}\mathbf{x}^T\mathbf{A}\} = \mathbf{x}^T\mathbf{A}\mathbf{x}$.

Using the same process in (9.6), $||\mathbf{x}_{m'} - \mathbf{x}_{m',n}||_2^2$ in $\xi_2$ of (9.5) can be written into

$$\begin{aligned}&||\mathbf{x}_{m'} - \mathbf{x}_{m',n}||_2^2 = (\mathbf{x}_{m'} - \mathbf{x}_{m',n})^T(\mathbf{x}_{m'} - \mathbf{x}_{m',n}) \\ &= \begin{bmatrix} \mathbf{0}_{1\times d} & 1 & -1 \end{bmatrix} \begin{bmatrix} \mathbf{I}_d & \mathbf{x}_{m'} & \mathbf{x}_{m',n} \\ \mathbf{x}_{m'}^T & \mathbf{Y}_{m'} & \mathbf{Y}_{m',n} \\ \mathbf{x}_{m',n}^T & \mathbf{Y}_{n,m'} & \mathbf{Y}_{n,n} \end{bmatrix} \begin{bmatrix} \mathbf{0}_{d\times 1} \\ 1 \\ -1 \end{bmatrix} \\ &= \text{tr}\left\{ \begin{bmatrix} \mathbf{0}_{d\times 1} \\ 1 \\ -1 \end{bmatrix} \begin{bmatrix} \mathbf{0}_{d\times 1} \\ 1 \\ -1 \end{bmatrix}^T \begin{bmatrix} \mathbf{I}_d & \mathbf{x}_{m'} & \mathbf{x}_{m',n} \\ \mathbf{x}_{m'}^T & \mathbf{Y}_{m'} & \mathbf{Y}_{m',n} \\ \mathbf{x}_{m',n}^T & \mathbf{Y}_{n,m'} & \mathbf{Y}_{n,n} \end{bmatrix} \right\} \\ &= \mathbf{Y}_{m'} - \mathbf{Y}_{m',n} - \mathbf{Y}_{n,m'} + \mathbf{Y}_{n,n}\end{aligned} \tag{9.7}$$

where $\mathbf{Y}_{m'} = \mathbf{x}_{m'}^T\mathbf{x}_{m'}$, $\mathbf{Y}_{n,n} = \mathbf{x}_{m',n}^T\mathbf{x}_{m',n}$, $\mathbf{Y}_{m',n} = \mathbf{x}_{m'}^T\mathbf{x}_{m',n}$.

The forms of (9.6) are convex, but the equality constraints of $\mathbf{Y} = \mathbf{y}^T\mathbf{y}$ are nonconvex. Using semidefinite relaxation, these equalities can be relaxed to inequality constraints of $\mathbf{Y} \succeq \mathbf{y}\mathbf{y}^T$. The matrix forms can be written as

$$\begin{bmatrix} \mathbf{Y} & \mathbf{y} \\ \mathbf{y}^T & 1 \end{bmatrix} \succeq 0 \tag{9.8}$$

where $\succeq$ means a positive definite (semidefinite) matrix, which is different from $\geq$.

For the constraint of (9.7), equality constraints of $\mathbf{Y}_{m'} = \mathbf{x}_{m'}^T\mathbf{x}_{m'}$, $\mathbf{Y}_{n,n} = \mathbf{x}_{m',n}^T\mathbf{x}_{m',n}$, $\mathbf{Y}_{m',n} = \mathbf{x}_{m'}^T\mathbf{x}_{m',n}$ are nonconvex. In (9.7), $\mathbf{Y}_{m'}$, $\mathbf{Y}_{n,n}$, and $\mathbf{Y}_{m',n}$ are coupled together. Thus, the matrix forms of the SDP relaxation for the constraint $\xi_2$ are

$$\begin{bmatrix} \mathbf{I}_d & \mathbf{x}_{m'} & \mathbf{x}_{m',n} \\ \mathbf{x}_{m'}^T & \mathbf{Y}_{m'} & \mathbf{Y}_{m',n} \\ \mathbf{x}_{m',n}^T & \mathbf{Y}_{n,m'} & \mathbf{Y}_{n,n} \end{bmatrix} \succeq 0 \tag{9.9}$$

Using the form of (9.6), (9.7), (9.8), and (9.9), the initial problem of (9.5) can be formulated to a semidefinite programming form. The unknown parameter vector could be summarized as $\theta = [\mathbf{y}, \mathbf{x}_{m'}, \mathbf{Y}_{m'}, \mathbf{Y}_{n,n}, \mathbf{Y}_{m',n}]$. (9.5) can be equivalently reformulated as

$$
\begin{aligned}
&\min\nolimits_{\theta} \epsilon \\
\text{s.t.} \quad & -\epsilon < \mathrm{tr}\left\{\begin{bmatrix} \mathbf{Y} & \mathbf{y} \\ \mathbf{y}^T & 1 \end{bmatrix}\begin{bmatrix} \mathbf{I}_d & -\mathbf{x}_m \\ -\mathbf{x}_m^T & \mathbf{x}_m^T\mathbf{x}_m \end{bmatrix}\right\} - \hat{\mathrm{r}}_m^2 < \epsilon, \\
& -\epsilon < \mathrm{tr}\left\{\begin{bmatrix} \mathbf{Y} & \mathbf{y} \\ \mathbf{y}^T & 1 \end{bmatrix}\begin{bmatrix} \mathbf{I}_d & -\mathbf{x}_{m'} \\ -\mathbf{x}_{m'}^T & \mathbf{x}_{m'}^T\mathbf{x}_{m'} \end{bmatrix}\right\} - \hat{\mathrm{r}}_{m'}^2 < \epsilon, \\
& -\epsilon < \mathbf{Y}_{m'} - \mathbf{Y}_{m',n} - \mathbf{Y}_{n,m'} + \mathbf{Y}_{n,n} - \hat{\mathrm{r}}_{m',n}^2 < \epsilon \\
& \begin{bmatrix} \mathbf{Y} & \mathbf{y} \\ \mathbf{y}^T & 1 \end{bmatrix} \succeq 0, \quad \begin{bmatrix} \mathbf{I}_d & \mathbf{x}_{m'} & \mathbf{x}_{m',n} \\ \mathbf{x}_{m'}^T & \mathbf{Y}_{m'} & \mathbf{Y}_{m',n} \\ \mathbf{x}_{m',n}^T & \mathbf{Y}_{n,m'} & \mathbf{Y}_{n,n} \end{bmatrix} \succeq 0
\end{aligned}
\tag{9.10}
$$

where $m \in \mathbb{R}_a^d$, $m' \in \mathbb{R}^d \cap \bar{\mathbb{R}}_a^d$, and $n \in \mathbb{R}_{a,m'}^d$. The location of the lost child (wearable tag) $\mathbf{y}$ can be extracted from the solution of $\theta = [\mathbf{y}, \mathbf{x}_{m'}, \mathbf{Y}_{m'}, \mathbf{Y}_{n,n}, \mathbf{Y}_{m',n}]$, where $\mathbf{y}$ is optimal in terms of all the available measurements.

### 9.4.5 *Virtual Anchor Assistance*

During the SDP optimization of (9.10), the location of $a_{m'}$ is used as an unknown parameter in $\theta$ as $\mathbf{x}_{m'}$. To utilize the feature that $\mathbf{x}_{m'}$ is accurate and reliable when $N_a \geq 3$, the estimated result of $\hat{\mathbf{x}}_{m'}$ from (9.10) could be used as a *virtual anchor* that optimizes the location result of $\hat{\mathbf{y}}$, and obtains an optimized value of $\hat{\mathbf{y}}_{op}$.

After the process of (9.10), the obtained $\hat{\mathbf{x}}_{m'}$ could be used to construct a new vector of nearby anchor points as $\mathbf{z}_{m_a} = [\mathbf{x}_m \quad \hat{\mathbf{x}}_{m'}]$, where $m_a = 1, \cdots, M_a + 1$. The ranging measurement vector could be reconstructed as $\mathbf{r}_{m_a} = [\hat{\mathbf{r}}_m \quad ||\hat{\mathbf{y}} - \hat{\mathbf{x}}_{m'}||_2]$. Another step of optimization via the newly constructed anchor vector $\mathbf{z}_{m_a}$ could be executed by relying on the minimization of residual error $\epsilon$ with the unknown parameter as $\theta = [\mathbf{y}_{op} \quad \mathbf{Y}_{op}]$, where $\mathbf{Y}_{op} = \mathbf{y}_{op}^T\mathbf{y}_{op}$. The constraint of this refinement could be written as

$$
-\epsilon < \mathrm{tr}\left\{\begin{bmatrix} \mathbf{Y}_{op} & \mathbf{y}_{op} \\ \mathbf{y}_{op}^T & 1 \end{bmatrix}\begin{bmatrix} \mathbf{I}_d & -\mathbf{z}_{m_a} \\ -\mathbf{z}_{m_a}^T & \mathbf{z}_{m_a}^T\mathbf{z}_{m_a} \end{bmatrix}\right\} - \mathrm{r}_{m_a}^2 < \epsilon \tag{9.11}
$$

For some cases, the refined result of $\mathbf{y}_{op}$ may suffer performance degradation than the initial result of $\hat{\mathbf{y}}$ due to the large error of the *virtual anchor*. Thus, a threshold-based detection process needs to be applied to mitigate the *virtual anchor* with low-confidence.

## 9.5 Evaluation

To illustrate the effectiveness of our proposed approach, we compare our proposed SDP-based cooperative location optimization proposed in (9.10) ("SDP-C") with a conventional LS-based approach ("Initial"). Approaches that using *virtual anchor* via LS and SDP are denoted as "SDP-C-VAL" and "SDP-C-VAS," respectively. We use average location error (ALE) as the performance metric. Simulation and experimental evaluation are conducted.

### 9.5.1 Simulation

**Sufficient for trilateration.** When the number of nearby participators with GPS location (anchors) is sufficient for trilateration, i.e., $M_a \geq 3$, we conduct simulation to evaluate the performance improvement contributed by the one-hop assistance. Fig. 9.5(b) shows the simulation results when the anchor error is added by $\sigma^2 = 2.25m$. In this case, the performance improvement contributed by the one-hop assistance is limited. Therefore, if $M_a \geq 3$, we could directly utilize the location of participators without leveraging multi-hops.

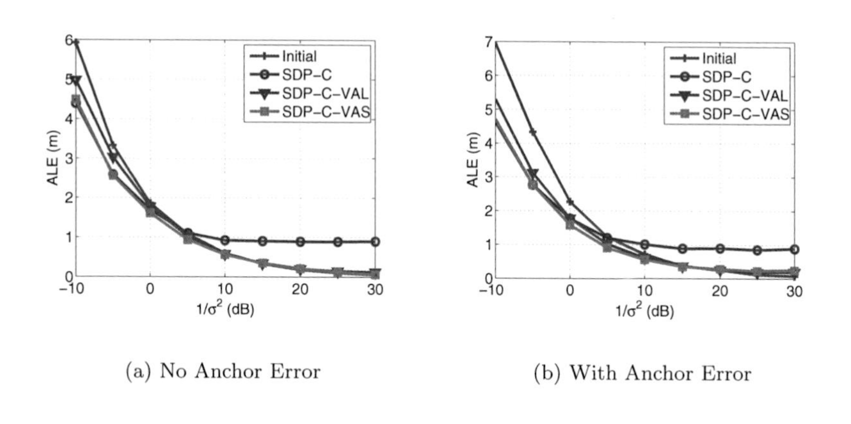


(a) No Anchor Error

(b) With Anchor Error

**Figure 9.5: Localization accuracy via one-hop when $M_a = 3$ and $N_a = 3$ for cases (a) without anchor error; (b) with anchor error ($\sigma^2 = 2.25m$).**

**Insufficient for trilateration.** When the number of anchors near the BLE tag is sufficient for trilateration, i.e., $M_a < 3$, the location of the BLE tag cannot be determined without ambiguity as in the case of "Initial" shown in Fig. 9.6 and Fig. 9.7. No matter how accurate the ranging result is, the accuracy shows no improvement due to the ambiguity in determining the location. After using one-hop assistance, the location of the BLE tag could be determined and the accuracy improves with better ranging results.

In Fig. 9.6, where $M_a = 2$ and $N_a = 3$, the performance improvement of our proposed SDP-based approaches over the "Initial" is significant. When $N_a = 3$, the location estimation result for the assistant node is reliable, and it could be utilized as a *virtual anchor* for further performance improvement.

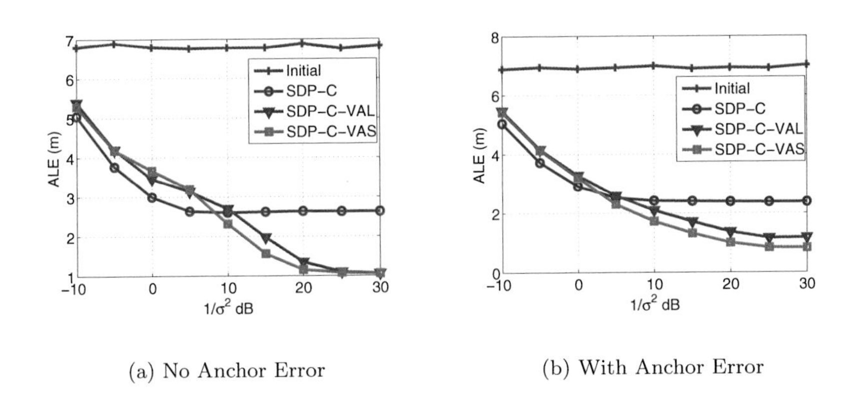


(a) No Anchor Error

(b) With Anchor Error

**Figure 9.6: Localization accuracy via one-hop when $M_a = 2$ and $N_a = 3$ for cases (a) without anchor error; (b) with anchor error ($\sigma^2 = 2.25m$).**

**No virtual anchor available.** When the number of accessed anchors of the assistant $m'$ is $N_a < 3$, i.e., insufficient for trilateration. The performance improvement is only contributed by leveraging all the measurements in SDP optimization. Fig. 9.8 shows the localization accuracy under different cases via one-hop assistance with anchor location errors. The achieved improvement over the "Initial" case is huge, and in turn demonstrates the effectiveness of SDP optimization even without *virtual anchor* assistance.

## 9.5.2 System Implementation

The fundamental requirement of crowd sensing is the sufficient user base. Incorporating the crowd sensing functionality into the mobile social network could be a feasible solution to improve the number of participators. We implement the FindingNemo functionality into the mission-based mobile social network ToGathor. As shown in Fig. 9.9, we implemented the social functions of: 1) add nearby wearable tag or smartphone; 2) friend lost and find friend again; 3) location-aware social groups; 4) track group members and sense nearby. Adding these modules on top of the existing functions of a mobile social network, e.g., basic user profile/account, message, could push the crowd sensing/sourcing task to the large scale societal level.

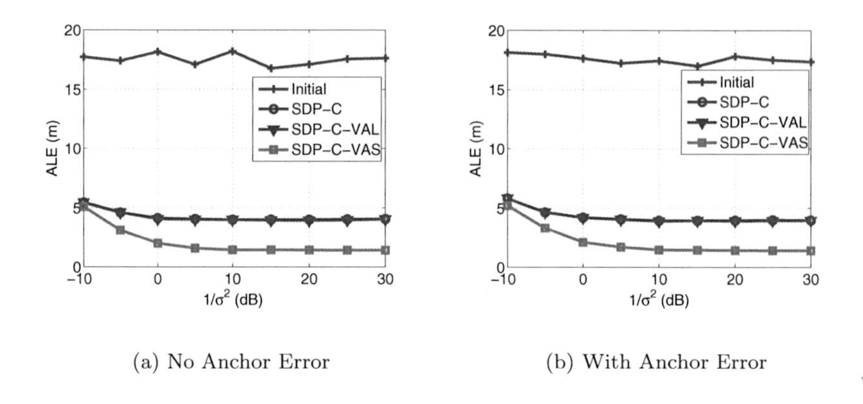


(a) No Anchor Error

(b) With Anchor Error

**Figure 9.7: Localization accuracy via one-hop when $M_a = 1$ and $N_a = 3$ for cases (a) without anchor error; (b) with anchor error ($\sigma^2 = 2.25m$).**

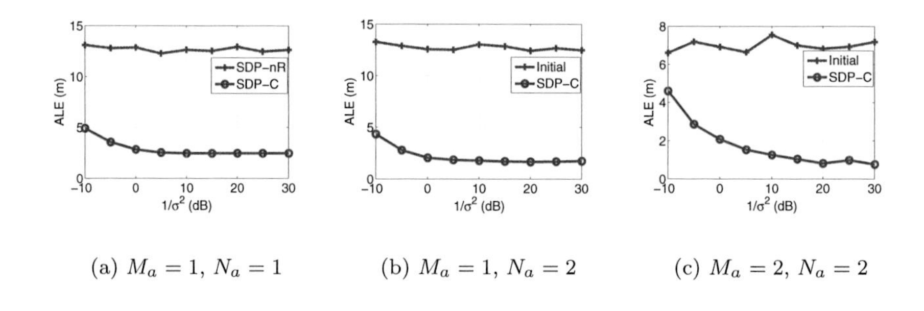


(a) $M_a = 1$, $N_a = 1$

(b) $M_a = 1$, $N_a = 2$

(c) $M_a = 2$, $N_a = 2$

**Figure 9.8: Localization accuracy via one-hop with anchor location error ($\sigma^2 = 2.25m$) for three cases.**

To improve the incentive for users' participation, one possible direction is to lower the user participating cost and improve the convenience. We propose to realize transparent participation in that all the tasks are done when the smartphone is in the background. Performing a background BLE scan and tracking in an app is lightweight. The latest version of iOS can handle this process via the OS, i.e., through iBeacon API. Evenif the app is switched off from the background, the OS could still launch the app automatically when the BLE notification is received. Google also released this nearby notification functionality in 2014 Google I/O. The native support of the nearby notification from modern mobile OS provides an essential foundation for transparent mobile crowd sensing. When the BLE connection is lost, the background app could be waked up and send notification to the group leader. This native OS support also enables transparent crowd sensing for locating the child. If the

**Figure 9.9: Basic functions of FindingNemo App.**

child is lost, nearby peers' smartphones installed with "FindingNemo," even in sleep mode, could be notified when the *SOS* alert is nearby. The app will be launched automatically after receiving this beacon, and the opportunistic communication and ranging via BLE will be performed in the background. After finishing the crowd sensing process, the participating smartphone will sleep again to save the battery life.

### 9.5.3 Experiment

**Experiment Setting.** Similar to the configurations in a simulation evaluation, we conducted an experiment to evaluate the system performance with different connectivity of nearby participators in different environments, as shown in Fig. 9.10. The location accuracy of participators is measured in different locations with different smartphones. As shown in Fig. 9.11(a), the obtained location error ranges from 2 meters to 15.4 meters with heteroge-

**Figure 9.10: Developed system and experiment setting.**

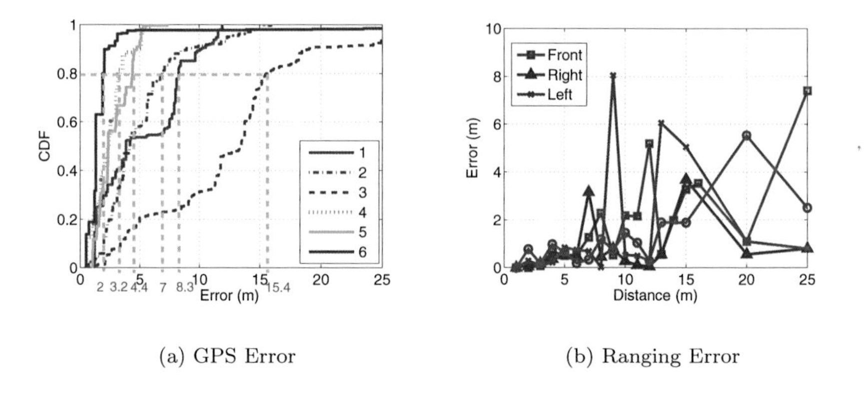


(a) GPS Error

(b) Ranging Error

**Figure 9.11: 1) The CDF of the GPS error; 2) the ranging error with distance.**

neous error distribution. The BLE ranging error versus the distance is shown in Fig. 9.11(b). We put the wearable tag in different angle directions, i.e., front, right, and left, for comparing the ranging performance. The overall location and ranging error is very large, and insufficient to locate the lost child if it only relies on one participator.

To emulate the situation of finding the lost child via multiple participators, we let one volunteer wearing the wearable tag perform random walking. Other volunteers acted as participators, and randomly moved around. We defined the number of participators with one-hop to the wearable tag as $M_a$; the number of participators with two-hop communication to the wearable tag via assistant as $N_a$. Various experiment configurations of $M_a$ and $N_a$ were conducted.

The BLE RSS ranging is performed similar to [56], where the accuracy is in the meter-level. The key problem of our proposed "FindingNemo" is relying on these inaccurate ranging results to localize the "unlocalizable" target, or solving the location ambiguity problem. Our goal is to achieve large accuracy improvement rather than struggling to slightly improve the ranging accuracy, since several meters of ranging difference does not matter too much when searching for the child.

**Sufficient for Trilateration.** When the number of nearby anchors is sufficient for trilateration, i.e., $M_a \geq 3$, the experimental results under different cases are as shown in Fig. 9.13. Similar to the conclusion obtained in simulation, performance improvement in this case is minimal. But it is still beneficial to utilize the one-hop assistance when $M_a \geq 3$. When the number of $N_a$ is reduced from 6 to 3, the performance difference is not apparent when $N_a \geq 3$.

**Insufficient for Trilateration.** When $M_a < 3$, we cannot localize the BLE tag without ambiguity via conventional methods. In real applications

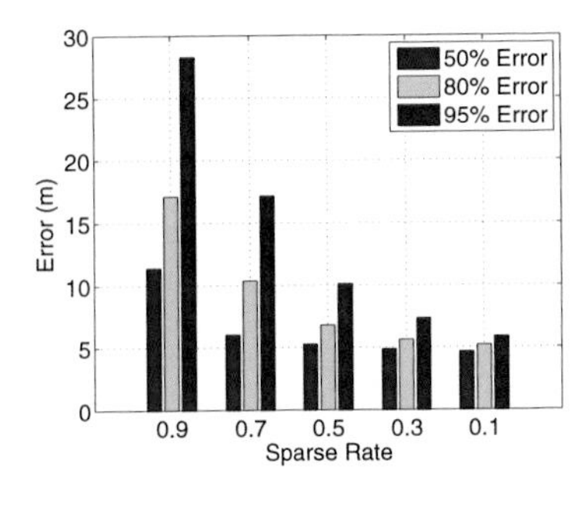


(a) Conventional SDP

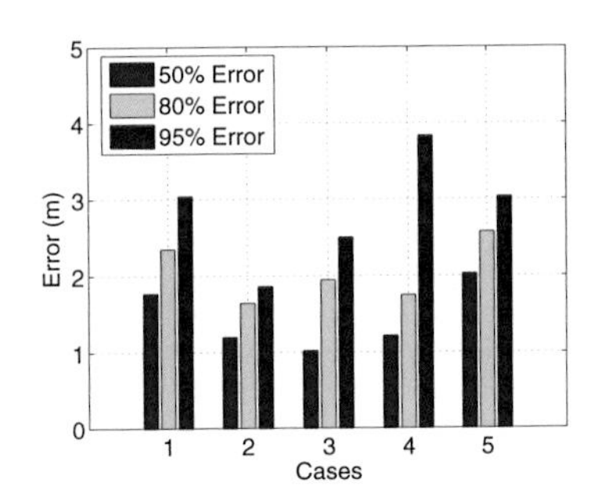


(b) Proposed Approach

**Figure 9.12: Localization error for: 1) conventional SDP with different ranging sparse rates; 2) proposed SDP approach for crowd sensing in different cases.**

of locating the lost child, $M_a$ is actually small for most of the cases. Leveraging the multi-hop and SDP-based optimization, we could make the target localizable and dramatically reduce the location error.

Fig. 9.14 shows the CDF of the ALE results when $M_a = 2$, $N_a = 3$ for different cases when the *virtual anchor* is utilized. The proposed "SDP-C-VAS" achieves best performance over most cases.

Fig. 9.15 shows the CDF of the ALE results when $M_a = 2$, $N_a = 2$ for different cases, where the *virtual anchor* cannot be utilized. Even without a *virtual anchor*, the performance improvement of using SDP optimization is still apparent.

When the number of accessed anchors $M_a$ is reduced to 1, the error of "Initial" is very large, as shown in Fig. 9.16. When the number of $N_a$ increases from 1 to 3, we can clearly see the performance improvement of using one-hop assistance, especially when the *virtual anchor* can be utilized.

When $M_a = 0$, the location of the target cannot be determined via "Initial," i.e., "NaN" for the ALE as shown in Table. 9.1. When leveraging one-hop assistance, the target could be localized. Table. 9.1 shows the ALE results when the number $N_a$ is changed from 6 to 1. Due to very limited measurements, the achieved accuracy of around 15m to 4.9m is sufficient, and really helps when searching for the child.

## 9.6 Related Work

To prevent your child from getting lost or to find your lost child if already separated, lots of systems and approaches have been developed. Conventional

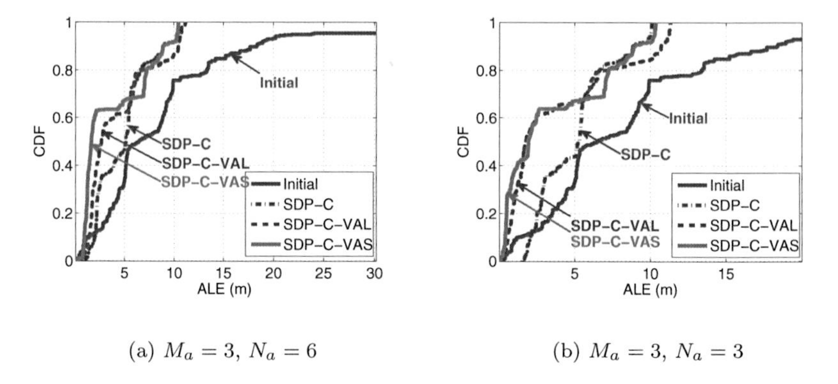


**Figure 9.13: Experimental localization accuracy via one-hop for two cases.**

**Table 9.1: Experimental localization accuracy via one-hop when $M_a = 0$**

| | Initial | SDP-C (m) | SDP-C-VAL (m) | SDP-C-VAS (m) |
|---|---|---|---|---|
| $N_a = 6$ | NaN | 6.9105 | 6.9105 | 4.9471 |
| $N_a = 3$ | NaN | 7.0643 | 7.0643 | 4.9710 |
| $N_a = 2$ | NaN | 9.7185 | 9.7185 | 9.7185 |
| $N_a = 1$ | NaN | 15.0772 | 15.0772 | 15.0772 |

approaches include guarding your child carefully or continuously keeping an eye on your child. But none of us could make sure that we would never get distracted, especially in places full of attractions. If the child is already lost, calling 911 or manually searching is labor-intensive. High-tech approaches include purchasing a GPS locator, and subscribing to a cellular service, e.g., Amber Alert GPS, PocketFinder, or AT&T Family Locator [21]. However, GPS and cellular are high cost in terms of energy consumption and hardware/service investment. Low-cost peer-to-peer communication devices with transmit and receive pairs are utilized to prevent child or pet loss, e.g., Toddler Tag Child Locator, Keeper 4.0, Chipolo [17]. However, all these approaches work only when the tag is within the communication range. If the tag is already lost, these approaches fail to provide any location or direction information to find the lost tag.

For open places without infrastructure, using crowds of mobile smartphones as the virtual localization infrastructure could be a feasible approach for finding the lost child. Crowd sensing is suitable for tasks that are hard, costly, or infeasible to finish without collaboration [46, 27]. When extended to mobile areas, the sensor-rich personal smartphone becomes the central of future MCS applications. Unleashing the potential of large-scale sensing, researchers propose solutions in terms of system architecture and algorithms to

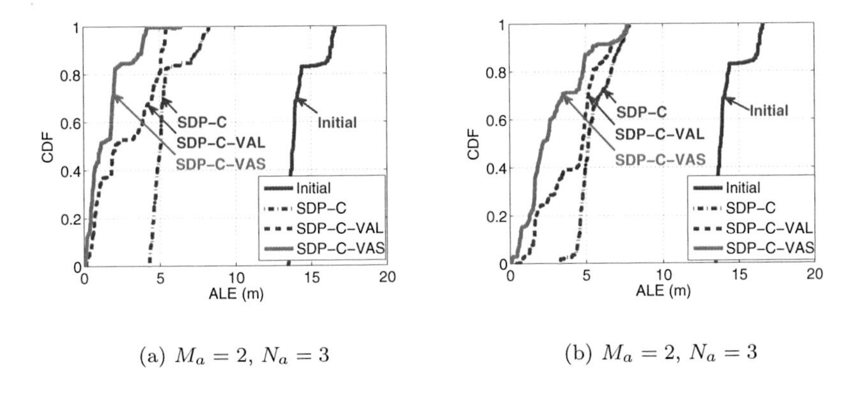


(a) $M_a = 2$, $N_a = 3$ (b) $M_a = 2$, $N_a = 3$

**Figure 9.14: Experimental localization accuracy via one-hop for two cases.**

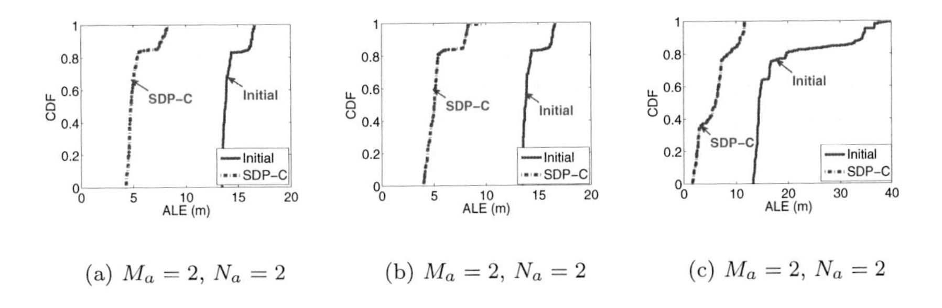


(a) $M_a = 2$, $N_a = 2$ (b) $M_a = 2$, $N_a = 2$ (c) $M_a = 2$, $N_a = 2$

**Figure 9.15: Experimental localization accuracy via one-hop for three cases.**

enable various specific applications. For example, mCrowd [126] is a system architecture for continuously sensing with high energy efficiency; the authors in [61] balance the performance needs of the application and the resource demands of continuous sensing on the phone. Wang [116] proposed a framework for an energy-efficient sensing strategy to recognize user states as well as to detect state transitions by powering only a minimum set of sensors. Crowd sensing-based applications are also emerging, e.g., the authors in [135] developed an application to predict the bus arrival time with mobile phone-based participatory sensing. Sensing the user's activity or surroundings via accelerometer, microphone- and GPS sensors becomes a hot topic of crowd sensing in [60, 69]. However, all these proposed sensing tasks are individual-based monitoring and loosely coupled between different participators.

Locating the lost child via MCS requires high coupling and collaboration between participators, in which the peer-to-peer measurements are key to the

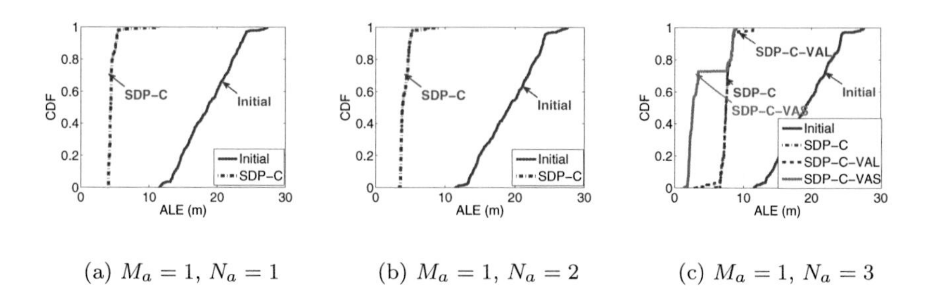


(a) $M_a = 1$, $N_a = 1$ (b) $M_a = 1$, $N_a = 2$ (c) $M_a = 1$, $N_a = 3$

**Figure 9.16: Experimental localization accuracy via one-hop for three cases.**

success and high accuracy of localization. Moreover, continuous sensing applications on a smartphone is challenging because of the high resource demands and limited battery capacity. Our proposed approach leverages highly efficient BLE notification for starting the sensing task, and lasts only a few seconds for demand-based transparent participation with low cost.

## 9.7 Conclusion

The ubiquitous availability, mobility, efficient computing and communication capacities of sensor-rich smartphones make mobile crowd sensing (MCS) a much more flexible and cost-effective localization and sensing method at scale. We designed "FindingNemo" for locating family members via MCS. The application requirements, incentive schemes, and design considerations are elaborated. We propose SDP-based cooperative location optimization via one-hop or multi-hop assistance to cover more participators. The proposed solution could locate the "unlocalizable" target with an ambiguous location, and significantly improve the location accuracy over conventional approaches. Further optimization via *virtual anchors* is proposed to leverage assistants with sufficient measurements for trilateration. The different configurations and accessibility of participators are analyzed via simulations and experiments, along with the comparison of performance improvement. The flexibility and accuracy of these proposed approaches may boost the rapid emergence of a consumer-centric participatory MCS.

# *Chapter 10*

# Improving Location Services via Social Collaboration

**Kaikai Liu**
*Assistant Professor, San Jose State University*

**Qiuyuan Huang**
*Ph.D. Student, University of Florida*

**Jiecong Wang**
*Master Student, University of Florida*

**Xiaolin Li**
*Associate Professor, University of Florida*

**Dapeng Oliver Wu**
*Professor, University of Florida*

## CONTENTS

## 10.1 Introduction

Most new models of smartphones have built-in global positioning system (GPS) receivers. The GPS on board enables a host of location-aware applications. According to a study published by the Pew Internet and American Life Project [1], more than 55% of smartphone owners use their phones to find directions, recommendations, or other information related to their present locations. In addition, geo-social "check-in" services such as Foursquare or Gowalla are very popular among young adults [1]. New digital cameras or smartphones are equipped with geo-tagging features [111], making it easy to group photos by location or track the user's footprint.

Obtaining the GPS position information incurs a high cost; the whole process includes many complex calculations, e.g., correlation, demodulation, tracking, ranging, and positioning. Moreover, satellite signals are hard to access, especially in indoor and harsh environments, due to the strong attenuation of the radio caused by building materials. The process of constantly searching and capturing the very weak beacon signal consumes a lot of power, and the estimated position is often inaccurate or even unavailable.

The rapid growth in people-centric mobile computing applications and location-based services has called for improved localization techniques. Energy efficiency and accuracy are the two main objectives of such improvements. Authors in [20, 44, 50] have paid attention to tradeoff between energy and location accuracy. They try to use low-power WiFi/GSM-based schemes to lower the frequency of GPS startups, but at the expense of lower accuracy and update rates. Other approaches utilize dedicated devices for localization when a GPS signal is unavailable [85, 81, 54, 57]. However, a lot of anchor nodes need to be placed at a very high density with known coordinates.

One compelling technique for improving the performance of localization is cooperative localization [80]. By exchanging the anchor node information and performing relative ranging between nodes, the position estimation for each node becomes possible and more accurate [49, 115]. However, existing cooperative localization techniques [79, 49, 115] require access to raw GPS ranging measurements. The GPS/WiFi position is the only information accessible by a user/application in a commercial smartpone. Therefore, the existing cooperative localization techniques [79, 49, 115] are not directly applicable to GPS-enabled smartphones. To deal with this inconvenience, the authors in [52, 74] propose practical approaches for optimizing the smartphone location results by leveraging the inter-node distance estimation. H. Liu et al. [52] map users' locations jointly against a WiFi signature map subject to ranging constraints, but show significant delay ($> 7s$) caused by TDMA-based peer-to-peer ranging and WiFi scanning. Nandakumar et al. [74] utilize the acoustic signal transmitted by desktop to assist the WiFi localization; however, not considering the mobile situation would limit its application in a smartphone.

In this chapter, we propose a cooperative location optimization (*Coloc*) scheme in a social network via crowd sensing. Unlike conventional cooperative localization that utilizes physical-layer information fusion, our proposed social-aided location optimization performs data fusion for nearby crowds of smartphones when coarse positions are already known. This social-aided fusion can achieve practical performance improvement at a lower cost with minimum added complexity. The rationale is that when a group of people in a common location all carry smartphones with GPS capability, the accuracy of localization can be significantly improved by fusing the GPS positions of the smartphones in this group. Theoretically, this performance gain is ascribed to the law of large numbers and location diversity. Adjacent samples of GPS results have high correlation with limited new information, while the location from peers contributes to diversity gains. Thus, GPS update rate as well as the power consumption could be lowered without sacrificing the accuracy if peer-assisted information is available. To further improve the accuracy, we perform peer-to-peer ranging via acoustic time-of-arrival (TOA) measurements. The achieved high-accuracy ranging results could be applied as distance constraints to optimize the overall group location accuracy. For utilizing the ranging information effectively, we derived the *necessary condition* and eliminated unnecessary ranging measurements. Two algorithms, i.e., sparse steepest descent optimization and polar optimization, are proposed to improve the overall location accuracy in a mobile environment. We want to emphasize that the driving forces of cooperative crowd localization are fast-improving smartphone technology and people-centric pervasive social computing.

The rest of this chapter is organized as follows. Section 10.2 summarizes our proposed use cases and previous work on smartphone-based ranging. Section 10.3 discusses our system design. Section 10.4 describes mathematical models, cooperative protocol, and the necessary condition for inter-node ranging. Section 10.5 presents the *Coloc* scheme for the case that relative distances between smartphones in a neighborhood are known. Section 10.6 and Section 10.7 present numerical results and experimental results, respectively. Section 10.8 summarizes related works. Section 10.9 concludes this chapter.

## 10.2 Preliminary

### *10.2.1 Use Cases*

Our proposed social-aided crowd localization scheme can be applied to two scenarios. The first scenario is that when you go outside with your friends, you and your friends are in the same route or bus, and many of you use smartphones to take geo-tagging photos occasionally. Without cooperation, every time you open your camera application, the smartphone starts up the GPS to perform localization. Even if your group is in the same place, anyone who wants to take photos needs to start up the GPS again, which slows down the response of the camera and drains the battery. Moreover, using one snapshot GPS measurement for localization is inaccurate. You may have experienced that some of your photos in your album show some places that you had never been before due to inaccurate GPS results. To reduce energy consumption and improve position accuracy, your group only needs to perform localization the first time. Using everyone's position result to do re-positioning cooperatively, a more precise position result is obtained and shared among your group members. When you are moving or stay stationary for a period of time, only one smartphone in your group needs to perform localization and update your group position automatically. Other smartphones in the group can just use the shared position, which is more accurate and power efficient than direct calculation by themselves.

The second scenario is journaling your location or moving traces when jogging with your friends. Running or jogging with friend is full of fun and can make you motivated to run. Journaling or logging workout locations, traces, or calories using a fitness app becomes a habit for many users. However, recording your GPS location continuously in the background could kill your smartphone battery quickly. If you only record significant changes of location to save energy, you could hardly see your complete moving traces. Only several dots of locations recorded in your path cannot satisfy your needs in recording your workout. Collaborating with your friend in journaling the location could save your smartphone energy and improve the location accuracy if needed. It could eliminate redundancy measurements and improve the location measurement rate in critical paths.

### 10.2.2 Relative Ranging

In cooperative localization, the relative distances between peers are the additional information input to optimize the overall location accuracy. Realizing relative ranging on a smartphone is crucial for the *Coloc* scheme. Using acoustic time-of-arrival (TOA)-based ranging has been demonstrated to have significantly better accuracy than received-signal-strength (RSS)-based ranging using WiFi/GSM/Bluetooth signals [52, 74, 53, 57].

Transmitting a simple acoustic beep and measuring its flight delay is a practical way to implement the accurate ranging on a smartphone [52, 74]. However, using a simple acoustic signal may cause the problem that there is no way to tell which smartphone emitted which signal, i.e., it causes ambiguity due to using an un-modulated signal. Resolving the problem by performing time-division multiple access and using a radio signal for assistance would increase the overall delay, which is especially serious for the acoustic signal (low transmission speed), e.g., for $N$ peers, total $N(N-1)/2$ relative distances need to be measured. For tracking users when they are walking around, sufficient ranging rate is required.

Based on our prior work [53, 57], we performed 2-PAM modulation for the acoustic signal and combine ranging and information bit transmission at the same time. With the information bits directly available in the ranging signal, we could identify the smartphone after signal demodulation. When one smartphone broadcasts its ranging beacon, other peers could all identify this beacon. Instead of performing transmit and reply for each ranging pair, we could achieve pair-wise ranging through one transmit and multiple replies. Thus, the significant amount of time used in round-robin ranging could be reduced. Moreover, we apply a cluster-based ranging approach to only estimate the user clusters with sufficient distance, i.e., the *necessary condition* for ranging presented in Section 10.4. This way, only several ranging measurements need to be performed, and the ranging delay could be minimized for tracking moving targets.

## 10.3 System Design

Fig. 10.1 illustrates the *Coloc* system architecture and major functional components. In this section, we sketch an overview of the design consideration, then elaborate on some important components in the system.

### 10.3.1 Design Consideration

In terms of accuracy, GPS is preferred over its alternatives, especially in outdoor environments, e.g., GSM/WiFi based approaches. However, GPS is extremely power hungry due to the inevitable complex computations. When the location information is demanded less, reducing the location update rate is one possible way to provide accurate position information while spending

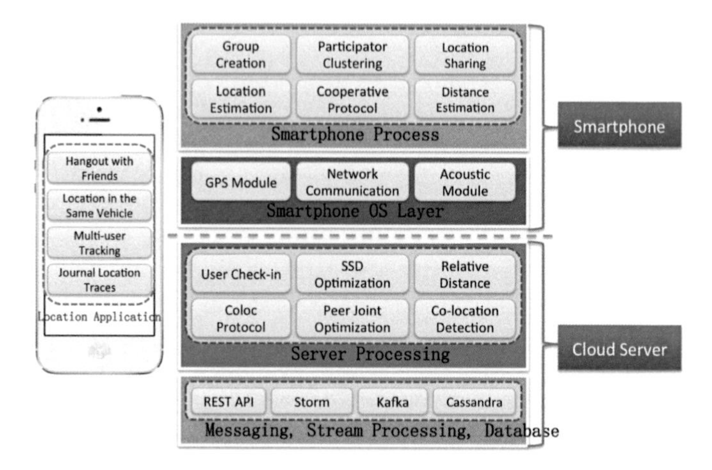


**Figure 10.1: System architecture.**

minimal energy. As the basic way to save energy, we also use reduced update rate for the GPS module, but we focus on balancing the GPS consumption to the network cost by applying the constraint that sufficient update rate is needed for tracking a moving target.

Using relative ranging information to improve the overall localization accuracy is the basic idea of our proposed *Coloc* scheme [52, 74]. The problem lies in how and when to utilize the ranging information. If the localization error surface of two peers is larger than their relative distance, the performance improvement contributed by this ranging measurement would be limited. Utilizing the ranging information when needed is essential in designing a cooperative scheme for location optimization.

## 10.3.2 System Overview

To realize the *Coloc* scheme, we proposed approaches for the smartphone to collect GPS data, report to the server, and use the refined results calculated by the server. The *Coloc* scheme consists of the following three key components: ***Coloc* Software Middleware in a Smartphone**: Each smartphone obtains position by its own GPS receiver during the start-up period. Three basic modules in a smartphone are utilized: the GPS module for coarse location estimation; the network module for communicating with the server and coarse-grained co-location detection; the acoustic module for peer detection and ranging. On top of the smartphone software middleware, cooperative location-based applications are supported, e.g., recording or tagging GPS trajectories when hanging out with friends; obtaining optimized location when multiple smartphones are in the same vehicle; tracking multi-users with a high accuracy and reliability requirement.

**Server Processing for Position Optimization**: The server receives all the GPS location information from all the users that checked in via our services. According to their coarse locations (same WiFi/BLE coverage), users could be divided into groups. Only the users in the same group could cooperate with each other for location optimization, where the size of the group is constrained by the maximum ranging distance. In each group we apply our *Coloc* scheme with relative ranging. Users in one group could also be clustered into small clusters. Widely-used clustering algorithms include K-means, unnormalized spectral clustering, the G-cut algorithm, and the normalized cuts algorithm. The reason that we perform clustering is that peer-to-peer ranging could be eliminated within one cluster to minimize the overall ranging cost and delay.

The server will send ranging coordination beacons to users for relative distance estimation. With all the information available, the server invokes the position optimization algorithm (i.e., neighborhood-based weighted least-squares estimation algorithm) to refine the position of each user by utilizing users' (coarse) GPS position information and the relative distances obtained in an iterative mode.

# 10.4 Crowd Sensing and Ranging Condition

## 10.4.1 *Crowd Cooperative Setting and Protocol*

The whole system setting is shown in Fig. 10.2. To realize crowd cooperative localization via social collaboration, we proposed a *Coloc* scheme for the smartphone to collect data and use the refined results calculated by the server. The *Coloc* scheme consists of the following five steps:

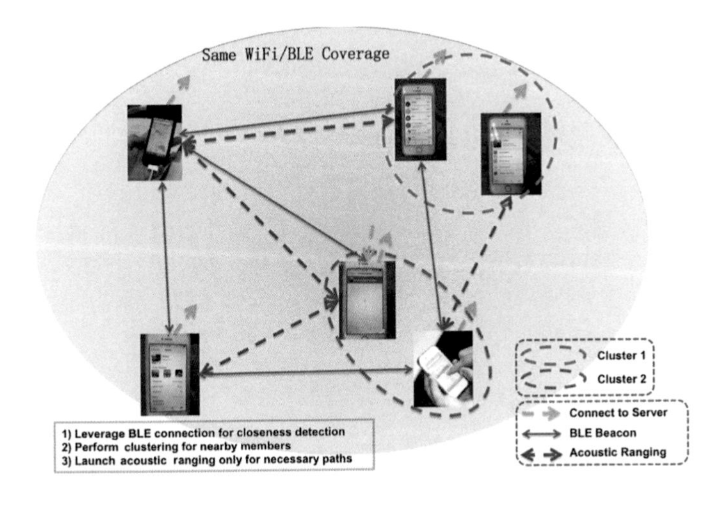


**Figure 10.2: System setting and protocol.**

1. **Group Creation**:

   When a user wants to hang out with their friends, they can create a group for all the group members and privately share their location information in this group. Then, the cooperative localization scheme is performed within this created group with privacy preserving. Another mode is crowd cooperation with nearby unknown participators. Users report their location and nearby network identification to the server; the server will perform co-location detection and automatically form a group for co-located users. In the following analysis, the meaning of "group" could be private or public.

2. **Initialization**:

   Each smartphone obtains its position by accessing its own OS location framework during the start-up period. Modern OS fuses GPS location and WiFi/Cellular location to the users when available. When we mention the initial location of the smartphone, we mean the obtained fused location results provided by OS. Assuming that there are WiFi access points near each smartphone or the group leader could broadcast BLE beacon to its members, this information (wireless ID and radio signal strengths (RSS)) could be used as coarse-grained co-location information. Then, each smartphone reports the following to the server: its initial position, IDs of its nearby WiFi or BLE beacon, and the corresponding RSS.

3. **Clustering and Co-Location Detection**:

   The server gets all the reported information and creates a group for crowded sensing results according to the co-location information. When the group is created, the server will further perform clustering to form clusters of co-located smartphones as shown in Fig. 10.2, i.e., partition all the smartphones into micro-groups according to their positions and Radio RSS. These micro-groups/clusters could be used as an closeness indication and evaluate the necessary condition for ranging.

4. **Distance Estimation**:

   When the user requests better location accuracy, relative ranging would be performed to optimize the overall localization accuracy. We estimate the relative distance between smartphones via acoustic TOA ranging. Not all peer-to-peer ranging is necessary, considering the balance of contribution, relative measurement delay, and energy cost.

   If the acoustic ranging is not available, the relative distance between any two smartphones is approximated from the GPS positions of the two smartphones and the BLE RSS if available.

5. **Position Optimization**:

   The server invokes the position optimization algorithm (i.e., neighborhood-based weighted least-squares estimation algorithm) to refine the position of each user by utilizing users' co-location information, the (coarse) GPS position information obtained in step 2, and the relative distances obtained in step 4. The position refinement works for the case without relative distances. If the acoustic TOA ranging result is available, higher positioning accuracy can be achieved.

6. **Position Updating**:

   The server sends the refined position back to each smartphone, and each smartphone updates its position with the received value. When one user wants to be a free-rider, and he is co-located with other participators, the other's location can be shared anomalously. In this way, significant power saving can be achieved since this user does not need to calculate the position value, at least not all the time.

### 10.4.2 Geo-Coordinate

We considered a social network consisting of $m$ collaborators in $\mathcal{R}^d$, where $d$ is the coordinate dimension, i.e., $d = 3$ for ellipsoidal space; $d = 2$ for the Cartesian space. Let $\mathcal{N}_g = 1, 2, \ldots, m$ denote the set of collaborators.

Assume the ground truth position of each collaborator is $\mathbf{p}_i$, $i \in \mathcal{N}_g$. With ellipsoidal coordinates, $\mathbf{p}_i$ can be written as a form of latitudes (radians), E longitudes (radians), and heights (m), i.e., $\mathbf{p}_i = (lat_i, lon_i, h_i)^T$. To simplify the process, we can change the ellipsoidal coordinates to the Cartesian coordinate under the standard of Geodetic Reference System 1980 (GRS80) by function $\mathbf{p}_i(x, y, z) = f_{ell}(\mathbf{p}_i(lat, lon, h))$ as

$$\begin{aligned} v &= a/\sqrt{(1 - e(\sin(lat))^2)} \\ x &= (v + h)\cos(lat)\cos(lon) \\ y &= (v + h)\cos(lat)\sin(lon) \\ z &= (v(1 - e) + h)\sin(lat) \end{aligned} \tag{10.1}$$

where $a$ and $e$ are the references of ellipsoid major semi-axis and eccentricity squared parameters defined in GRS80.

For small-scale geographic space, we can focus on the 2D Cartesian coordinate without the heights ($h$) information. By subtracting a pre-defined reference point $\mathbf{p}_{ref} = (x_f, y_f)^T$, a local coordinate obtained by the GPS module in the smartphone is $\hat{\mathbf{p}}_i = \hat{\mathbf{p}}_i - \hat{\mathbf{p}}_{ref} = (\hat{x}_i, \hat{y}_i)^T$, $i \in \mathcal{N}_g$ for plane-coordinate. The estimation error is $\mathbf{e}_i = |\hat{\mathbf{p}}_i - \mathbf{p}_i|$. Assuming that the position estimate is unbiased, $\mathbf{e}_i$ follows a zero-mean Gaussian distribution as $\mathbf{e}_i \sim \mathcal{N}(0, \mathbf{\Sigma}_i)$, where $\mathbf{\Sigma}_i$ is the error covariance matrix and is assumed to be a diagonal matrix with diagonal entries of $(\sigma_i^x)^2$ and $(\sigma_i^y)^2$. Then, the position matrix of each collaborator can be written as $\hat{\mathbf{P}} = [\hat{\mathbf{p}}_1 \quad \hat{\mathbf{p}}_2 \quad \ldots \quad \hat{\mathbf{p}}_m] \in \mathcal{R}^{d \times m}$.

### 10.4.3 Mathematical Modeling

The estimation error of the location can be written as $\mathbf{e}_i = |\hat{\mathbf{p}}_i - \mathbf{p}_i|$. Assuming that the position estimated is unbiased, $\mathbf{e}_i$ follows a zero-mean Gaussian distribution as $\mathbf{e}_i \sim \mathcal{N}(0, \mathbf{\Sigma}_i)$. So, the probability density function of $\hat{\mathbf{p}}_i$ can be written as

$$f(\hat{\mathbf{p}}_i) = \frac{1}{\sqrt{2\pi \det(\mathbf{\Sigma}_i)}} \exp\left(-\frac{\mathbf{D}^T \mathbf{\Sigma}_i^{-1} \mathbf{D}}{2}\right) \tag{10.2}$$

where $\mathbf{D} = (\hat{\mathbf{p}}_i - \mathbf{p}_i)$, and $\det(\mathbf{\Sigma}_i)$ calculates the determinant of $\mathbf{\Sigma}_i$. $\mathbf{\Sigma}_i$ is the error covariance matrix and is assumed to be a diagonal matrix with diagonal entries of $(\sigma_i^x)^2$ and $(\sigma_i^y)^2$. Then, the position matrix of each collaborator can be written as $\hat{\mathbf{P}} = [\hat{\mathbf{p}}_1 \quad \hat{\mathbf{p}}_2 \quad \ldots \quad \hat{\mathbf{p}}_m] \in \mathcal{R}^{d \times m}$.

The problem of social-aided cooperative localization can be modeled so as to refine the estimated positions ($\hat{\mathbf{p}}_i$) obtained by GPS. The additional information that we utilize to optimize the accuracy of GPS position are the co-location or relative distances ($\mathbf{D}$) between collaborators.

For a pair of collaborators $\mathbf{p}_i$ and $\mathbf{p}_j$, their Euclidean distance can be denoted as $d_{ij} = ||\mathbf{p}_i - \mathbf{p}_j|| = \sqrt{(x_i - x_j)^2 + (y_i - y_j)^2}$, where $||\cdot||$ is the 2-norm of the vector. Considering the measurement error, the estimated distance between collaborators is the noised version of $d_{ij}$ as $\hat{d}_{ij} = d_{ij} + n_{ij}$, where $n_{ij}$ is a Gaussian noise component with $n_{ij} \sim \mathcal{N}(b_{ij}, \sigma_{ij}^2)$. The term of $b_{ij}$ is a range bias induced by non-line-of-sight (NLOS) propagation, and $b_{ij} = 0$ when the measurement is in line-of-sight (LOS) condition. In real situations, the inter-node distance information is not fully available, i.e., some of the measurements are missing or unavailable. To deal with such conditions, we define the distance measurement matrix as $\mathbf{D} = \{d_{ij} : (i,j) \in \mathcal{N}_g\}$ with $d_{ij} = 0$ representing the unavailable measurements. The matrix $\mathbf{D}$ is a sparse matrix with sparse rate $\gamma$ defined as the number of $d_{ij} = 0$ terms divided by the total number of $m(m-1)/2$.

Fisher information $J$ (the reciprocal of CRLB) is often used as a metric to assess the accuracy of a particular position estimation. Hence, parameters to be estimated are the collaborator's refined position $\hat{\mathbf{p}}_k = (\hat{x}_k, \hat{y}_k)^T, k \in \mathcal{N}_g$ by using their initial position and relative distance. For notational convenience, we denote the unknown parameter as $\theta = [\hat{p}_k]$, where $1 \le k \le N_g$. Let $\hat{\theta}$ denote an estimation of the parameter $\theta$. The error covariance matrix of $\hat{\theta}$ satisfies information inequality as

$$\mathbb{E}_r\{(\hat{\theta} - \theta)(\hat{\theta} - \theta)^T\} \ge \mathbf{J}_\theta^{-1} \tag{10.3}$$

where $\mathbf{J}_\theta$ is the Fisher information matrix (FIM) of non-random parameter $\theta$.

The joint likelihood ratio of the discrete random vector $\mathbf{r}$ of the received signal and random parameter $\theta$ can be shown as $f(\mathbf{r}, \theta) = f(\mathbf{r}|\theta) \cdot g(\theta)$, where $f(\mathbf{r}|\theta)$ is the conditional pdf, and $g(\theta)$ is the *a priori* probability density function of $\theta$. The generalized Fisher Information Matrix (FIM) for $\theta$ is given by

$$\mathbf{J}_\theta \triangleq \mathbb{E}_{\mathbf{r},\theta}\{[\frac{\partial}{\partial\theta} ln f(\mathbf{r}, \theta)] \cdot [\frac{\partial}{\partial\theta} ln f(\mathbf{r}, \theta)]^T\} \tag{10.4}$$

(10.4) can be further decomposed,

$$\mathbf{J}_{\theta} = \underbrace{\mathbf{J}_{f(\mathbf{r},\theta)|j=i}}_{\text{GPS position info}} + \underbrace{\mathbf{J}_{f(\mathbf{r},\theta)|j\neq i}}_{\text{Info. from cooperation}} + \underbrace{\mathbf{J}_{g(\theta)}}_{\text{Prior Infor.}} \tag{10.5}$$

where the first term indicates the position information from a collaborator using GPS; the second term indicates the inter-ranging information between collaborator $i$ and $j$; and the third term denotes *a priori* information on $\theta$.

From (10.5), we know that the cooperative localization contributes to the second term; the resulting FIM can be much better than conventional localization methods that just use *a prior* information and $j = i$ term. By using the initial GPS position result and inter-note information as a prerequisite, performing post-decision optimization can obtain a more accurate position result, $\hat{p}_k = (\hat{x}_k, \hat{y}_k)^T, k \in \mathcal{N}_g$.

### 10.4.4 Crowd Clustering and Co-Location Detection

In the step of grouping, we use a distributed clustering algorithm to form clusters, i.e., partition all the co-located smartphones (within the same WiFi/BLE coverage) into clusters. Widely-used clustering algorithms include K-means, un-normalized spectral clustering, the G-cut algorithm, and the normalized cuts algorithm.

The K-means algorithm, a.k.a. the Lloyd algorithm is based on the nearest neighbor criterion and the centroid criterion. However, the K-means algorithm cannot be used for grouping smartphones if the GPS positions of the smartphones are not available due to significant signal attenuation in indoor environments, high rise building environments, or dense forest.

Even if the GPS position is available, it is still not accurate enough. We propose the following method to obtain distances or affinity measures between any pair of smartphones. Suppose the network under consideration consists of $N$ smartphones. When a smartphone (say group leader, Node $i$) has Internet access, it sends a BLE beacon signal. Any smartphone (say, Node $j$ ($j \neq i$)) that is able to detect the beacon signal records the power $P_{ij}$ of the received BLE beacon, and $P_{ij}$ will be regarded as an affinity measure between Node $i$ and Node $j$; for any node $k$ that is not able to detect the beacon signal, we set $P_{ik} = 0$. Note that given a path loss model and transmission power $P_i^{(t)}$, we can convert $P_{ij}$ to a rough distance $d_{ij}$ between Node $i$ and Node $j$, up to a constant scaling factor. For example, assume the path loss is proportional to the $n$-th power of distance, i.e., $P_i^{(t)}/P_{ij} = \kappa \times d_{ij}^n$, where $\kappa$ is a constant; then $d_{ij} = \kappa^{-\frac{1}{n}} \times (P_i^{(t)}/P_{ij})^{\frac{1}{n}}$.

Once other smartphones also get a chance to send a beacon signal, we can obtain an affinity matrix $S$ (where $S \in \mathbb{R}^{N\times N}$) and the entry at the $i$-th row and the $j$-th column of $S$ is $P_{ij}$. We assume that the transmission power $P_i^{(t)}$ is the same for all $i$. In the case that $P_i^{(t)}$ are different for different $i$, the affinity matrix $S$ will not be symmetric; then we will use $S_{sym} = S + S^T$

as the affinity matrix, which is symmetric. Un-normalized spectral clustering and the normalized cuts algorithm are able to group smartphones, given an affinity matrix.

The following shows the procedure of the normalized cuts algorithm.

Input: Affinity matrix $S$ (where $S \in \mathbb{R}^{N \times N}$), number $K$ of clusters to construct.

1. Compute the unnormalized Laplacian $L$.
2. Compute the first $K$ generalized eigenvectors $u_1, \cdots, u_K$ of the generalized eigenproblem $Lu = \lambda Du$.
3. Let $U$ ($U \in \mathbb{R}^{N \times K}$) be the matrix containing the vectors $u_1, \cdots, u_K$ as columns.
4. For $i = 1, \cdots, N$, let $y_i$ ($y_i \in \mathbb{R}^K$) be the vector corresponding to the $i$-th row of $U$.
5. Cluster the $N$ points $\{y_i \ (i = 1, \cdots, N)\}$ in $\mathbb{R}^K$ with the K-means algorithm into clusters $C_1, \cdots, C_K$.
6. Run the above procedure 50 times with different randomly permuted matrix $S$.
7. Choose the clusters $C_1, \cdots, C_K$ with minimum distortion as output.

Output: Clusters $A_1, \cdots, A_K$ with $A_i = \{j | y_j \in C_i\}$. Using 60 smartphone locations as an example, the input affinity matrix and the clustering result is shown in Fig. 10.3(a) and Fig. 10.3(b), respectively.

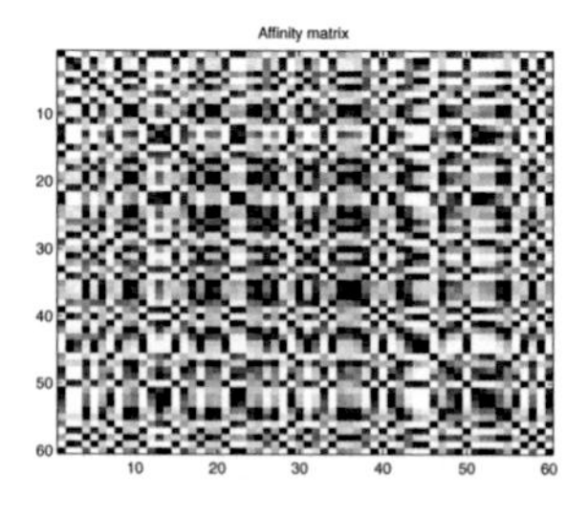


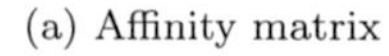
(a) Affinity matrix

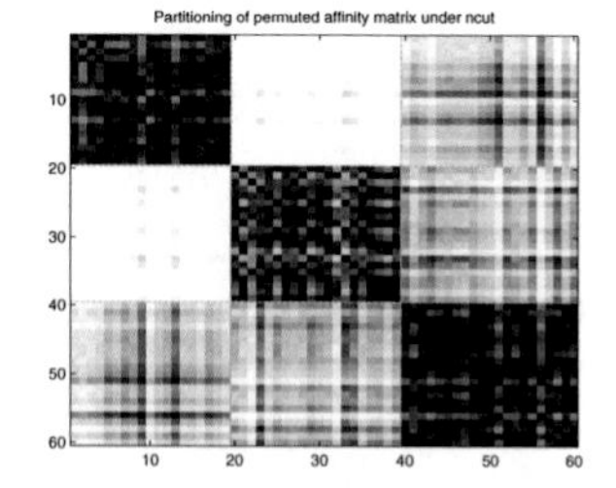


(b) Clustering result

**Figure 10.3: 1. The input affinity matrix of the clustering algorithms for 60 smartphone locations with three clusters; 2. the output of normalized cuts clustering algorithm.**

### 10.4.5 Necessary Conditions for Relative Ranging

Performing pair-wise ranging for a large amount of peers may cause substantial energy consumption and delay. In reality, some of these ranging pairs are unnecessary or only contribute to limited performance improvement. Selecting the ranging pairs that are necessary could be an effective solution to balance the performance improvement and ranging cost. In this subsection, we derive the necessary condition for ranging based on the error probability distribution. The rationale is that we analyze the performance gains contributed by directly fusing the location of co-location users, while these performance gains would be decreasing for larger relative distance. By analyzing the maximum allowable distance for performance gains of location fusion without using distance information, we can set this maximum allowable distance as the necessary condition of ranging. If the pair-wise distance is within the maximum allowable distance, directly fusing the co-located users could also improve the location accuracy, with no need for a costly ranging process. Consider the extreme case first: If all the collaborators are co-located in the same place, this co-location information of collaborators can be utilized to improve the overall localization accuracy due to the correlation between different estimated positions. For the co-location clusters $C_1, \ldots, C_K$, the mixture of the position information of cluster $C_k$ can be written as

$$\hat{\mathbf{p}}_k = \frac{1}{N_k} \sum_{i \in C_k} \gamma_i \hat{\mathbf{p}}_i \tag{10.6}$$

where $\gamma_i$ is the weighting coefficient of initial location for users in cluster $C_k$, and can be calculated by the historical position variance of user $i$.

To illustrate the performance gains with regard to the maximum allowable distance, we focus on the location fusion of a two-users case with $\hat{\mathbf{p}}_{i,j} = \gamma_i \hat{\mathbf{p}}_i + \gamma_j \hat{\mathbf{p}}_j$. The probability density function of the mixed random variable $\hat{\mathbf{p}}_{i,j}$ is $f(\hat{\mathbf{p}}_{i,j}) = \gamma_i f(\hat{\mathbf{p}}_i) + \gamma_j f(\hat{\mathbf{p}}_j)$. If the equal weighting method is used for information fusion, the coefficients are $\gamma_i = \gamma_j = \frac{1}{2}$. The location estimation result $\hat{\mathbf{p}}_{i,j}$ still follows a Gaussian distribution as $(\hat{\mathbf{p}}_i + \hat{\mathbf{p}}_j)/2 \sim \mathcal{N}((\mathbf{p}_i + \mathbf{p}_j)/2, \mathbf{\Sigma}_{i,j})$, where $\mathbf{\Sigma}_{i,j}$ is a diagonal matrix with diagonal entries of $((\sigma_i^x)^2 + (\sigma_j^x)^2)/4$ and $((\sigma_i^y)^2 + (\sigma_j^y)^2)/4$.

The mean square error (MSE) is often used as a characteristic metric to illustrate the accuracy of the result. The MSE of the estimation of $\hat{\mathbf{p}}_i$ is $MSE_i = (\sigma_i^x)^2 + (\sigma_i^y)^2$. Define $(\sigma_i^p)^2 = (\sigma_i^x)^2 + (\sigma_i^y)^2$. The MSE of $\hat{\mathbf{p}}_{i,j}$ is

$$\begin{aligned} M\hat{S}E_i &= \mathrm{E}[||\mathbf{p}_i - \hat{\mathbf{p}}_{i,j}||^2] \\ &= \mathrm{E}[||\mathbf{p}_i - (\hat{\mathbf{p}}_i + \hat{\mathbf{p}}_j)/2||^2] \\ &= \frac{1}{4}||\mathbf{p}_j - \mathbf{p}_i||^2 + \frac{1}{4}(\sigma_i^p)^2 + \frac{1}{4}(\sigma_j^p)^2 \end{aligned} \tag{10.7}$$

where $||\mathbf{p}_j - \mathbf{p}_i||^2$ is the 2-norm of the distance difference, i.e., the biased value of the estimator. The MSE for the initial position estimation result is

$MSE_i = (\sigma_i^p)^2$. The $\hat{\mathbf{p}}_{i,j}$ can be defined as the difference of the MSE value, as

$$\begin{aligned} \delta MSE_i &= \frac{1}{4}||\mathbf{p}_j - \mathbf{p}_i||^2 + \frac{1}{4}(\sigma_i^p)^2 + \frac{1}{4}(\sigma_j^p)^2 - (\sigma_i^p)^2 \\ &= \frac{1}{4}||\mathbf{p}_j - \mathbf{p}_i||^2 + \frac{1}{4}(\sigma_j^p)^2 - \frac{3}{4}(\sigma_i^p)^2 \end{aligned} \tag{10.8}$$

In order to achieve performance gains for user $i$ when using the position of user $j$ for information fusion, the condition $\delta MSE_i < 0$ should be satisfied. Define the performance gain of user $i$ using the position information from $i$ and $j$ as $G_i(i,j) = -\delta MSE_i$. The maximum allowable distance constraint can be shown as

$$||\mathbf{p}_j - \mathbf{p}_i||^2 < 3(\sigma_i^p)^2 - (\sigma_j^p)^2 \tag{10.9}$$

(10.9) means the condition that the performance gains can be achieved by using co-location information fusion. Only if the condition (10.9) is satisfied can two users be called "co-location." If the initial measurement variance of users $i$ and $j$ are approximately the same, i.e., $\sigma_i^p = \sigma_j^p = \sigma^p$, then (10.9) can be simplified as $d_{ij}^p < \sqrt{2}\sigma^p$, where $d_{ij}^p = ||\mathbf{p}_j - \mathbf{p}_i||$ is the relative distance calculated by using the measured GPS position. Since $\sigma^p = \sqrt{(\sigma^x)^2 + (\sigma^y)^2}$, if $\sigma^x = \sigma^y = \sigma$, then $d_{ij}^p < \sqrt{2}\sqrt{2\sigma^2} = 2\sigma$.

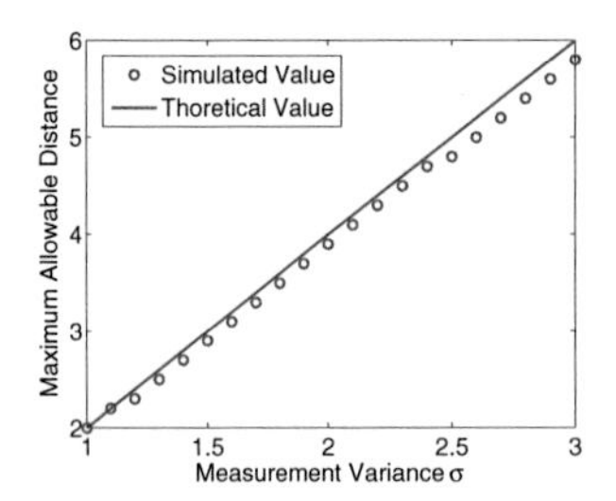


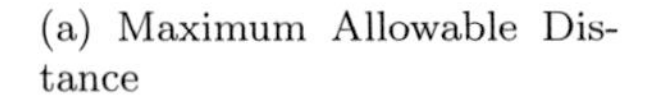
(a) Maximum Allowable Distance

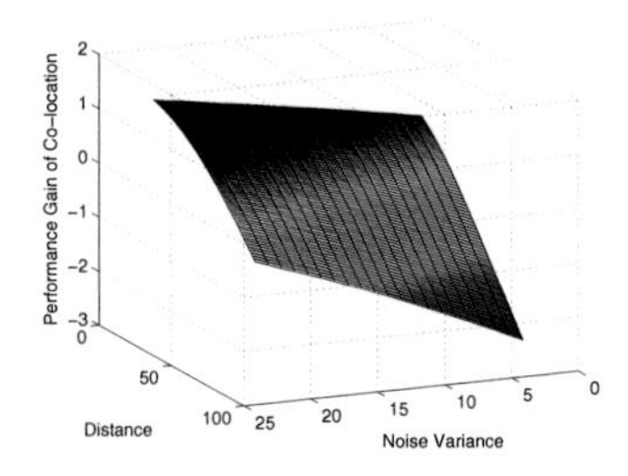


(b) Performance Gains

**Figure 10.4: (a) The relation between measurement variance and maximum allowable distance (peer-to-peer ranging is not necessary). (b) The performance gains with regard to the relative distance and variance.**

The relation between measurement variance and maximum allowable distance when peer-to-peer ranging is not necessary is shown in Fig. 10.4(a); the performance gains with regard to the relative distance and variance is shown in Fig. 10.4(b). Note that $d_{ij}^p$ is different from the ranging measurement $\hat{d}_{ij}$;

$d_{ij}^p$ is obtained by fusing GPS positions of smartphones, while $\hat{d}_{ij}$ is obtained by inter-user ranging. Using the initial measured coarse GPS location information, $d_{ij}^p$ can be estimated. In addition, $d_{ij}$ is the unknown true distance between Node $i$ and Node $j$.

If $d_{ij}^p$ does not meet the constraint of (10.9), then we can call it a *necessary condition* for ranging, since the pair-wise ranging is needed for improving the location accuracy.

## 10.5 Cooperative Location Optimization

If the estimated $d_{ij}^p$ violates (10.9), i.e., meets the *necessary condition* for ranging, then pair-wise ranging should be conducted. To improve the positioning accuracy in this condition, we develop a cooperative localization scheme that leverages relative distances among the smartphones.

### *10.5.1 Sparse Steepest Descent Optimization*

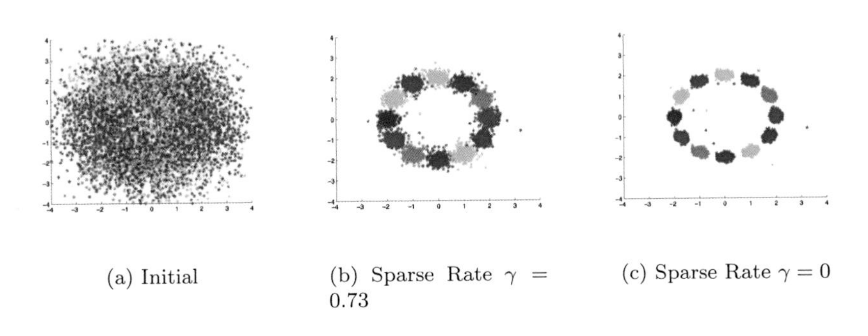

(a) Initial

(b) Sparse Rate $\gamma = 0.73$

(c) Sparse Rate $\gamma = 0$

**Figure 10.5: Numerical results with 12 users under $\sigma = 1$ and $R = 2$: (a) initial positions, (b) refined positions obtained by SSD with $\gamma = 0.73$, and (c) refined positions obtained by SSD $\gamma = 0$.**

With two independent measurements $\hat{\mathbf{p}}_i$ and $\hat{d}_{ij}$ available, the problem can be described so as to refine the position $\hat{\mathbf{p}}_i$ by utilizing the relative ranging information $\hat{d}_{ij}$. Typically, the ranging accuracy of $\hat{d}_{ij}$ is more accurate than the GPS positioning accuracy due to the short distance between users. We use the following neighborhood-based weighted least-squares estimation to improve the positioning accuracy of $\hat{\mathbf{p}}_i, \forall i$, i.e., minimizing the squared error between the calculated distance and the measured distance:

$$\hat{\mathbf{P}} := \arg\min_{\hat{\mathbf{P}}} e(\hat{\mathbf{P}}) = \arg\min_{\hat{\mathbf{P}}} \sum_{(i,j)\in\mathcal{N}_g} \mu_{ij}(||\hat{\mathbf{p}}_i - \hat{\mathbf{p}}_j|| - \hat{d}_{ij})^2 \tag{10.10}$$

where $e(\hat{\mathbf{P}})$ is the total sum of distance errors between all the users, and $\mu_{ij}$ is a weight that is inversely proportional to the variance $\sigma_{ij}^d$. $\hat{\mathbf{P}}$ is a matrix whose columns are $\hat{\mathbf{p}}_i$, $i \in \mathcal{N}_g$, where $\mathcal{N}_g$ is the set of all the collaborators in a neighborhood.

The objective function of (10.10) achieves the minimum value when the total distance calculated by GPS position is equal to the measured distance, i.e., more accurate results of position are achieved at the level of the ranging accuracy. To solve the optimization problem of (10.10), we apply steepest descent method to reduce the error function and calculate the updated version of user position.

Performing the gradient operation $\nabla$ of the error function $e(\hat{\mathbf{P}}) = \sum_{(i,j)\in\mathcal{N}_g} \mu_{ij}(||\hat{\mathbf{p}}_i - \hat{\mathbf{p}}_j|| - \hat{d}_{ij})^2$ with respect to the user $i$ has

$$\nabla_i e(\hat{\mathbf{P}}) = 2 \sum_{(i,j)\in\mathcal{N}_g} \mu_{ij}(||\hat{\mathbf{p}}_i - \hat{\mathbf{p}}_j|| - \hat{d}_{ij})\nabla_i(||\hat{\mathbf{p}}_i - \hat{\mathbf{p}}_j|| - \hat{d}_{ij}) \tag{10.11}$$

where $\hat{d}_{ij}$ is a measurement value, $\nabla_i \hat{d}_{ij} = 0$. $||\hat{\mathbf{p}}_i - \hat{\mathbf{p}}_j||$ represents the distance from $\hat{\mathbf{p}}_i$ to $\hat{\mathbf{p}}_j$, i.e., $||\hat{\mathbf{p}}_i - \hat{\mathbf{p}}_j|| = \sqrt{(x_i - x_j)^2 + (y_i - y_j)^2}$. The gradient of such a distance can be written as $\nabla_i ||\hat{\mathbf{p}}_i - \hat{\mathbf{p}}_j|| = (\hat{\mathbf{p}}_i - \hat{\mathbf{p}}_j)/||\hat{\mathbf{p}}_i - \hat{\mathbf{p}}_j||$. Then (10.11) can be calculated as

$$\begin{aligned} \nabla_i e(\hat{\mathbf{P}}) &= 2 \sum_{(i,j)\in\mathcal{N}_g^+} \mu_{ij}(||\hat{\mathbf{p}}_i - \hat{\mathbf{p}}_j|| - \hat{d}_{ij}) \frac{\hat{\mathbf{p}}_i - \hat{\mathbf{p}}_j}{||\hat{\mathbf{p}}_i - \hat{\mathbf{p}}_j||} \\ &= 2 \sum_{(i,j)\in\mathcal{N}_g^+} \mu_{ij}(1 - \hat{\mathbf{d}}_{ij}^n)(\hat{\mathbf{p}}_i - \hat{\mathbf{p}}_j) \end{aligned} \tag{10.12}$$

where $\hat{\mathbf{d}}_{ij}^n = \hat{d}_{ij}/||\hat{\mathbf{p}}_i - \hat{\mathbf{p}}_j||$ is the normalized relative distance; it also characterizes the difference between measured distance and calculated distance from position. After optimization, $\hat{\mathbf{d}}_{ij}^n$ should approach 1. $\mathcal{N}_g^+$ represents the sparse set that $\hat{d}_{ij} \neq 0$. The relative ranging results between users are not fully available such that some measurements of $\hat{d}_{ij}$ are missing, i.e., $\hat{d}_{ij} = 0$. The sparse property of the distance matrix $\mathbf{D} = \{d_{ij} : (i,j) \in \mathcal{N}_g\}$ causes the performance gains contributed by distance restraint to be not fully available especially when the sparse rate $\gamma$ is high. However, such a sparse feature can be utilized to speed up the processing by using a sparse matrix operation.

After obtaining the gradient function of the error function $e(\hat{\mathbf{P}})$, the new position can be updated by using

$$\hat{\mathbf{P}} := \hat{\mathbf{P}} + \alpha \nabla_i e(\hat{\mathbf{P}}) \tag{10.13}$$

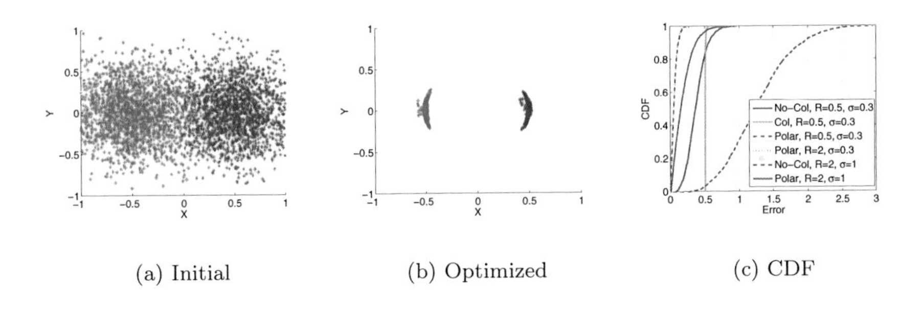


(a) Initial (b) Optimized (c) CDF

**Figure 10.6: The 2-users case of using polar optimization, initial measurement (a), after polar optimization (b) and CDF (c), $\sigma = 0.3$.**

where $\alpha$ is the iterative step size and $\alpha \in (0, 1]$. Eq. (10.13) should be interpreted column-wisely as $\hat{\mathbf{p}}_i := \hat{\mathbf{p}}_i + \alpha \nabla_i e(\hat{\mathbf{P}}), \forall i$ with $\hat{\mathbf{p}}_i = (\hat{x}_i, \hat{y}_i)^T$.

The steepest descent approach is a local optimization method with strong requirement of the initial value selection. However, for our application where GPS position results can be used as the initial value, the overall performance of steepest descent can be guaranteed to provide an optimized value.

### 10.5.2 Weighting Center-Based Polar Optimization

In the previous subsection, the optimized position results are achieved by minimizing the error between $||\hat{\mathbf{p}}_i - \hat{\mathbf{p}}_j||$ and measured distance $\hat{d}_{ij}$. The optimization process utilizes the gradient iteration. Another feasible approach is to assume the measured distance is accurate and replace the true relative distance with $\hat{d}_{ij}$. The weighting center between two users' positions is more accurate than the individual results. Then, update $\hat{\mathbf{p}}_i := f(\hat{\mathbf{p}}_i, \hat{d}_{ij})$ with the relative distance and weighting center.

The relation to the positions of users $i$ and $j$ can be expressed as $d \equiv ||\mathbf{p}_i - \mathbf{p}_j||$. For the measured relative distance $\hat{d}_{ij}$, $d$ can be replaced by $d \triangleq \hat{d}_{ij}$. The initial position measurement $\hat{\mathbf{p}}_i$ follows Gaussian distribution with mean value of $\mathbf{p}_i$. The weighting center of position $\hat{\mathbf{p}}_i$ and $\hat{\mathbf{p}}_j$ is theoretically more stable because random deviation can be canceled out with high probability. Denote the weighting center $\mathbf{p}_{ij}^w = (\hat{\mathbf{p}}_i + \hat{\mathbf{p}}_j)/2$, which can be viewed as more accurate than $\hat{\mathbf{p}}_i$ and $\hat{\mathbf{p}}_j$, where $\mathbf{p}_{ij}^w = (\hat{x}_{ij}^w, \hat{y}_{ij}^w)^T$, $\hat{\mathbf{p}}_i = (\hat{x}_i, \hat{y}_i)^T$, $\hat{\mathbf{p}}_j = (\hat{x}_j, \hat{y}_j)^T$. The angle from the position of node $i$ to node $j$ is estimated as

$$\hat{\theta} = \arctan(y_i - y_j)/(x_i - x_j) \tag{10.14}$$

With the weighting center and $\theta$ available, the node positions $i$ and $j$ can be re-estimated in the polar-coordinate domain. The position of user $j$ can be calculated by transferring the polar-coordinate to Cartesian coordinate by

$$\begin{aligned}\hat{x}_i &:= \hat{x}_{ij}^w + \mathbf{a}_x d/2 \cdot \cos(\hat{\theta}) \\ \hat{y}_i &:= \hat{y}_{ij}^w + \mathbf{a}_y d/2 \cdot \sin(\hat{\theta}) \\ \hat{x}_j &:= \hat{x}_{ij}^w - \mathbf{a}_x d/2 \cdot \cos(\hat{\theta}) \\ \hat{y}_j &:= \hat{y}_{ij}^w - \mathbf{a}_y d/2 \cdot \sin(\hat{\theta})\end{aligned} \tag{10.15}$$

where $\mathbf{a}_x$ and $\mathbf{a}_y$ are the unit vector from the direction of node $i$ to $j$, with the equation as $\mathbf{a}_x = (x_i - x_j)/|x_i - x_j|$ and $\mathbf{a}_y = (y_i - y_j)/|y_i - y_j|$.

For every iteration process, we need to use the measured position results of $\hat{\mathbf{p}}_i$ and $\hat{\mathbf{p}}_j$ to update the weighting centering $\mathbf{p}_w$ and $\theta$. The coefficient of updating is chosen as $(W_m + n - 1)/(W_m + n)$, where $W_m$ is the window length, $n$ is the iteration step. Then, substituting $\mathbf{p}_{ij}^w$ and $\theta$ in (10.15) with new estimated, the optimized position results for node $i$ and $j$ are obtained.

Different from the calculation of (10.11) that performs over all the available nodes of $\sum_{(i,j)\in\mathcal{N}_g^+}$, (10.15) only processes for two users, i.e., users $i$ and $j$. Through performing such pair-wise optimization over the whole sparse set $\mathcal{N}_g^+$, the positions for all the users can be optimized.

## 10.6 Numerical Results

To illustrate the performance gains contributed by the *Coloc* scheme, we conduct Monte Carlo simulation to calculate the error cumulative distribution funnction (CDF) by changing the noise variance of initial position results. The (x, y) coordinates of the positions of twelve users (smartphones) are shown as a scatter figure in Fig. 10.7(a); the positions of each user follow the same two-dimensional Gaussian distribution and are shown by different colors.

The mean CDF curves for twelve users of various approaches and different sparse rates are shown in Fig. 10.8(a) with initial measurement variance of $\sigma = 0.3$. The "MA" represents the conventional moving average method used for the initial measurements, while "SSD" represents our proposed sparse steepest descent optimization approach. Even when the ranging sparse rate is very high ($\gamma = 0.73$), i.e., only several ranging pair measurements are utilized, the performance superiority over "MA" is still sufficient. Another interesting point lies in there being no apparent performance degradation when sparse rate is lower than $\gamma = 0.4$. Such a property can help reduce the overall ranging costs and delay while maintaining desired performance gains.

The performance of the *Coloc* scheme using ranging information can even be improved when our proposed sparse steepest descent optimization and weighting center-based polar optimization are combined together. Since polar-based optimization is performed for two users, i.e., in a local way, we execute the polar method after the global SSD approach. The measurement results are shown in Fig. 10.7. "Initial" is the initial position measurement; the "SSD"

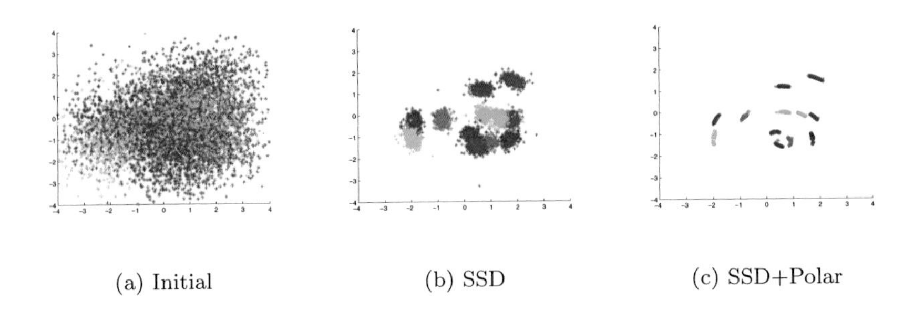

(a) Initial (b) SSD (c) SSD+Polar

**Figure 10.7: Numerical results with 12 users under $\sigma = 1$ and $R = 2$ (x and y-coordinates are in meters): (a) initial positions, (b) refined positions obtained by SSD, and (c) refined positions obtained by SSD+Polar.**

case uses our proposed sparse steepest descent optimization; "SSD+Polar" uses the polar optimization after the SSD processing. The CDF figure is shown in Fig. 10.8(b). We can know that using polar and SSD optimization, the performance gains are larger than when using the conventional moving average method. When combining SSD and polar together, the performance can even be improved, as shown in Fig. 10.8(b).

# 10.7 Experimental Validation

## *10.7.1 System Implementation*

We designed a mobile social network from the ground up to enable social crowd sensing, to improve the location services as shown in Fig. 10.9. With a better-designed mobile social network app or framework that could be embedded in other apps, users could easily invite their friends to hang out and enable crowd sensing for localization. The dedicated mobile social network make the social collaboration process more convenient. In terms of usage, we make the cooperative sensing process transparent to users, and perform all the tasks in the background when necessary.

As shown in Fig. 10.9, users can create location groups and invite friends to join. Nearby friends could be scanned via BLE beacon and invited to cooperative localization, tracking, and sharing locations. Only friends that accepted your request could share locations with you and let you do tracking and free-ride. The BLE RSS is continuously monitored between you and your friends.

Using location journaling application as an example. Fig. 10.10 is the log trace analysis result using Apple Xcode Instrument for a fifteen-minute workout. Before cooperation, recording location trace continuously consumes a lot of energy and CPU cycle, i.e., the energy usage and CPU activity are high.

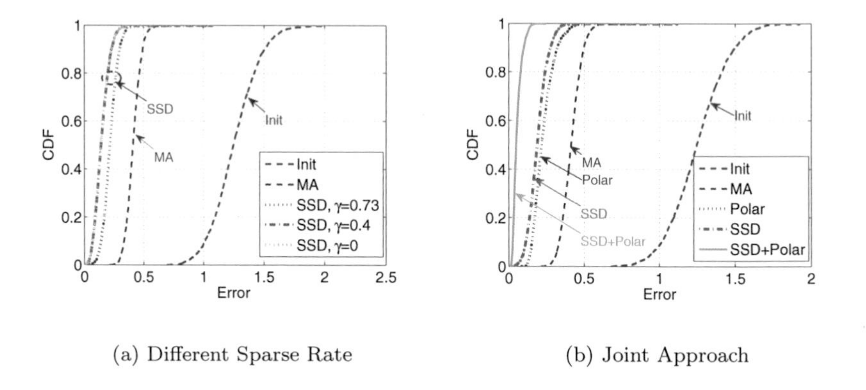


(a) Different Sparse Rate

(b) Joint Approach

**Figure 10.8: The CDF of location accuracy under various processing types: 1. Using SSD with different sparse rate of ranging; 2. using joint optimization approaches of SSD and Polar.**

When performing cooperative localization, the GPS startup time period could be significantly reduced, and only launches GPS when necessary. Under similar location trace accuracy, the energy usage and CPU activity after cooperation is reduced, as shown in Fig. 10.10. If recording the location for longer time, the improvement in terms of energy savings is more apparent.

## *10.7.2 Experiment Setup*

We conducted experiments by using smartphones to collect location data, and validated our proposed cooperative localization technique by using these real measured results. The data is collected by using Apple iOS smartphones (iPhone4, iPhone4S, and iPhone5 are used in the experiment). The location is obtained via the default module in iOS location management, which combines the GPS and WiFi location results. Two cases of situations are tested: stationary situation for accuracy test; moving situation for tracking and dynamic performance test.

## *10.7.3 Case Study I: Stationary Users*

To evaluate the performance of our proposed cooperative location optimization approach for multi-users in real environments, we conducted measurements for nine users with random positions in a campus environment. The initial measurement results are shown in Fig. 10.11(a). Different from the simulation results, the obtained GPS results show strong correlation among adjacent measurements. That's also why lowering the GPS update rate is possible, to save energy without sacrificing the accuracy. For convenience, we

**Figure 10.9: The designed mobile social network.**

denote "Init" as the initial position results; "Col" is the result obtained by only utilizing the co-location information without ranging; "Polar," "SSD," and "Polar+SSD" are our proposed schemes that uses the ranging-based information for collaboration. We follow the same terms/notations used in Section 10.6.

We applied the normalized cuts algorithm to the affinity matrix corresponding to Fig. 10.11(a), and obtained the clustering results of four clusters as shown in Fig. 10.12(a), i.e., the positions with large similarity measures are grouped together. By clustering nine users into four clusters with co-location, we can perform location fusion without relative ranging. This approach is labeled as "Col." The CDF results of using different algorithms are shown in Fig. 10.12(b). We observe that the conventional moving average "MA" approach does not show performance improvement over the initial position results due to the dependency between adjacent measurements. By clustering nine users into four clusters with co-location, the location accuracy of "Col" is much better than "MA" as shown in Fig. 10.12(b).

If the relative distance information can be obtained, the accuracy can even be improved by using "Polar" and "SSD" approaches. The scatter figure of using "SSD" is shown in Fig. 10.11(b). After performing joint optimization of "SSD+Polar," the more accurate results are shown in Fig. 10.11(c). By

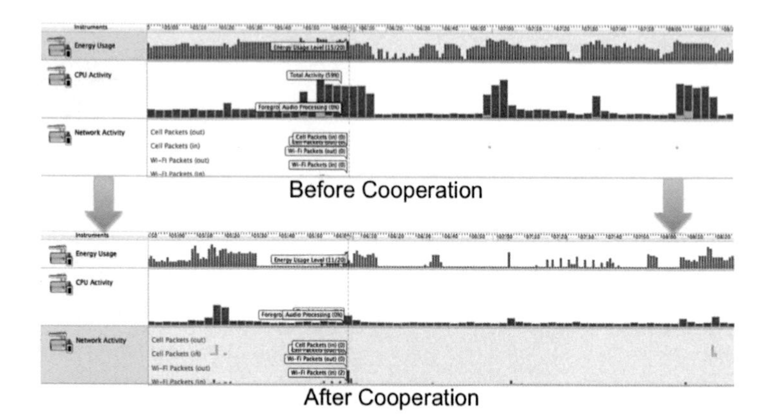


**Figure 10.10: The comparison of energy usage, CPU activity, and network activity before and after cooperation.**

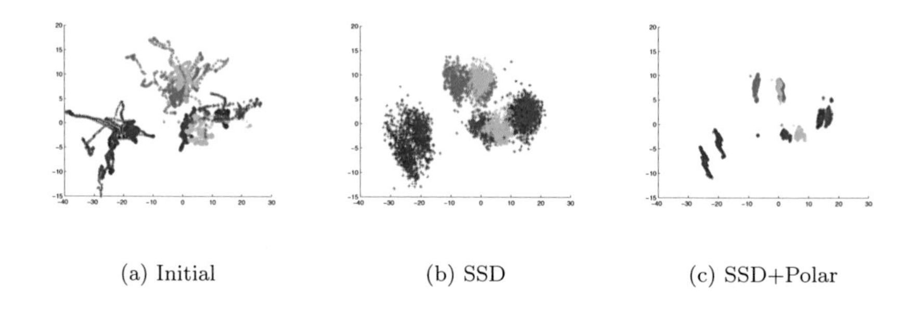

(a) Initial (b) SSD (c) SSD+Polar

**Figure 10.11: Experimental results with 9 users (x and y-coordinate is in meters): (a) initial positions obtained by GPS, (b) refined positions obtained by SSD, and (c) refined positions obtained by SSD+Polar.**

comparing the results to the initial measurement results in Fig. 10.11(a), the performance improvement of using "SSD+Polar" is significant.

From the statistical results of Fig. 10.12(b), and using 80% probability as an example, the initial GPS accuracy is around 5m. After fusing the co-located users without ranging, the achieved accuracy is about 4m. When using our proposed ranging-based optimization approach "SSD+Polar," the positioning accuracy is approximately 1.2m.

### *10.7.4 Case Study II: Moving Users*

To evaluate the performance improvement of the *Coloc* scheme in moving scenarios, we conducted experiments for moving users. Users carry GPS-enabled

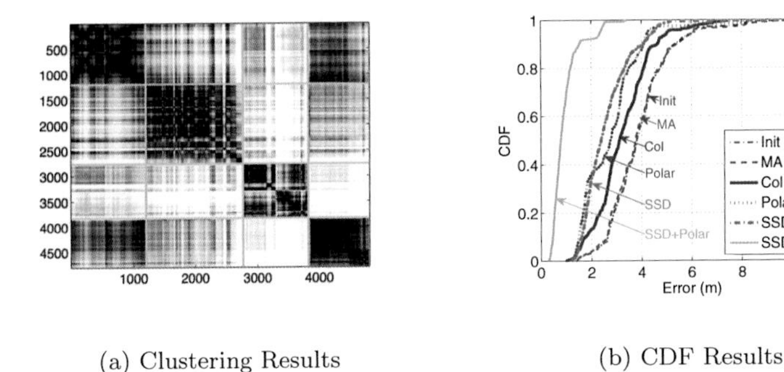


(a) Clustering Results

(b) CDF Results

**Figure 10.12: 1. The clustering results for the 9 users; 2. The CDF of location accuracy under various processing types.**

smartphones and perform cooperative localization with peers when walking in a campus parking lot. The GPS update time interval is $t_G$; we use lower $t_G$ when applying the *Coloc* scheme to save energy. Fig. 10.13(a) shows the initial measurement of a GPS trajectory of 4 users when walking around a parking lot, where $t_G = 0.997s$. Using low-update GPS data ($t_G = 1.994s$), and after performing our proposed "SSD" approach, the devision of the GPS trajectory has been greatly suppressed as shown in Fig. 10.13(b). After apply the "Polar" approach in addition to "SSD," the trajectory is more smooth as shown in Fig. 10.13(c), which is much better than the initial high update rate data. To test the effectiveness of the *Coloc* scheme when users are walking in two separate groups with certain distance, we conduct an experiment by letting three users form a group and walk in parallel with another user. The walking traces of these four users are shown in Fig. 10.13(a). After "SSD" and joint "SSD" and "Polar" optimization, the accuracy of walking traces improved significantly as shown in Fig. 10.13(b) and Fig. 10.13(c). These results demonstrate the energy efficiency and accuracy of our proposed *Coloc* scheme.

## 10.8 Related Work

Optimizing the GPS localization goes a long way back — more than one decade, from improving the RF component design, signal processing, ranging, and localization algorithm, to a differential GPS system, and assisted-GPS [40, 115]. However, the inherent complexity of the localization problem makes the further improvement of the GPS system hard to achieve.

When smartphones became an important personal companion, researchers proposed to use other auxiliary sensors embedded in a smartphone to improve the accuracy of GPS. Hybrid approaches have been proposed to balance the

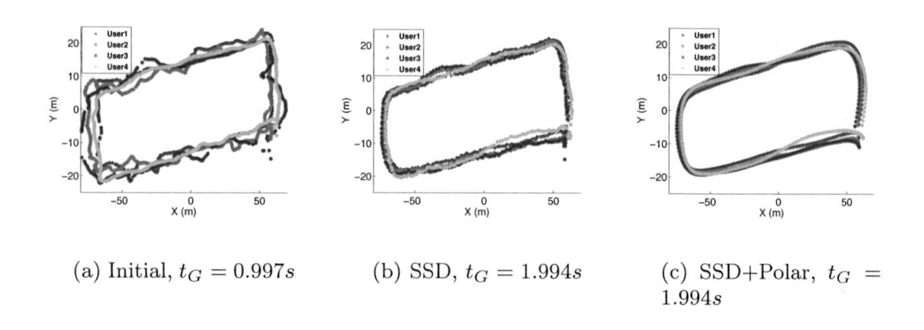


(a) Initial, $t_G = 0.997s$ (b) SSD, $t_G = 1.994s$ (c) SSD+Polar, $t_G = 1.994s$

**Figure 10.13: Experiment results of 4 user's GPS trajectory when walking around a parking lot.**

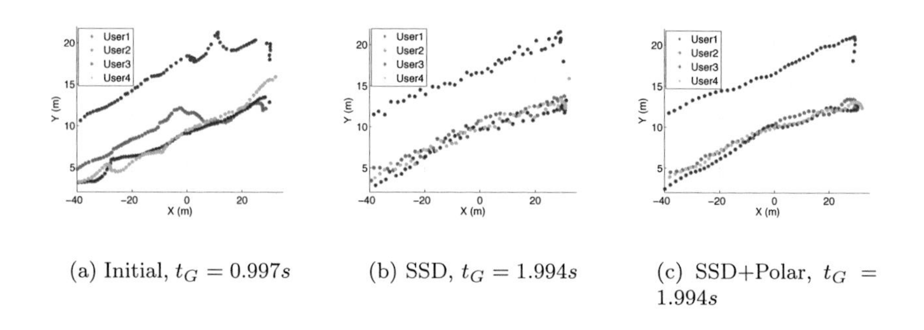


(a) Initial, $t_G = 0.997s$ (b) SSD, $t_G = 1.994s$ (c) SSD+Polar, $t_G = 1.994s$

**Figure 10.14: Experiment results of 4 user's GPS trajectory when walking along a line.**

power and accuracy of GPS, e.g., using WiFi fingerprinting, or an accelerometer [50, 20].

Recent approaches that use the microphone sensor in a smartphone for accurate ranging demonstrate a practical way for achieving accurate auxiliary measurements [52, 74]. H. Liu et al. [52] improved the accuracy of WiFi-based localization subject to ranging constraints. The problem is that the error of WiFi is even larger than the maximum ranging distance of the acoustic signal; the performance gains contributed by peer-wise ranging would be limited. The CDF results demonstrated in [52] only show improvement in overall error (most contributed by reduced bias), but the slope (determines the resolution) remains the same after their peer-assisted localization approach. Nandakumar et al. [74] utilized the acoustic signal transmitted by desktop to assist the WiFi localization, however, the authors do not consider the mobile situation, which would limit their application in real scenarios. These two approaches also suffer slow update time for the localization ($> 7s$) due to the time-divided multiple pair-wise ranging and the inherent low transmission speed of the acoustic signal. For $N$ peers, a total of $N(N-1)/2$ ranging pairs need to be measured, and resulting at least $N(N-1)$ times acoustic signal transmission for two-way ranging mode. Reducing the ranging complexity and improving the performance gains of the location optimization algorithm are the two key challenges.

## 10.9 Conclusions

To address the positioning inaccuracy and power inefficiency of current smartphone localization, in this chapter we proposed a social-aided *Coloc* scheme. Specifically, we use neighborhood-based weighted least-squares estimation when relative distances between smartphones are available. The energy efficiency is achieved by sharing location information among co-located users and lower the GPS update rate. Numerical and experimental results conclusively demonstrate that our proposed cooperative localization schemes can achieve considerable performance gain in both indoor and outdoor environments. In the experiments of nine users with random positions, the positioning accuracy of our scheme was 1.2m with a confidence level of 80%. In contrast, a regular GPS receiver has an accuracy of 4.7m. The optimized GPS trajectory also demonstrates the effectiveness of the *Coloc* scheme for tracking moving targets.

# *Chapter 11*

# Hiding Media Data via Shaders: Enabling Private Sharing in the Clouds

**Kaikai Liu**

*Assistant Professor, San Jose State University*

**Xiaolin Li**

*Associate Professor, University of Florida*

## CONTENTS

## 11.1 Introduction

In the era of cloud and big data, the sharing and storing of massive quantities of various types of data is prevalent. Our smartphones can store the transmitted/received media messages in ever-expanding social networks, e.g., Facebook, WhatsApp, and iMessage. Our photos, videos and other media data are stored and synced in media storage services like Flickr and YouTube. When enjoying the disruptive reform of cloud services via all kinds of free storage and message channels, you may be aware that your private data are assaulted from all sides. Attracted by the big opportunities of monetizing your data, the Internet's big brothers have hoarded your personal data and sold it for billions. Even with some cloud service providers who really respect users' privacy, eavesdropping or hacking may still happen, e.g., 2014 celebrity photo leakage from Apple iCloud. While some service providers claim they have deleted your data, e.g., Snapchat, hackers can still steal 100,000 sensitive photos in "the Snappening." Due to the potential leakage of sensitive information, many users are reluctant to share/store their data to clouds.

Pressure from users' privacy awareness following these leakage events are encouraging many more privacy-preserving behaviors, products, and solutions. Using the law to sue these service providers is one approach. There are also technical approaches that could help you to prevent the leakage. The direct solution is encryption, i.e., encrypt all of our data. The advanced encryption standard (AES) is widely expected to become the *de facto* standard for encrypting all forms of electronic data. However, encryption could not solve all the problems due to its inherent limitations.

Most importantly, encryption is typically incompatible with existing clouds and mobile clients; e.g., YouTube, Flickr, Facebook, and Google+ do not allow you to upload media in an encrypted format. Alternatively, format-compliant and low-complexity encryption techniques have become popular for media data. Authors in [88, 105] proposed encryption approaches along with the standard JPEG compression. The basic idea is to perform encryption and JPEG compression jointly, so that the photo is perceptually encrypted but still in JPEG format. Thus, the encrypted photo is acceptable to existing social networks or cloud services, while the true content is invisible to eavesdroppers. However, joint encryption with compression has the drawback of format limitation, e.g., it is only suitable for the JPEG format. Another practical problem is that these approaches need to break down the existing hardware-accelerated codecs for compression and decompression, which is hard to access and inefficient to re-implement via software. Encryption after compression via selective binary processing could maintain the format [119]. However, this scheme requires fairly deep parsing into the bit stream to identify necessary parts of the bit stream to be encrypted [51], resulting in significant processing overheads [136].

Placing the encryption before the compression stage can be inherently compliant to the syntax format without bit stream identification because the image/video is processed before the compression process. Encryption approaches via chaotic mapping belong to this category [124, 122]. However, one significant challenge is the loss of correlation between continuous video frames when disordering the inherent redundancy, resulting in lower compression ratios with high communication cost. Further, a small amount of pixel error of encrypted image during transmission would make the data undecryptable even with the same key.

We propose a low-complexity privacy-preserving scheme for big media data with the following salient features: 1) Format-compliant and compression-independent, putting the encryption before compression without modifying the existing efficient hardware-accelerated codecs; 2) correlation-preserving and transmission error robustness, minimizing the communication overhead and risk; 3) low-complexity and easy programming interface, leveraging hardware-acceleration without significant modification of existing software stacks.

## 11.2 System Model

### 11.2.1 Problem Statement

Media data sharing is becoming popular on mobile devices due to multitudes of advantages of the cloud services, such as the cloud storage and flexible multimedia message. However, this paradigm shift results in the loss of control over media data as well as new security and privacy issues. Users' privacy fully depends on the provider's reliability and security guarantees. Cloud storage providers may assure the data privacy via database encryption. Mobile device providers may guarantee the security level of local data on disk.

However, cloud storage, service, communication network, and mobile device are provided separately by different vendors. Existing protection approaches have not covered the full path from end to end. When the privacy leakage happens, it is hard to know which part cause the problem. This is in opposition to the data protection requirement that customers know where and what happens to their data.

### 11.2.2 Threat Model and Assumption

In this paper, we focus on three types of threats to privacy that result from sharing users' media data via the cloud. 1) The fragility of local software and hardware on mobile clients. Users' devices can be stolen or accessed. The local data on disk can be accessed by others via the software bugs or device jail-breaking. 2) The eavesdropping on the communication links, e.g., unprotected WiFi. The communication path is very complex from the mobile device to the cloud server with multiple independent physical links and vendors. Security risks that threaten the communication include eavesdropping, interception, man-in-the-middle attacks, and DNS spoofing. 3) The untrustable cloud service providers. They may leak users' privacy, either by commission or omission. Inadequate storage protection, attacking or hacking the server, and unauthorized server access are all possible ways of privacy breach.

### 11.2.3 Compatible with Existing Cloud Services

The key point we keep in mind is that we cannot overthrow the existing cloud backend and mobile client, which is the essential part of the current media-rich social networks. Few will like privacy-preserving products with less functionality; few will try a new product that is completely incompatible with their existing software or subscribed services. For example, users are lured to use free terabyte-level photo storage services like Flickr, although they are still concerned with the privacy of their posted photos even with a "lock" icon; users are attracted to sharing their videos and photos with their families or friends, but still concerned about eavesdropping or privacy leakage.

The goal of this chapter is to design a full privacy protection scheme that enables users to protect the privacy of their media data in a transparent way, while still benefiting from the existing cloud services, e.g., Flickr, Facebook, YouTube. One possible solution is to keep the media format with perceptual encryption, i.e., format-compliant.

### 11.2.4 Format-Compliant and Compression-Independent Solutions

Privacy-persevering algorithms should not affect the operation of image/video codecs, and the need to preserve the format. Mobile devices usually apply a standardized video codec with hardware acceleration for image/video compression and decompression, which is often packed into a stand-alone module and not accessible to developers. Standardized video codec technologies like MPEG-1, MPEG-2 (ISO/IEC, 2000), MPEG-4, and H.26X AVC (advanced video coding) are widely applied in current mobile devices for efficiently transmitting/storing videos over bandwidth-limited networks via compression. Algorithms with joint compression and encryption are proposed to achieve security/privacy and compression rate at the same time [88]. However, these approaches need to break down the existing hardware-accelerated codecs, which are hard to access, and it is inefficient to realize all the standard formats in current mobile devices. Thus, *compression-independent* algorithms are preferred to joint compression and encryption approaches, with no need to modify the existing codec.

## 11.3 System Design

### *11.3.1 Design Objective*

Compared with text or small binary data, media data is characterized by a number of peculiarities, such as large data size, various resolution/rate versions, real-time requirements for video, the use of standardized hardware-accelerated codecs, high compression-rate requirement, standardized data formats, and user-specific privacy requirements. These peculiarities raise a couple of specific requirements for privacy-persevering techniques.

**Hardware Acceleration:** Leveraging the existing cloud and mobile solutions, we propose approaches that perform chained media transform via the OpenGL Shader. The superior of the Shader is its full parallel computation architecture leveraging GPU. We utilize the Fragment Shaders in the OpenGL rendering pipeline (after Rasterizer) for the pixel manipulation required in our proposed algorithm. The size covered by a fragment is related to the pixel area. The current Shader program could perform pixel-level computation at full frame rate.

**Keep Existing Code Intact:** Although GPU is faster, it is still not compatible with existing solutions in the mobile client. Most of the apps are built upon normal picture and video framework that does not have OpenGL canvas. Our solution is to build our transform-based media cipher as an image filter. Developers only need minimum change to their existing code base, i.e., add one filter to the photo or video.

### *11.3.2 Proposed Approach*

The proposed privacy protection architecture is shown in Fig. 11.1. In terms of encryption algorithm, our objective is to jointly achieve format-compliance, compression-independence, and correlation-preserving. However, it seems that these goals conflict with each other, and there is no solution yet available that could meet all these criteria for our application demand. Putting the encryption before compression, e.g., image scrambling and permutation approaches via chaotic map, could be format-compliant and compression-independent, but inherently conflict with correlation-preserving.

We propose a privacy-preserving scheme based on encryption before compression, and solving its inherent problems of correlation loss and transmission error sensitiveness. We also propose an additional feature that allows users to choose arbitrary photos as the encryption keys instead of remembering a long password. The algorithm detail will be illustrated in Section 11.4.

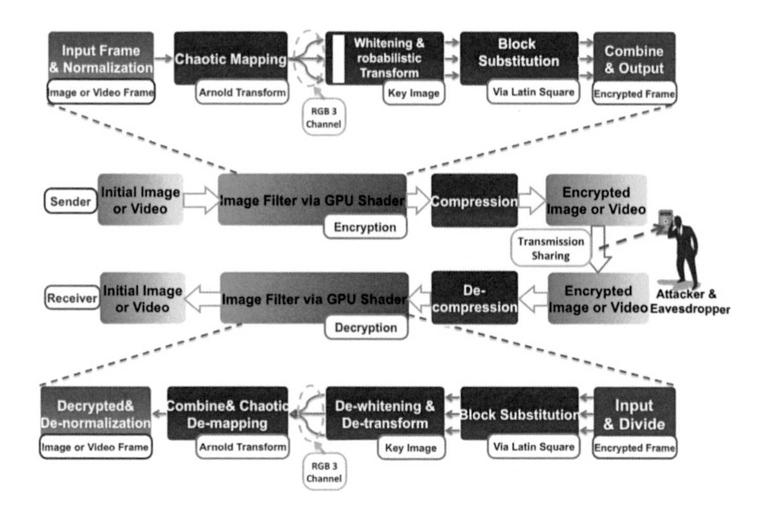


**Figure 11.1: System Architecture.**

# 11.4 Privacy Protection Algorithm Design

## *11.4.1 Design Principles*

The proposed privacy-preserving scheme puts encryption before compression, and jointly achieve format-compliance, compression-independence, and correlation-perserving via three core steps: 1) Chaotic Mapping; 2) Whitening; 3) Block Substitution.

Unlike traditional Chaotic Mapping approaches [122], we utilize generalized low-order chaotic mapping to minimize the correlation loss, and make the decryption process insensitive to the transmission noise of the encrypted media. The reduced security level is compensated by the Whitening and Block Substitution stages.

The Whitening process is different from the traditional random noise-based approach. Instead of using a random bit map for XOR operation [51], we allow users to select any photo as the random key map. The XOR process is generalized from the bit operation to image pixel operation. This step enhances the security level by using the users' own photo as a perceptual key instead of a long password.

The Block Substitution stage further enhances the security level via random-like pre-defined patterns.

## 11.4.2 Chaotic Mapping

### *11.4.2.1 Pros and Cons of Chaotic Mapping*

Among existing encryption methods, the chaos map-based image encryption method is a family of methods that are good for encryption purposes due to its high sensitivities to its parameters and initial values, the mixing property and the ergodicity [122]. Chaos-based image encryption methods have been researched for years since Fridrich's first approach for digital image encryption [26]. Many existing chaotic image encryption algorithms have good cryptographical properties, but they also have defects when compared with normal cryptosystems. The most significant problem when applying the chaotic mapping in our proposed application is its round-off error, which may make the decrypted image/video blurred especially after multiple rounds or higher orders. The nonreversible problem is caused by the real number definition of chaotic systems, i.e., infinite scale, while a normal cryptosystem is defined on finite numbers. The round-off error in real number quantization is small but intolerable for current high definition image/video requirements.

To deal with the round-off problem, several nonchaotic image encryption methods were researched by using various random-like patterns [124]. Although these nonchaotic approaches eliminate round-off errors by using random-like patterns, their confusion and diffusion properties are not good enough. To overcome the round-off problem of chaotic image encryption and improve the confusion and diffusion properties of the nonchaotic approach, we propose a chained solution under the guideline of the Markov cipher by using generalized low-order chaotic mapping and multi-channel block substitution via a nonchaotic random pattern.

### *11.4.2.2 Generalized Low-Order Chaotic Mapping*

Define the chaotic mapping as $\Gamma$. Using $\Gamma$ for image encryption, it should be invertible. There are lots of chaotic maps available, e.g., Arnold's cat map, Baker's map, logistic map, tent map. Using the Arnold's cat map as an example, the transform of $\Gamma$ could be written as $\Gamma : (x, y) \rightarrow (2x+y, x+y) \bmod N$, where $N$ is the pixel dimension. The initial coefficient of the Arnold mapping is $a_r = [2 \quad 1; 1 \quad 1]$; its higher order $O$-th could be written as $a_r^O$. We encrypt the image frame with different orders of Arnold transform. The encrypted images are shown in Fig. 11.2. It is clear to see that higher order contributes a better encryption property with higher randomization. However, higher orders show pixel blur due to the amplification of the round-off error as stated before.

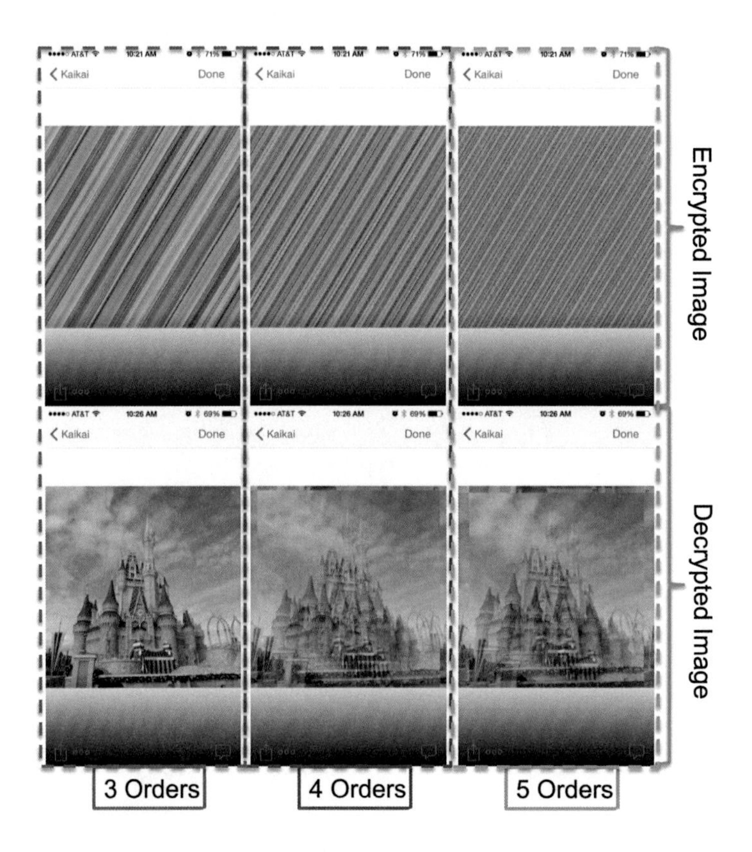


**Figure 11.2: Encrypted and decrypted image under different orders of chaotic map.**

To minimize the round-off error, a low-order chaotic map should be utilized. However, a low-order chaotic map has a poor encryption property. Another problem for existing chaotic maps is the limited unknown parameters, i.e., coefficients for a chaotic map is well defined. Using the coefficients and initial values as the encryption key is not secure enough.

Our solution is to generalize the chaotic map by maintaining its basic properties, and eliminate other unused properties, e.g., periodic property (returning to its original state after a number of steps). The basic requirement for the chaotic mapping $\Gamma$ is $\det(\Gamma) = \pm 1$ and therefore its inverse has integer entries, where $det\cdot$ calculates the determinant of the matrix. The initial coefficient of the Arnold mapping is $[2 \quad 1; 1 \quad 1]$, which is well known and less secure. By generalizing the Arnold mapping, we could have $\Gamma = [a \quad b; c \quad d]$, where $ad - bc = \pm 1$. The first few orders of $\Gamma$ also satisfy our requirements.

### 11.4.3 Whitening and Probabilistic Transformation

In cryptography, key whitening is a common technique to increase the security level of a cipher by mixing the data with the key, e.g., in DES and AES [124]. The most common form of whitening is using XOR (exclusive or) bit operation between a plaintext message and a key. For image data where each pixel is bytes instead of the bit, typical solutions could include iteratively applying the XOR to each bit from the most significant bit (MSB) to the least significant bit (LSB). However, such XOR operation via multi-iteration becomes inefficient for a color image.

As stated in [124], the XOR operation could be generalized to transposition in the finite field by $y = (x + s) \bmod 255$, where $x$ is the input, $y$ is the output, and $s$ is the corresponding byte in the key matrix. Instead of using generated matrix as the key, we provide a scheme that could allow users to define the key matrix by just selecting their own image/photo. Using the user's own image data could result in a very long key and help users to remember their own passwords easily. Motivated by using a personal photo as a key, we have the pixel-level encryption process as

$$C(i,j,k) = (RF(I(i,j,k)) + RF(\hat{S}(i,j,k))) \bmod 255 \tag{11.1}$$

where $RF(I(i,j,k))$ and $RF(\hat{S}(i,j,k))$ are the transformed input image and key image, respectively. The transformation process depends on the value of $k = 1, 2, 3$ for the color space. $RF(\cdot)$ could flip the row up to down, flip the column left to right, or perform matrix transposition. The selection of all these transformations is determined by a random probability $\rho \in [0, 1]$. $RF(\cdot)$ could be written as

$$RF(I(i,j,k)) = \tag{11.2}$$
$$\begin{cases} I^T(i,j,1) + F_x(I(i,j,2)) + F_y(I(i,j,2)) & \text{if } \rho \in [0, 0.1] \\ F_y(I(i,j,1)) + F_x(I(i,j,2)) + I^T(i,j,3) & \text{if } \rho \in (0.1, 0.2] \\ \cdots & \text{if } \rho \in (0.2, 0.3] \end{cases}$$

Where $F_x(\cdot)$ is flip the column left to right; $F_y(\cdot)$ is flip the row up to down. The determined $\rho$ is also one part of the key and should be transmitted to the receiver side. The added transformation of $RF(\cdot)$ is to improve the randomness of image whitening, and make it perceptually unrecognizable.

### 11.4.4 Multi-Channel Block Substitution via Latin Square

Non-chaotic mapping approaches via random-like pattern in the integer domain could eliminate the round-off error in chaotic approaches. However, the drawback lies in its weak random properties that would not guarantee a sufficient security level, i.e., do not have good confusion and diffusion.

In this subsection, we propose to utilize the Latin square as the random-like pattern, and enhance its security level via these approaches: 1) multi-channel processing in the color domain; 2) normalized block substitution for scale-independence.

**Latin Square.** For dimension $N$, define the symbol set as $\mathcal{S} = \{S_0, S_1, \cdots, S_k, \cdots, S_{N-1}\}$, the indicator function of Latin square is $f_{Latin}(i,j,k)$. A Latin square $L(i,j)$ is a square matrix that each symbol in $\mathcal{S}$ appears exactly once in each row and each column, where the indicator function is $f_{Latin}(i,j,k) = 1$ when $L(i,j) = S_k$; otherwise $f_{Latin}(i,j,k) = 0$. For each row, we have $\sum_{i=0}^{N-1} f_{Latin}(i,j,k) = 1$; for each column, $\sum_{j=0}^{N-1} f_{Latin}(i,j,k) = 1$, which implies that symbol $S_k$ is occurred once in each row and column.

**Multi-channel and Normalized Block Substitution.** In general, the permutation procedure is to find a bijective mapping for bit/pixel. The property of the Latin square matrix is a permutation of the integer number sequence of length $N$, which naturally becomes a one-to-one bijective mapping for pixel locations. Assume the input image frame is $I(i,j)$ with size as $N_r \times N_c$. The Latin square size is $N \times N$. To make the proposed approach suitable for various frame sizes, we perform normalization for the input frame $I(i,j)$ via $\mathcal{N}\{I(i,j)\}$. For pixel $((i,j))$ in $I(i,j)$, the new indexes for the row and column after normalization are $\bar{i} = i/N_r \in [0,1]$ and $\bar{j} = j/N_c \in [0,1]$, respectively.

For color images, each pixel of $I(i,j)$ or $I(\bar{i},\bar{j})$ is a tuple (red, green, blue) on a 0 to 255 scale. For better randomization, we divide the tuple into three independent channels as $I_{red}(\bar{i},\bar{j})$, $I_{green}(\bar{i},\bar{j})$, and $I_{blue}(\bar{i},\bar{j})$, Latin square substitution could be performed for each color space with different Latin squares ($L_{red}$, $L_{green}$, $L_{blue}$) as keys.

For each substitution via Latin square, the row and column could be performed independently. Using column substitution for the red channel as an example, the new obtained pixel position could be written as

$$\bar{j}^s_{red} = f^S_j(L_{red}, \bar{i}, \bar{j}) = L_{red}(\lfloor \bar{i} \times N \rfloor, \lfloor \bar{j} \times N \rfloor) + \bar{j} - \lfloor \bar{j} \times N \rfloor \tag{11.3}$$

where $\lfloor \bar{i} \times N \rfloor$ and $\lfloor \bar{j} \times N \rfloor$ calculate the block position in the Latin square matrix ($N$) of the current normalized pixel $(\bar{i},\bar{j})$; $\bar{j} - \lfloor \bar{j} \times N \rfloor$ calculates the remainder pixel position within the block for column. The row substitution could be written as

$$\bar{i}^s_{red} = f^S_i(L_{red}, \bar{i}, \bar{j}) = L_{red}(\lfloor \bar{i} \times N \rfloor, \lfloor \bar{j} \times N \rfloor) + \bar{i} - \lfloor \bar{i} \times N \rfloor \tag{11.4}$$

(11.3) and (11.4) combined and performed for the three RGB channels with different Latin squares could be the forward encryption process via random-like pattern substitution. The input pixel $(\bar{i}, \bar{j})$ of (11.3) and (11.4) constructs the plaintext of the input image frame in red space, where the output $(\bar{i}^s_{red}, \bar{j}^s_{red})$ is the pixel of the ciphertext in red space.

The inverse process for $(\bar{i}^s_{red}, \bar{j}^s_{red})$ in the row and column direction, i.e., the encryption, could be written as

$$\bar{i} = \arg \max_{z=\lfloor \bar{i}^s_{red} \times N \rfloor} (f_{Latin}(\bar{i}^s_{red}, z, \bar{j}^s_{red})) + \bar{i}^s_{red} - \lfloor \bar{i}^s_{red} \times N \rfloor \tag{11.5}$$

$$\bar{j} = \arg \max_{z=\lfloor \bar{j}^s_{red} \times N \rfloor} (f_{Latin}(z, \bar{j}^s_{red}, \bar{i}^s_{red})) + \bar{j}^s_{red} - \lfloor \bar{j}^s_{red} \times N \rfloor$$

where $f_{Latin}(\cdot)$ is the indicator function of Latin square $L_{red}$. The element of $f_{Latin}(i, j, k)$ is most zero with only one '1' element ($f_{Latin}(i, j, k) = 1$ when $L(i, j) = S_k$), i.e., the maximum is equal to 1. This process implements the inverse searching over the Latin Square and decrypts the image pixel row and column location, and recovers the plaintext from the ciphertext.

### 11.4.5 Security Analysis

To ensure security level, the cipher scheme should be robust enough for a brute-force attack, or exhaustive key searching. Hackers may crack any encrypted image file by searching all possible keys in the key space until the right key is found. The key length plays an important role in an encryption system, and determines the feasibility of attack. Bigger key space means more difficulties in terms of time complexity.

The proposed encryption process contains three parts: 1) Chaotic Mapping, where the total available mapping number determines the security level; 2) Whitening and Probability Transformation, where the key image and the number of transformations are the key for the high security; 3) Latin Square, where the size of the Latin square matters. The low-order chaotic mapping number is in the range of $p_1 = 100$; the number of transformations is in the range of $p_2 = 10$, while the size of the Latin square is $256 \times 256$. The total number of these spaces is near $p_t = 256! \times 256! \times 100! \times 10!$. In addition to the encryption key space, we also utilized the custom image as the key with a size of $256 \times 256$. Combing the $p_t$ and the custom image key, the key space becomes very large and cannot be easily attacked via brute force.

## 11.5 Evaluation

### 11.5.1 System Evaluation

#### 11.5.1.1 Media Sharing Process

Fig. 11.3 shows our designed secure media sharing process to open social media channels. We leverage the image key in addition to the normal key for better security. To meet the design objective of easy-to-use and low-complexity computation, we integrate our proposed privacy-preserving techniques into one customized image filter. The image filter works in the raw image domain and does not require image format compliance. A highly integrated block could simplify the integration process to existing code. To improve the efficiency for the pixel-wise computation in our approach, we design and implement this customized image filter in the GPU via the OpenGL Shader. Shader is a program designed to run on some stage of a graphics processor, and written in the OpenGL Shading Language. We utilize the Fragment Shaders in the OpenGL rendering pipeline (after Rasterizer) for the pixel manipulation required in our proposed algorithm. The size covered by a fragment is related to the pixel area. Thus, the computationally intensive pixel-by-pixel operation could be converted to fragment processing with highly paralleled implementation in GPU. The reason that we utilize normalization and block-based processing in algorithm design is to fit the GPU Shader processing framework for high computation efficiency.

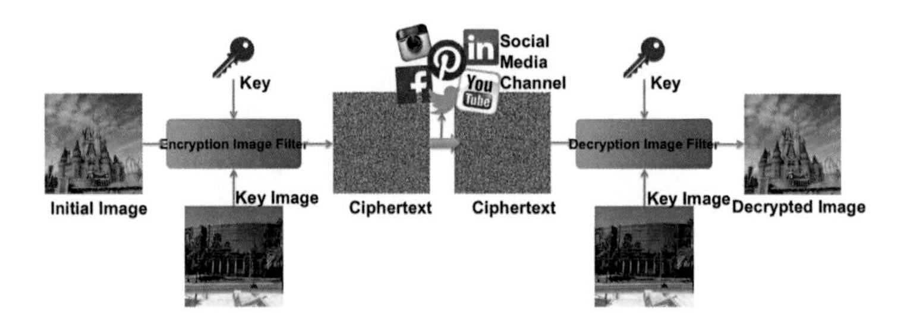


**Figure 11.3: The secure media sharing process.**

The encrypted image/video file, i.e., the ciphertext, can be shared with open media channels, e.g., Facebook, Twitter. The eavesdropper or other unauthorized users cannot get the real meaning of the ciphertext. On the receiver-side, the encrypted image can be decrypted via the key image and the normal key.

### *11.5.1.2 Computational Complexity*

To evaluate the computational efficiency and execution time, we compare the combination of our proposed approaches with standard AES 256 cipher and joint JPEG encryption and compression approaches [88, 105]. Since we do not have the implementation of the joint JPEG encryption and compression, we utilize the software module of JPEG compression to emulate the approaches in [88, 105]. The rationale is: 1) Adding encryption to the JPEG compression module should take longer computation time; 2) Changing the JPEG compression requires the use of software implementation, which cannot utilize the existing hardware accelerated version.

In Fig. 11.4, we compare 1) Chaotic Mapping; 2) Chaotic Mapping + Whitening; 3) Chaotic Mapping + Whitening+ Block Substitution; 4) Hardware-Accelerated AES256; 5) Software-Based Compression. All these cases include the complete process from reading the data, encryption, and compression, whereas cases 1–4 utilize the standard hardware-accelerated compression module. The results are obtained from the Apple iPhone5S via different image/video sizes. The three test cases are: 1) 48KB image; 2) 4.2MB image; 3) 32.9MB video. From Fig. 11.4, we observe at least threefold improvement in the execution time when compared with the standard highly optimized AES 256 cipher. The software solution of compression is even slower than the AES cipher. Adding more steps to our Shader-based approach does not increase the execution time significantly, i.e., cases 1–3 do not differ too much.

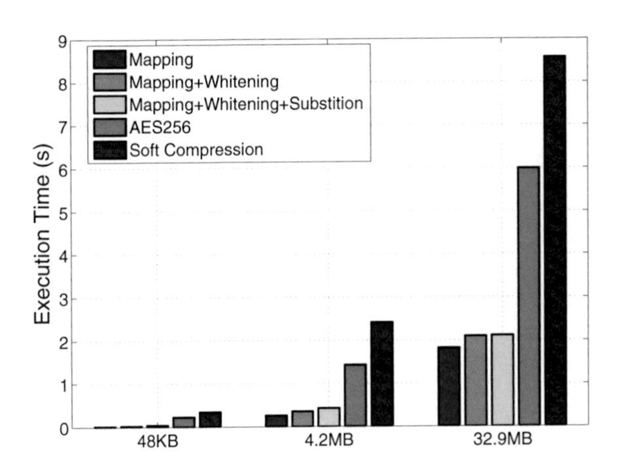


**Figure 11.4: The execution time of different approaches.**

## 11.5.2 Security Analysis

### *11.5.2.1 Whitening and Probabilistic Transformation*

Fig. 11.5(a) shows the utilized sample key image in the whitening process. After performing the operation in (11.1), the XOR mixed ciphertext without probabilistic transformation (PT) is shown in Fig. 11.5(b). One problem of the normal XOR process is the potential leakage of the key image if the input image is monochrome. As shown in the lower part of Fig. 11.5(b), we could see the coarse outline of the key image. This problem could be solved by leveraging the probabilistic transformation as shown in Fig. 11.5(c), which adopts different transforms in the three color channels.

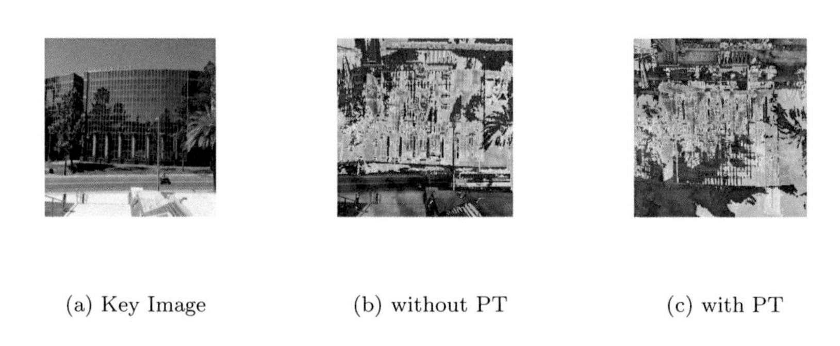

(a) Key Image (b) without PT (c) with PT

**Figure 11.5: Whitening via key image: 1) Key image; 2) Homogeneous whitening via XOR mixing without PT; 3) non-homogenous whitening via XOR mixing with PT.**

#### *11.5.2.2 Multi-Channel Block Substitution*

Fig. 11.6(a) shows the row substitution result with the only color channel. The obtained noisy image is still perceptually visible. Even for the two-channels result Fig. 11.6(b), it is still insufficient for concealing the details. After using three channel substitution, Fig. 11.6(c) shows the final encrypted image with high randomness.

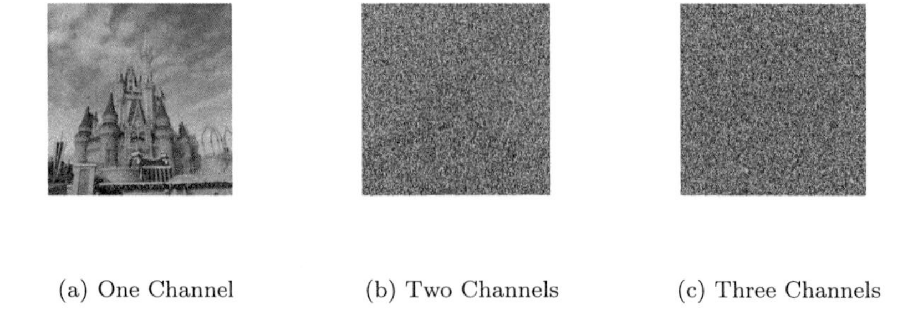

(a) One Channel (b) Two Channels (c) Three Channels

**Figure 11.6: Color image after substition via different color channels: 1) red channel; 2) red and green channel; 3) all three channels.**

The image histogram for the encrypted image in Fig. 11.6(c) is shown in Fig. 11.7(b). Compared with the initial plaintext in Fig. 11.7(a), the histogram of the ciphertext is flattened via the multi-channel block substitution process, which implies a high security level.

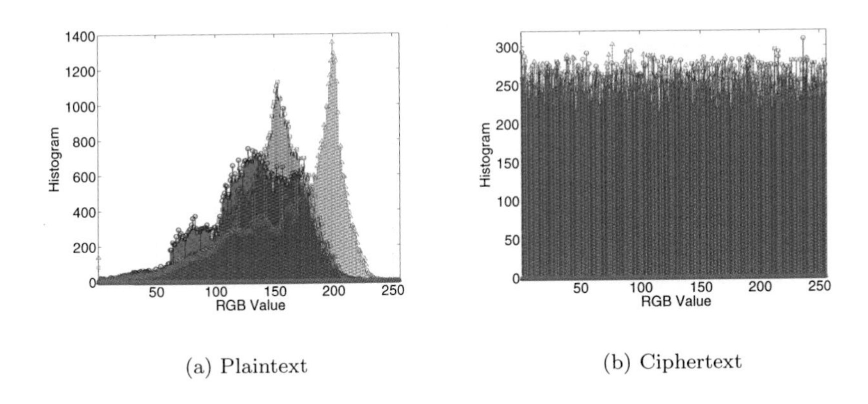


(a) Plaintext (b) Ciphertext

**Figure 11.7: The color histogram for (a) plaintext; (b) ciphertext.**

#### *11.5.2.3 Overall Security Level*

We propose approaches to preserve the correlation via low-order chaotic mapping. The reduced security level of chaotic mapping is compensated by the whitening and block substitution processes. The homogeneous problem of block substitution is also solved by the non-homogeneous transform in the chaotic mapping process and the whitening process. Fig. 11.8(a) shows the correlation result between the ciphertext and plaintext; the resulting correlation is almost zero in most areas, i.e., higher randomness and better security level. Fig. 11.8(b) shows the correlation result between the decrypted image and plaintext with a very high peak correlation, i.e., perfect recovery. This chained solution balances complexity and security level, and meets our objective for correlation-preserving.

### 11.5.3 Correlation Preserving and Noise Robustness

Our proposed privacy-preserving scheme preserves the input format, independent of the existing compression module. For video data, the correlation between frames is also preserved. The rationale is as follows: 1) The utilized chaotic mapping is low-order and only performs one iteration, making the image frame with the same context to produce similar output rather than the

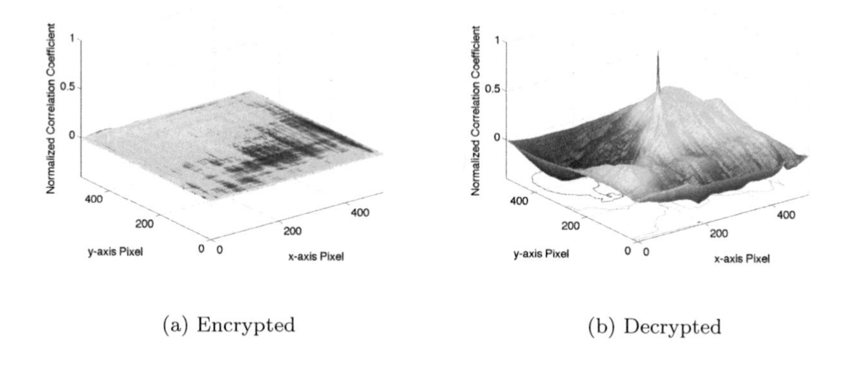


(a) Encrypted (b) Decrypted

**Figure 11.8: The normalized correlation coefficient for (a) encrypted image; (b) decrypted image.**

random output in true chaotic mapping. 2) The whitening and block substitution process are invertible, consistent according to the key, and no random noise is added.

To quantify the correlation between different image frames, we utilize the maximum normalized correlation coefficient (MNCC) as a metric [91]. We randomly change the pixel of the initial image/video frame in terms of pixel difference in percentage (the x-axis of Fig. 11.9(a)). The calculated MNCC of the initial frame is shown in Fig. 11.9(a) with decreasing trends. From Fig. 11.9(a), the normal encryption process via chaotic mapping results in a very low correlation, which is intentional in terms of protecting the plaintext. However, we need to preserve the correlation between adjacent image frames for better compression ratio to save the communication cost. As shown in Fig. 11.9(a), our proposed chained approaches could preserve the correlation when the pixel difference is low.

Another important requirement for our cipher system is the noise robustness when sharing and transmitting the media over the lossy channel. To evaluate the noise robustness, we randomly add noise to the encrypted media data. Fig. 11.9(b) shows the MNCC with regard to the different percentage of pixel error when the noise is added. When the noise becomes strong, i.e., more pixels are affected, the MNCC of our proposed approach decreases slowly, which demonstrates strong robustness over the transmission error. However, the normal approach shows very quick decreasing of MNCC due to the high-order chaotic mapping.

To get a perceptual feeling of the noise robustness, we set the same pixel error for the encrypted image. Fig. 11.10(a) shows the decrypted image via the normal approach; Fig. 11.10(b) shows the decrypted image using our proposed

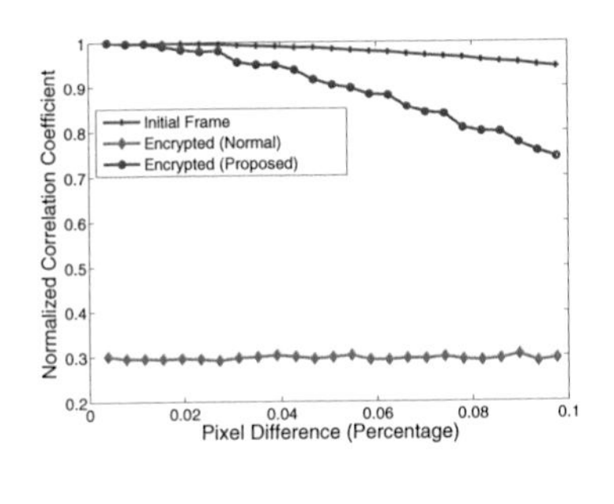


(a) Correlation Preserving

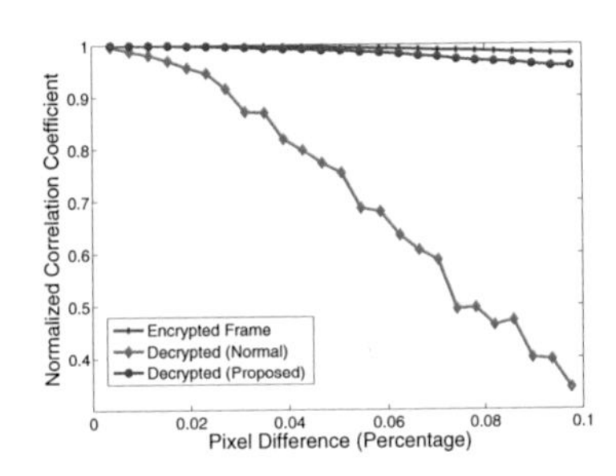


(b) Robust to Noise in Transmission

**Figure 11.9: The maximum normalized correlation coefficient.**

chained approach. Apparently, our proposed approach achieves better image quality under the same transmission error.

(a) Normal Approach

(b) Proposed Approach

**Figure 11.10: Robust to noise in transmission.**

## 11.6 Related Work

Protecting media data privacy without sacrificing the services provided by cloud service providers requires format-compliant encryption solutions. The format-compliant encryption technique can be divided into three different parts according to the relative position of the encryption and compression module.

**Encryption after Compression.** Putting the encryption after compression could minimize the processing data size. Authors in [119] proposed solutions to place it after compression and maintain the format compliance. The core idea is to only encrypt the information-carrying fields, and leave the syntax part unencrypted. The selective encryption scheme could also be applied according to the various privacy requirements. However, the encrypted code words may not be the valid code words defined in the MPEG standard [45]. This scheme also requires fairly deep parsing into the bit stream to identify the parts of the bit stream to be encrypted [51]. This incurs a significant processing overhead. Zhu et al. [136] state that this scheme is even slower than naive algorithm.

**Joint Compression and Encryption.** Joint compression and encryption could solve this problem and be inherently compliant with the syntax format without bit stream identification because the syntax is formatted before the compression process. Authors in [88] proposed solutions that threshold the coefficient of DCT transform of photos into two parts. The main part is stored in a JPEG-compliant format to leverage the cloud services; the other part is encrypted and transmitted via different service providers, e.g., Dropbox. However, this process only suits the DCT-based photo compression format, e.g., JPEG, not other formats. The transmission of two parts per-photo is not convenient. Another practical problem is that the solution needs to break down the photo DCT and compression loop. Re-implementing the whole process is not efficient and fast enough compared to the native hardware-accelerated versions.

**Encryption before Compression.** Putting the encryption before compression and making it compression-independent could comply with the format syntax. Most image scrambling and permutation approaches belong to this category. For example, using chaotic mapping could be a good format-compliant cipher [122, 124, 123]. One significant challenge for these compression-independent approaches is the correlation loss between continuous video frames [124, 122]. Encryption algorithms disorder the inherent redundancy of the input plaintext using cryptographic operations, and make it independent for two plaintexts even with the same key. As a result, the following compression ratio will be very low since there is much less inter-frame redundancy to compress. Another problem is the pixel error sensitivity. When some transmission error happens, the initial image cannot be decrypted even with the same key.

## 11.7 Conclusion

In this chapter, we aimed to overhaul cloud media sharing by letting users assure themselves that no one can eavesdrop or understand what they are watching, posting, and communicating. To meet the objective of designing a format-compliant, compression-independent, and correlation-preserving cipher, we proposed chained approaches via chaotic mapping, image-based key whitening, and Latin square pattern-based substitution. Users can choose their favorite image as the key instead of remembering a long password. To lower the computational cost and encryption delay, we propose to utilize the GPU Shader for parallel pixel processing. We integrate all the proposed approaches into a customized image filter for easy use without modifying the existing code base. Experimental results demonstrate a sufficient security level for cloud media data.

# *Chapter 12*

# CONCLUSION AND FUTURE DIRECTIONS

**Kaikai Liu**

*Assistant Professor, San Jose State University*

**Xiaolin Li**

*Associate Professor, University of Florida*

## CONTENTS



## 12.1 Book Summary

In this book, we proposed a complete location ecosystem for future mobile smart life application, including hardwares (anchor nodes, wearable tags), firmwares (DSP in RTOS, Zigbee/BLE/Audio), algorithms (fine-grained localization, social cooperative localization), Smartphone Apps (crowd sensing, mobile social networks), and cloud servers (algorithm offloading, streaming processing, messaging, and datastore).

In the first part, we proposed a practical and accurate multi-modal step-by-step navigation eco-system, codenamded Guoguo, that assists the blind and visually impaired towards an independent and dignified lifestyle when navigating indoors (e.g., navigating to the bathroom without assistance). The prototype eco-system utilizes various software and hardware components to fill the long-lasting gap of indoor localization. Guoguo consists of an anchor network with a coordination protocol to transmit modulated localization beacons using high-band acoustic signals. To address the challenges of utilizing the audible-band acoustic signal in smartphone localization, i.e., strong attenuation, interference-richness, high sound disturbance, and difficulty in synchronization, we proposed comprehensive schemes to improve the localization accuracy and extend coverage without sound disturbance. For the first time, we can locate a smartphone user at the centimeter-level, which has significant implications for potential indoor location services and applications compared with existing meter-level localization solutions. We further propose approaches to improve its coverage, accuracy, and location update rate with low-power consumption. Experimental results demonstrated that the achieved average localization accuracy is about $3 \sim 10$cm in a typical indoor environments. Guoguo represents a leap in progress in smartphone-based indoor localization, opening enormous new opportunities for indoor location-based services, positioning and navigation systems, and other commercial, educational, or entertainment applications. The location approaches described here are the prototype eco-system that will eventually fuse several complementary assistance softwares together towards accessibility assistance and navigation for the blind, disabled, and ultimately the elderly.

In the second part, we focused on the application of accessing the information automatically via context-awareness. Based on the location estimation approach in the first part, we pushed location awareness to context-aware augmented reality (AR). AR aims to render the world that users see and overlay information that reflects the real physical dynamics. The digital view could be potentially projected near the Point-of-Interest (POI) in a way that makes the virtual view attached to the POI even when the camera moves. Achieving smooth support for movements is a subject of extensive studies. One of the key problems is where the augmented information should be added to the field of vision in real time. Existing solutions either leverage GPS location for rendering outdoor AR views (hundreds of kilometers away) or rely on image markers for small-scale presentation (only for the marker region). To realize AR applications under various scales and dynamics, we proposed a suite of algorithms for fine-grained AR view tracking to improve the accuracy of attitude and displacement estimation, reduce the drift, eliminate the marker, and lower the computation cost. Instead of requiring extremely highly accurate absolute locations, we proposed multi-modal solutions according to mobility levels without additional hardware requirements. Experimental results demonstrated significantly less error in projecting and tracking the AR view. These results are expected to make users excited to explore their surroundings with enriched content.

In the previous two parts, we focused on the infrastructure-based fine-grained approach. However, in some applications, e.g., locating lost children or other group members in an open and uncontrolled area, infrastructure is not always available. To solve this problem, we move from infrastructure-based solutions to crowd sensing in the third part, and further push the location sensing solution for broader impact via participatory social networks. For applications like guarding a child or pet in crowded places full of attractions, conventional location tracking approaches require fixed anchor networks, fingerprinting points as references, or bulky devices with GPS and cellular connections. The newly proposed miniature wearable devices are attractive in terms of cost, size, and portability. However, locating a lost child/pet in open and uncontrolled areas without the high-cost cellular and GPS trackers is mission impossible due to the limited communication coverage of the miniature wearable devices. To overcome this limitation, we proposed mobile crowd sourcing/sensing (MCS)-based collaborative localization via nearby opportunistically connected participators with smartphones. To obtain sufficient measurements for better location resolution, we utilized one-hop and multi-hop assistants to reach more participators for transparent sensing assistance. To overcome the bias and unsolvable problem caused by insufficient measurements in crowd sensing, we proposed global optimization approaches based on semidefinite programming (SDP) via sparse measurement constraints. Multi-hop opportunistic ranging and coarse-grained location information are leveraged to jointly optimize the location of the wearable tag and assistant with unknown locations. We conducted extensive experiments and simulations in various scenarios. Compared with other classic algorithms, our proposed approach achieves significant accuracy improvement and could locate the "unlocalizable" child. Utilizing the ubiquitousness of "crowds" of sensor-rich smartphones, our proposed approach has enormous potential to truly unleash the power of collaborative locating and searching at a societal scale.

## 12.2 Future Directions

The new generation of smartphones has almost realized the early vision of Mark Weiser's vision/dream from a human-computer interaction perspective: computing everywhere, personal computing, the social dimension of computing, and privacy implications [96]. Multiple technical approaches are moving towards these envisioned mobile smart life applications. One of the key approaches is context-awareness that facilitates real-world interaction of computing, whereas the location plays the key role in context configurations.

In this book, we proposed enabling techniques for three envisioned mobile smart life applications: 1) helping the blind or other disabled to live independently via step-by-step navigation; 2) accessing the information automatically via context-awareness; 3) locating lost children or other group members via mobile crowd sensing. The next step is facilitating the real-world impact of more complex location and context recognition, moving towards next-generation opportunistic context-awareness and large-scale ensembles of crowd sensing interacting with social entities. For example, with fine-grained location/pose/activity sensing, context-aware information can be pushed to you automatically without any keyword typing in explicit search engines. Your mobile device, e.g., smartphone or wearable tag, could be your assistant covering both online and offline. Your group members will not only include online "friends" but also entities with whom a person has a relationship (e.g., children, elders, pets) with a wearable tag; they can interact with each other in a transparent, reactive, and proactive manner beyond the conventional online social networks (where users perform common interactions such as manually typing, searching, and posting).

The results of the techniques studied in this book, with both their limitations and their promises, are moving towards these environed smart life scenarios. However, there are many research directions waiting to be explored.

**Rehabilitation Technology for the Blind.** People who are visually impaired face a multitude of challenges every day that can prevent them from getting where they want to go, even giving up places that are essential to their life, e.g., school, clinic, retail store, and city facilities. The journey to public places is very daunting and leaves them stressed and anxious. We proposed a system architecture and enabling techniques for smartphone based step-by-step navigation for the blind. To fully enable the freedom of mobility for the blind, significant efforts in the areas of sensing, voice interaction, and social assistance is needed. The blind's navigation system should start by establishing their location within the building, and on the map. It will require voice input for the destination or purpose, and then determines the best route or action to get them done, and guides them along it via verbal cues and feedbacks. This is because visually impaired people would have difficulty using the smartphone's display to understand where they are located, and what's nearby. Everything should be done verbally. Speech recognition approaches via the smartphones or any other wearable device's microphone are necessary. The speech recognition engine should not only understand what they need, but also needs to perform related actions. However, indoor navigation and interaction is far more complex than outdoor GPS navigation via voice assistance, where the road is clear and predefined. There are so many obstacles, blockages, floors, and routes in indoor places, e.g., door, staircase, elevator, and restricted areas. The voice interpretation is inaccurate and not rich enough when compared with normal people's vision. Existing speech recognition and interaction engines highly rely on their knowledge set or database, whereas special commands and voice-related map interactions should be defined specifically for complex indoor navigation purposes.

**Context-Awareness and Augmented Reality.** Augmented reality (AR) has been in science fiction for decades [33]. Thanks to the power of smartphones, the technology is almost here to make it happen, e.g., existing GPS location-based AR for nearby Points-of-Interest (POI), image marker-based solutions for rendering additional information. Mobile tech keeps moving toward AR, but it is still far from perfect in terms of object recognition and AR view tracking with various mobilities. We proposed one approach via sensor fusion to achieve efficient AR view tracking, and this gives us an insight into the exciting future possibilities for AR applications. To really achieve the dreamed of AR potentials, lots of barriers must be overcome. Possible extensions include energy efficient vision sensing; highly reliable attitude and location sensing; sensor fusion via INS and computer vision; realtime processing back for recognition; machine learning back-end for context recommendation. All these techniques are expected to have high impact in the future smart life applications and our daily activities.

**Wearable and Cyber-Physical Social Networking.** The recent transition to smart wearables opens up a slew of new opportunities for vendors, developers, researchers, and manufactrures. Growth in the smart wearables points to an emerging battleground for future smart life applications. In this dissertation, we proposed solutions for wearable-based group member tracking and localization via mobile crowd sensing. The demand for other innovative applications related to security or other areas has been absolutely astounding, e.g., fitness, identification, notification. These will raise the expectations of what a smart wearable can do. We are not there yet, but we are seeing the building blocks of what is to come. Several key challenges associated with smart wearables include portable hardware design with longer battery life, energy efficient localization approaches, context-awareness, and wireless connectivity via multi-hop assistance. When the wearable devices are ubiquitously available for everyone, a new way of social interaction could be designed via nearby sensing and interaction that mimics the real physical encounters. Participatory crowd missions could be conducted in a transparent way without disturbing the users. Incentive mechanism design is the key for these mobile crowd applications.

# References


[1] Geosocial and location-based services on smartphones. 2011.

[2] AccessMagazine. Fixing broken sidewalks, 2015.

[3] ADA. Americans with Disabilities ACT: Guide for places of lodging, 2015.

[4] Apple. Cmattitude class reference, 2014.

[5] Apple. ibeacon, 2014.

[6] M. Azizyan, I. Constandache, and R. Roy Choudhury. Surroundsense: mobile phone localization via ambience fingerprinting. In *Proceedings of the 15th annual international conference on mobile computing and networking*, pages 261–272. ACM, 2009.

[7] Ronald Azuma, Yohan Baillot, Reinhold Behringer, Steven Feiner, Simon Julier, and Blair MacIntyre. Recent advances in augmented reality. *Computer Graphics and Applications, IEEE*, 21(6):34–47, 2001.

[8] P. Bahl and V.N. Padmanabhan. Radar: An in-building rf-based user location and tracking system. In *INFOCOM 2000. Nineteenth Annual Joint Conference of the IEEE Computer and Communications Societies. Proceedings. IEEE*, volume 2, pages 775–784. IEEE, 2000.

[9] P. Biswas, T.C. Liang, K.C. Toh, Y. Ye, and T.C. Wang. Semidefinite programming approaches for sensor network localization with noisy distance measurements. *Automation Science and Engineering, IEEE Transactions on*, 3(4):360–371, 2006.

[10] Gaetano Borriello, Alan Liu, Tony Offer, Christopher Palistrant, and Richard Sharp. Walrus: wireless acoustic location with room-level resolution using ultrasound. In *Proceedings of the 3rd international con-*

*ference on mobile systems, applications, and services*, pages 191–203. ACM, 2005.

[11] S.P. Boyd and L. Vandenberghe. *Convex optimization*. Cambridge University Press, 2004.

[12] BSD. Opencv, 2014.

[13] Cassandra. The apache cassandra project, 2014.

[14] Manmohan Krishna Chandraker, Christoph Stock, and Axel Pinz. *Real-time camera pose in a room*. Springer, 2003.

[15] Y. Chen, D. Lymberopoulos, J. Liu, and B. Priyantha. Fm-based indoor localization. In *Proceedings of the 10th international conference on mobile systems, applications, and services*, pages 169–182. ACM, 2012.

[16] K.W. Cheung, H.C. So, W.K. Ma, and Y.T. Chan. Least squares algorithms for time-of-arrival-based mobile location. *Signal Processing, IEEE Transactions on*, 52(4):1121–1130, 2004.

[17] Chipolo. Chipolo :: Nothing is lost, 2014.

[18] I. Constandache, X. Bao, M. Azizyan, and R.R. Choudhury. Did you see bob?: human localization using mobile phones. In *Proceedings of the sixteenth annual international conference on mobile computing and networking*, pages 149–160. ACM, 2010.

[19] I. Constandache, R.R. Choudhury, and I. Rhee. Towards mobile phone localization without war-driving. In *INFOCOM, 2010 Proceedings IEEE*, pages 1–9. IEEE, 2010.

[20] I. Constandache, S. Gaonkar, M. Sayler, R.R. Choudhury, and L. Cox. Enloc: Energy-efficient localization for mobile phones. In *INFOCOM 2009, IEEE*, pages 2716–2720. IEEE, 2009.

[21] Cozi. Child tracking: What parents should know, 2014.

[22] D. Dardari, C.C. Chong, and M. Win. Threshold-based time-of-arrival estimators in UWB dense multipath channels. *IEEE Transactions on Communications*, 56(8):1366–1378, 2008.

[23] Bas des Bouvrie. *Improving rgbd indoor mapping with imu data*. Ph.D. thesis, Masters thesis, Delft University of Technology, 2011.

[24] Estimote. Estimote beacons real world context for your apps.

[25] Friedrich Fraundorfer and Davide Scaramuzza. Visual odometry: Part ii: Matching, robustness, optimization, and applications. *Robotics & Automation Magazine, IEEE*, 19(2):78–90, 2012.

[26] Jiri Fridrich. Image encryption based on chaotic maps. In *Systems, Man, and Cybernetics, 1997. Computational Cybernetics and Simulation, 1997 IEEE International Conference on*, volume 2, pages 1105–1110. IEEE, 1997.

[27] Raghu K. Ganti, Fan Ye, and Hui Lei. Mobile crowdsensing: current state and future challenges. *Communications Magazine, IEEE*, 49(11):32–39, 2011.

[28] M. Grant and S. Boyd. Cvx: Matlab software for disciplined convex programming. *Available at httpstanford edu boydcvx*, 2008.

[29] Blippar group. Layar, 2014.

[30] I. Guvenc, C.C. Chong, and F. Watanabe. Analysis of a linear least-squares localization technique in los and nlos environments. In *Vehicular Technology Conference, 2007. VTC2007-Spring. IEEE 65th*, pages 1886–1890. IEEE, 2007.

[31] I. Guvenc and Z. Sahinoglu. Threshold-based TOA estimation for impulse radio UWB systems. In *IEEE International Conference on Ultra-Wideband*, pages 420–425, 2005.

[32] Richard Hartley and Andrew Zisserman. *Multiple view geometry in computer vision.* Cambridge University Press, 2003.

[33] Simon Hill. Get past the gimmicks and gaze upon the future of augmented reality apps, 2014.

[34] Fabian Hoflinger, Rui Zhang, Joachim Hoppe, Amir Bannoura, Leonhard M. Reindl, Johannes Wendeberg, Manuel Buhrer, and Christian Schindelhauer. Acoustic self-calibrating system for indoor smartphone tracking (assist). In *Indoor Positioning and Indoor Navigation (IPIN), 2012 International Conference on*, pages 1–9. IEEE, 2012.

[35] J.D. Hol. Sensor fusion and calibration of inertial sensors, vision, ultra-wideband and gps. *Linköping Studies in Science and Technology. Dissertations*, (1368), 2011.

[36] Humanware. Trekker Breeze+ handheld talking gps, 2015.

[37] Myung Hwangbo, Jun-Sik Kim, and Takeo Kanade. Inertial-aided klt feature tracking for a moving camera. In *Intelligent Robots and Systems, 2009. IROS 2009. IEEE/RSJ International Conference on*, pages 1909–1916. IEEE, 2009.

[38] G. Jin, X. Lu, and M.S. Park. An indoor localization mechanism using active rfid tag. In *Sensor Networks, Ubiquitous, and Trustworthy Computing, 2006. IEEE International Conference on*, volume 1, pages 4–pp. IEEE, 2006.

[39] Kafka. Apache kafka: A high-throughput distributed messaging system, 2014.

[40] E.D. Kaplan and C.J. Hegarty. *Understanding GPS: principles and applications*. Artech House Publishers, 2006.

[41] Hirokazu Kato. Artoolkit: library for vision-based augmented reality. *IEICE, PRMU*, pages 79–86, 2002.

[42] S.M. Kay. *Fundamentals of Statistical Signal Processing, Volume 2: Detection Theory*. Prentice Hall PTR, 1998.

[43] Bernd Kitt, Andreas Geiger, and Henning Lategahn. Visual odometry based on stereo image sequences with ransac-based outlier rejection scheme. In *Intelligent Vehicles Symposium (IV), 2010 IEEE*, pages 486–492. IEEE, 2010.

[44] M.B. Kjærgaard, J. Langdal, T. Godsk, and T. Toftkjær. Entracked: energy-efficient robust position tracking for mobile devices. In *Proceedings of the 7th international conference on mobile systems, applications, and services*, pages 221–234. ACM, 2009.

[45] Rob Koenen. Overview of the mpeg-4 standard. *ISO/IEC JTC1/SC29/WG11 N*, 1730:11–13, 2002.

[46] Nicholas D. Lane, Emiliano Miluzzo, Hong Lu, Daniel Peebles, Tanzeem Choudhury, and Andrew T. Campbell. A survey of mobile phone sensing. *Communications Magazine, IEEE*, 48(9):140–150, 2010.

[47] Tobias Langlotz, Claus Degendorfer, Alessandro Mulloni, Gerhard Schall, Gerhard Reitmayr, and Dieter Schmalstieg. Robust detection and tracking of annotations for outdoor augmented reality browsing. *Computers & graphics*, 35(4):831–840, 2011.

[48] Patrick Lazik and Anthony Rowe. Indoor pseudo-ranging of mobile devices using ultrasonic chirps. In *Proceedings of the 10th ACM Conference on Embedded Network Sensor Systems*, pages 99–112. ACM, 2012.

[49] J.Y. Lee and R.A. Scholtz. Ranging in a dense multipath environment using an UWB radio link. *IEEE Journal on Selected Areas in Communications*, 20(9):1677–1683, 2002.

[50] K. Lin, A. Kansal, D. Lymberopoulos, and F. Zhao. Energy-accuracy trade-off for continuous mobile device location. In *Proceedings of the 8th international conference on mobile systems, applications, and services*, pages 285–298. ACM, 2010.

[51] Fuwen Liu and Hartmut Koenig. A survey of video encryption algorithms. *Computers & Security*, 29(1):3–15, 2010.

[52] Hongbo Liu, Yu Gan, Jie Yang, Simon Sidhom, Yan Wang, Yingying Chen, and Fan Ye. Push the limit of wifi based localization for smartphones. In *Proceedings of the 18th annual international conference on mobile computing and networking*, pages 305–316. ACM, 2012.

[53] K. Liu, X. Liu, and X. Li. Acoustic ranging and communication via microphone channel. In *Proc. IEEE Globecom '12*, Anaheim, California, USA, IEEE, 2012.

[54] K. Liu, H. Yin, and W. Chen. Low complexity tri-level sampling receiver design for uwb time-of-arrival estimation. In *Communications (ICC), 2011 IEEE International Conference on*, pages 1–5. IEEE, 2011.

[55] Kaikai Liu, Qiuyuan Huang, Jiecong Wang, Xiaolin Li, and Dapeng Wu. Improving gps service via social collaboration. In *Mobile Adhoc and Sensor Systems (MASS), 2013 IEEE 10th International Conference on.* IEEE, 2013.

[56] Kaikai Liu, Qiuyuan Huang, Jiecong Wang, Xiaolin Li, and Dapeng Oliver Wu. Improving gps service via social collaboration. In *Mobile Ad-Hoc and Sensor Systems (MASS), 2013 IEEE 10th International Conference on*, pages 393–401. IEEE, 2013.

[57] Kaikai Liu, Xinxin Liu, and Xiaolin Li. Guoguo: Enabling fine-grained indoor localization via smartphone. In *Proceedings of the 11th international conference on mobile systems, applications, and services.* ACM, 2013.

[58] Kaikai Liu, Xinxin Liu, and Xiaolin Li. Guoguo: Enabling fine-grained indoor localization via smartphone. In *Proceedings of the 11th annual international conference on mobile systems, applications, and services*, pages 235–248. ACM, 2013.

[59] Kaikai Liu, Xinxin Liu, Lulu Xie, and Xiaolin Li. Towards accurate acoustic localization on a smartphone. In *INFOCOM, 2013 Proceedings IEEE*, pages 495–499. IEEE, 2013.

[60] Hong Lu, Wei Pan, Nicholas D. Lane, Tanzeem Choudhury, and Andrew T. Campbell. Soundsense: scalable sound sensing for people-centric applications on mobile phones. In *Proceedings of the 7th international conference on mobile systems, applications, and services*, pages 165–178. ACM, 2009.

[61] Hong Lu, Jun Yang, Zhigang Liu, Nicholas D. Lane, Tanzeem Choudhury, and Andrew T. Campbell. The jigsaw continuous sensing engine for mobile phone applications. In *Proceedings of the 8th ACM Conference on Embedded Networked Sensor Systems*, pages 71–84. ACM, 2010.

[62] A. Mandal, C.V. Lopes, T. Givargis, A. Haghighat, R. Jurdak, and P. Baldi. Beep: 3d indoor positioning using audible sound. In *Consumer Communications and Networking Conference, 2005. CCNC. 2005 Second IEEE*, pages 348–353. IEEE, 2005.

[63] Justin Gregory Manweiler, Puneet Jain, and Romit Roy Choudhury. Satellites in our pockets: an object positioning system using smartphones. In *Proceedings of the 10th international conference on mobile systems, applications, and services*, pages 211–224. ACM, 2012.

[64] G. Mao, B. Fidan, and B. Anderson. Wireless sensor network localization techniques. *Computer Networks*, 51(10):2529–2553, 2007.

[65] Alex T. Mariakakis, Souvik Sen, Jeongkeun Lee, and Kyu-Han Kim. Sail: Single access point-based indoor localization. In *Proceedings of the 12th annual international conference on mobile systems, applications, and services*, pages 315–328. ACM, 2014.

[66] F. Landis Markley and John L. Crassidis. *Fundamentals of Spacecraft Attitude Determination and Control.* Springer, 2014.

[67] Lorenz Meier, Petri Tanskanen, Lionel Heng, Gim Hee Lee, Friedrich Fraundorfer, and Marc Pollefeys. Pixhawk: A micro aerial vehicle design for autonomous flight using onboard computer vision. *Autonomous Robots*, 33(1-2):21–39, 2012.

[68] C. Meng, Z. Ding, and S. Dasgupta. A semidefinite programming approach to source localization in wireless sensor networks. *Signal Processing Letters, IEEE*, 15:253–256, 2008.

[69] Emiliano Miluzzo, Nicholas D. Lane, Shane B. Eisenman, and Andrew T. Campbell. Cenceme–injecting sensing presence into social networking applications. In *Smart Sensing and Context*, pages 1–28. Springer, 2007.

[70] M. Minami, Y. Fukuju, K. Hirasawa, S. Yokoyama, M. Mizumachi, H. Morikawa, and T. Aoyama. Dolphin: a practical approach for implementing a fully distributed indoor ultrasonic positioning system. *UbiComp 2004: Ubiquitous Computing*, pages 347–365, 2004.

[71] Mixare. Mix augmented reality engine, 2011.

[72] Marko Modsching, Ronny Kramer, and Klaus ten Hagen. Field trial on gps accuracy in a medium size city: The influence of built-up. In *3rd Workshop on Positioning, Navigation and Communication*, pages 209–218, 2006.

[73] Alessandro Mulloni, Daniel Wagner, Istvan Barakonyi, and Dieter Schmalstieg. Indoor positioning and navigation with camera phones. *Pervasive Computing, IEEE*, 8(2):22–31, 2009.

[74] R. Nandakumar, K.K. Chintalapudi, and V.N. Padmanabhan. Centaur: locating devices in an office environment. In *Proceedings of the 18th annual international conference on mobile computing and networking*, pages 281–292. ACM, 2012.

[75] Rob Napier and Mugunth Kumar. *iOS 7 Programming Pushing the Limits*. John Wiley & Sons, 2014.

[76] Joseph Newman, David Ingram, and Andy Hopper. Augmented reality in a wide area sentient environment. In *Augmented Reality, 2001. Proceedings. IEEE and ACM International Symposium on*, pages 77–86. IEEE, 2001.

[77] NFB. National Federation of the Blind, 2015.

[78] L.M. Ni, Y. Liu, Y.C. Lau, and A.P. Patil. Landmarc: indoor location sensing using active rfid. *Wireless networks*, 10(6):701–710, 2004.

[79] A. Nosratinia, T.E. Hunter, and A. Hedayat. Cooperative communication in wireless networks. *Communications Magazine, IEEE*, 42(10):74–80, 2004.

[80] N. Patwari, J.N. Ash, S. Kyperountas, A.O. Hero III, R.L. Moses, and N.S. Correal. Locating the nodes: cooperative localization in wireless sensor networks. *Signal Processing Magazine, IEEE*, 22(4):54–69, 2005.

[81] C. Peng, G. Shen, Y. Zhang, Y. Li, and K. Tan. Beepbeep: a high accuracy acoustic ranging system using cots mobile devices. In *Proceedings of the 5th international conference on embedded networked sensor systems*, pages 1–14. ACM, 2007.

[82] Chunyi Peng, Guobin Shen, and Yongguang Zhang. Beepbeep: A high-accuracy acoustic-based system for ranging and localization using cots devices. *ACM Transactions on Embedded Computing Systems (TECS)*, 11(1):4, 2012.

[83] William Russell Pensyl, Daniel Keith Jernigan, Tran Cong Thien Qui, Hsin Pei Fang, and Lee Shang Ping. Large area robust hybrid tracking with life-size avatar in mixed reality environment: for cultural and historical installation. In *Proceedings of the 7th ACM SIGGRAPH International Conference on Virtual-Reality Continuum and Its Applications in Industry*, page 9. ACM, 2008.

[84] Ian R. Petersen and Andrey V. Savkin. *Robust Kalman filtering for signals and systems with large uncertainties*. Birkhäuser Boston, 1999.

[85] N.B. Priyantha, A. Chakraborty, and H. Balakrishnan. The cricket location-support system. In *Proceedings of the 6th annual international conference on mobile computing and networking*, pages 32–43. ACM, 2000.

[86] Qualcomm. Qualcomm gimbal proximity beacon.

[87] Qualcomm. Vuforia, 2015.

[88] Moo-Ryong Ra, Ramesh Govindan, and Antonio Ortega. P3: Toward privacy-preserving photo sharing. In *NSDI*, pages 515–528, 2013.

[89] Anshul Rai, Krishna Kant Chintalapudi, Venkata N. Padmanabhan, and Rijurekha Sen. Zee: Zero-effort crowdsourcing for indoor localization. In *Proceedings of the 18th annual international conference on mobile computing and networking*, pages 293–304. ACM, 2012.

[90] Pavel Rajmic. Method for real-time signal processing via wavelet transform. *Nonlinear Analyses and Algorithms for Speech Processing*, pages 368–378, 2005.

[91] K. Ramamohan Rao and Ping Yip. *Discrete cosine transform: algorithms, advantages, applications*. Academic Press, 2014.

[92] Redis. Redis.io, 2014.

[93] Miguel Ribo, Markus Brandner, and Axel Pinz. A flexible software architecture for hybrid tracking. *Journal of Robotic Systems*, 21(2):53–62, 2004.

[94] Ethan Rublee, Vincent Rabaud, Kurt Konolige, and Gary Bradski. Orb: an efficient alternative to sift or surf. In *Computer Vision (ICCV), 2011 IEEE International Conference on*, pages 2564–2571. IEEE, 2011.

[95] A.H. Sayed, A. Tarighat, and N. Khajehnouri. Network-based wireless location: challenges faced in developing techniques for accurate wireless location information. *Signal Processing Magazine, IEEE*, 22(4):24–40, 2005.

[96] Albrecht Schmidt, Bastian Pfleging, Florian Alt, Alireza Sahami Shirazi, and Geraldine Fitzpatrick. Interacting with 21st-century computers. *IEEE Pervasive Computing*, (1):22–31, 2011.

[97] Guobin Shen, Zhuo Chen, Peichao Zhang, Thomas Moscibroda, and Yongguang Zhang. Walkie-markie: indoor pathway mapping made easy. In *Proceedings of the 10th USENIX conference on Networked Systems Design and Implementation*, pages 85–98. USENIX Association, 2013.

[98] A.M.C. So and Y. Ye. Theory of semidefinite programming for sensor network localization. *Mathematical Programming*, 109(2):367–384, 2007.

[99] Sonitor. Sonitor system website. http://www.sonitor.com/, 2014.

[100] Kenneth Southall and Walter Wittich. Barriers to vision rehabilitation: a qualitative approach, 2008.

[101] Storm. Storm - the apache software foundation, 2014.

[102] Guo-Lin Sun and Wei Guo. Bootstrapping m-estimators for reducing errors due to non-line-of-sight (nlos) propagation. *Communications Letters, IEEE*, 8(8):509–510, 2004.

[103] S.P. Tarzia, P.A. Dinda, R.P. Dick, and G. Memik. Indoor localization without infrastructure using the acoustic background spectrum. In *Proceedings of the 9th international conference on mobile systems, applications, and services*, pages 155–168. ACM, 2011.

[104] J.D. Taylor. *Ultra-wideband radar technology*. CRC, 2001.

[105] Matt Tierney, Ian Spiro, Christoph Bregler, and Lakshminarayanan Subramanian. Cryptagram: Photo privacy for online social media. In *Proceedings of the first ACM conference on online social networks*, pages 75–88. ACM, 2013.

[106] Mostafa Uddin and Tamer Nadeem. Rf-beep: A light ranging scheme for smart devices. In *Pervasive Computing and Communications (PerCom), 2013 IEEE International Conference on*, pages 114–122. IEEE, 2013.

[107] Mostafa Uddin and Tamer Nadeem. Spyloc: a light weight localization system for smartphones. In *Proceedings of the 19th annual international conference on mobile computing & networking*, pages 223–226. ACM, 2013.

[108] J.M. Valin, F. Michaud, J. Rouat, and D. Létourneau. Robust sound source localization using a microphone array on a mobile robot. In *Intelligent Robots and Systems, 2003.(IROS 2003), Proceedings, 2003 IEEE/RSJ International Conference on*, volume 2, pages 1228–1233. IEEE, 2003.

[109] A. Van den Bos. *Parameter estimation for scientists and engineers*. Wiley Online Library, 2007.

[110] DWF Van Krevelen and R Poelman. A survey of augmented reality technologies, applications and limitations. *International Journal of Virtual Reality*, 9(2):1, 2010.

[111] W. Viana, J. Bringel Filho, J. Gensel, M. Villanova-Oliver, and H. Martin. Photomap: from location and time to context-aware photo annotations. *Journal of Location Based Services*, 2(3):211–235, 2008.

[112] Daniel Wagner, Gerhard Reitmayr, Alessandro Mulloni, Tom Drummond, and Dieter Schmalstieg. Real-time detection and tracking for augmented reality on mobile phones. *Visualization and Computer Graphics, IEEE Transactions on*, 16(3):355–368, 2010.

[113] R. Wand, A. Hopper, V. Falcao, and J. Gibbons. The active bat location system. *ACM Transactions on Information Systems*, pages 91–102, 1992.

[114] He Wang, Souvik Sen, Ahmed Elgohary, Moustafa Farid, Moustafa Youssef, and Romit Roy Choudhury. No need to war-drive: unsupervised indoor localization. In *Proceedings of the 10th international conference on mobile systems, applications, and services*, pages 197–210. ACM, 2012.

[115] Ning Wang and Liuqing Yang. Further results on cooperative localization via semidefinite programming. In *Information Sciences and Systems (CISS), 2011 45th Annual Conference on*, pages 1–6. IEEE, 2011.

[116] Yi Wang, Jialiu Lin, Murali Annavaram, Quinn A. Jacobson, Jason Hong, Bhaskar Krishnamachari, and Norman Sadeh. A framework of energy efficient mobile sensing for automatic user state recognition. In *Proceedings of the 7th international conference on mobile systems, applications, and services*, pages 179–192. ACM, 2009.

[117] Z. Wang, S. Zheng, Y. Ye, and S. Boyd. Further relaxations of the semidefinite programming approach to sensor network localization. *SIAM Journal on Optimization*, 19(2):655–673, 2008.

[118] Wayfinding. The future of blind navigation, 2015.

[119] Jiangtao Wen, Mike Severa, Wenjun Zeng, Max H. Luttrell, and Weiyin Jin. A format-compliant configurable encryption framework for access control of video. *Circuits and Systems for Video Technology, IEEE Transactions on*, 12(6):545–557, 2002.

[120] Zac White. iphone-ar-toolkit, 2012.

[121] Wikipedia. Gps for the visually impaired, 2015.

[122] Yue Wu, Gelan Yang, Huixia Jin, and Joseph P. Noonan. Image encryption using the two-dimensional logistic chaotic map. *Journal of Electronic Imaging*, 21(1):013014–1, 2012.

[123] Yue Wu, Yicong Zhou, Joseph P. Noonan, and Sos Agaian. Design of image cipher using latin squares. *Information Sciences*, 264:317–339, 2014.

[124] Yue Wu, Yicong Zhou, Joseph P. Noonan, Karen Panetta, and Sos Agaian. Image encryption using the sudoku matrix. In *SPIE Defense, Security, and Sensing*, pages 77080P–77080P. International Society for Optics and Photonics, 2010.

[125] Roni Yadlin. Attitude determination and bias estimation using kalman filtering, 2009.

[126] Tingxin Yan, Matt Marzilli, Ryan Holmes, Deepak Ganesan, and Mark Corner. mcrowd: a platform for mobile crowdsourcing. In *Proceedings of the 7th ACM Conference on Embedded Networked Sensor Systems*, pages 347–348. ACM, 2009.

[127] J. Yang and Y. Chen. Indoor localization using improved rss-based lateration methods. In *Global Telecommunications Conference, 2009. GLOBECOM 2009. IEEE*, pages 1–6. IEEE, 2009.

[128] J. Yang, S. Sidhom, G. Chandrasekaran, T. Vu, H. Liu, N. Cecan, Y. Chen, M. Gruteser, and R.P. Martin. Detecting driver phone use leveraging car speakers. In *Proceedings of the 17th annual international conference on mobile computing and networking*, pages 97–108. ACM, 2011.

[129] Zheng Yang, Chenshu Wu, and Yunhao Liu. Locating in fingerprint space: wireless indoor localization with little human intervention. In *Proceedings of the 18th annual international conference on mobile computing and networking*, pages 269–280. ACM, 2012.

[130] H. Yin, Z. Wang, L. Ke, and J. Wang. Monobit digital receivers: design, performance, and application to impulse radio. *Communications, IEEE Transactions on*, 58(6):1695–1704, 2010.

[131] YourStory. Wearable technology poised to surge: things all CEOs CIOs must know., 2015.

[132] Moustafa Youssef and Ashok Agrawala. The horus wlan location determination system. In *Proceedings of the 3rd international conference on mobile systems, applications, and services*, pages 205–218. ACM, 2005.

[133] Zengbin Zhang, David Chu, Xiaomeng Chen, and Thomas Moscibroda. Swordfight: enabling a new class of phone-to-phone action games on commodity phones. In *Proceedings of the 10th international conference on mobile systems, applications, and services*, pages 1–14. ACM, 2012.

[134] Pengfei Zhou, Mo Li, and Guobin Shen. Use it free: Instantly knowing your phone attitude. In *ACM MobiCom*, volume 2014, 2014.

[135] Pengfei Zhou, Yuanqing Zheng, and Mo Li. How long to wait?: predicting bus arrival time with mobile phone based participatory sensing. In *Proceedings of the 10th international conference on mobile systems, applications, and services*, pages 379–392. ACM, 2012.

[136] Bin B. Zhu, Chun Yuan, Yidong Wang, and Shipeng Li. Scalable protection for mpeg-4 fine granularity scalability. *Multimedia, IEEE Transactions on*, 7(2):222–233, 2005.

# Index

## M